EXTRAGALACTIC RADIO SOURCES

INTERNATIONAL ASTRONOMICAL UNION
UNION ASTRONOMIQUE INTERNATIONALE

SYMPOSIUM No. 97

ORGANIZED BY IAU IN COOPERATION WITH URSI,
HELD AT ALBUQUERQUE, U.S.A., AUGUST 3–7, 1981

EXTRAGALACTIC RADIO SOURCES

EDITED BY

DAVID S. HEESCHEN

and

CAMPBELL M. WADE

National Radio Astronomy Observatory, Charlottesville, VA, U.S.A.

D. REIDEL PUBLISHING COMPANY

DORDRECHT : HOLLAND / BOSTON : U.S.A. / LONDON : ENGLAND

Library of Congress Cataloging in Publication Data

Main entry under title:

Extragalactic radio sources.

Includes indexes.
1. Radio sources (Astronomy)–Congresses. 2. Galaxies Congresses. I. Heeschen, David S. II. Wade, Campbell M. III. International Astronomical Union. IV. International Union of Radio Science.
QB857.E94   522'.682   81-21163
ISBN 90-277-1384-7   AACR2
ISBN 90-277-1385-5 (pbk.)

Published on behalf of
the International Astronomical Union
by
D. Reidel Publishing Company, P.O. Box 17, 3300 AA Dordrecht, Holland

All Rights Reserved
Copyright © 1982 by the International Astronomical Union

Sold and distributed in the U.S.A. and Canada
by Kluwer Boston, Inc.,
190 Old Derby Street, Hingham, MA 02043, U.S.A.

In all other countries, sold and distributed
by Kluwer Academic Publishers Group,
P.O. Box 322, 3300 AH Dordrecht, Holland

D. Reidel Publishing Company is a member of the Kluwer Group

No part of the material protected by this copyright notice may be reproduced or utilized in any form or by any means, electronic or mechanical, including photocopying, recording or by any informational storage and retrieval system, without written permission from the publisher

Printed in The Netherlands

TABLE OF CONTENTS

Titles of invited papers are given in capital letters.

PREFACE	xi
ORGANIZING COMMITTEES	xii
LIST OF PARTICIPANTS	xiii
INTRODUCTORY LECTURE / J. Oort	1
LARGE SCALE EMISSION FROM RADIO GALAXIES / H. Andernach & R. Wielebinski	13
Evolutionary Tracks of Extended Radio Sources / J. Baldwin	21
Rotating Gas and the Shapes of Radio Sources / L. Sparke	25
Aperture Synthesis Observations of Cygnus A at 86.2 GHz / M. Birkinshaw & M. Wright	27
The Spectral Index Distribution of Cygnus A / P. Scott	29
Observations of Radio Galaxies and QSR with RATAN-600 / N. Soboleva & Y. Parijskij (poster)	33
A Possible Evolutionary Feature in the Spectra of Radio Galaxies / O. Slee (poster)	35
High Resolution VLA Observations of Quasars with Distorted Radio Structure / J. Stocke, W. Christiansen, & J. Burns (poster)	39
The Radio Spectrum Across Three Tailed Radio Galaxies / H. Andernach (poster)	41
VLA and Optical Mapping of the Quasar PKS 0812+020 / F. Ghigo, L. Rudnick, K. Johnston, P. Wehinger, & S. Wyckoff (poster)	43
What Bends Wide-Angle Tailed Radio Sources? / J. Burns, J. Eilek, & F. Owen	45
Nuclear Ejection--One Side at a Time / L. Rudnick	47
Time Dependent Energy Supply in Radio Sources and Morphology of Radio Lobes / W. Christiansen, A. Pacholczyk, & J. Scott (poster)	51
Multi-Frequency Polarization Studies of Radio Galaxies / G. Pooley	53
A Complete Sample of Radio Galaxies / P. Shaver, I. Danziger, R. Ekers, R. Fosbury, W. Goss, D. Malin, A. Moorwood, & J. Wall	55
Extended Structure in High-Redshift Radio Sources / P. Duffett-Smith & A. Purvis	59

Extended Optical Line Emission Associated with Radio Galaxies /
  W. van Breugel & T. Heckman.. 61
Extended Emission Lines in Radio Galaxies / R. Fosbury............. 65
Emission Lines: Sign of a New Energy Source? / D. De Young........ 69
Optical Inverse Compton Emission in Extragalactic Radio
  Sources / S. Okoye & O. Obinabo.. 71
Proton-Proton Collisions in Extragalactic Radio Sources /
  S. Okoye & P. Okeke (poster).. 75

A PRELIMINARY EXAMINATION OF THE EFFECT OF CLUSTER GAS ON
  TAILED RADIO GALAXIES / D. Harris.. 77

Radio-Optical Studies of a Complete Sample of Abell Clusters
  / B. Mills & R. Hunstead.. 85
Studies of a Complete Sample of Abell Clusters at 1400 MHz /
  R. White, F. Owen, & R. Hanisch (poster).................................... 87
Two Peculiar Radio Galaxies in A 1367 / G. Gavazzi & W. Jaffe
  (poster)... 89
Radio Observations at 1.4 GHz of Abell Clusters / C. Fanti,
  R. Fanti, L. Feretti, A. Ficarra, I. Gioia, G. Giovannini,
  L. Gregorini, F. Mantovani, B. Marano, L. Padrielli,
  P. Parma, P. Tomasi, & G. Vettolani (poster)................................ 91
The Radio Emission of Interacting Galaxies / E. Hummel,
  J. van der Hulst, J. van Gorkom, and C. Kotanyi (poster)................... 93
Stephan's Quintet Revisited / J. van der Hulst & A. Rots
  (poster)... 95

A MORPHOLOGICAL CLASSIFICATION OF CLUSTERS OF GALAXIES FROM
  EINSTEIN IMAGES / C. Jones & W. Forman....................................... 97

RADIO AND X-RAY STRUCTURE OF CENTAURUS A / E. Feigelson............ 107

Emission Regions in Centaurus A / R. Price & J. Graham............. 115
X-Ray Emission from Centaurus A / J. Terrell....................... 117
VLBI Observations of the Nucleus of Centaurus A / R. Preston,
  A. Wehrle, D. Morabito, D. Jauncey, M. Batty, R. Haynes,
  A. Wright, & G. Nicolson... 119

SYSTEMATICS OF LARGE-SCALE RADIO JETS / A. Bridle................. 121

Radio and X-Ray Observations of Large Scale Jets in Quasars /
  J. Wardle & R. Potash.. 129
4C 18.68: A QSO with Precessing Radio Jets? / A. Gower &
  J. Hutchings (poster).. 133
The Quasar Jet 4C 32.69 at 1.4 GHz / J. Dreher (poster)............ 135
Bent Jets in Radio Quasars / S. Neff (poster)...................... 137
The Jets in 3C 449 Revisited / T. Cornwell & R. Perley (poster).... 139
Recent WSRT and VLA Observations of the Jet Radio Galaxy
  NGC 6251 / A. Willis, R. Strom, R. Perley, & A. Bridle........ 141
NGC 4258: A Bent Jet in a Spiral Galaxy / R. Sanders.............. 145

KILOPARSEC SCALE STRUCTURE IN HIGH LUMINOSITY RADIO SOURCES
    OBSERVED WITH MTRLI / P. Wilkinson 149

A Suggested Classification and Explanation for Hotspots in
    Some Powerful Radio Sources / P. Kronberg & T. Jones 157
Hot-Spots in Luminous Extragalactic Radio Sources / R. Laing
    (poster) 161
Morphology and Power of Radio Sources / P. Scheuer 163
The Radio Jet of 3C 273 / R. Conway 167
Relativistic Beaming and Quasar Statistics / I. Browne & M. Orr ... 169
The Radio Core in 3C 236 / E. Fomalont, A. Bridle, & G. Miley
    (poster) 173
The Arcsecond Morphology of Compact Radio Sources / R. Perley 175
Highly Polarized Emission from the E-Hotspot in DA240 /
    S. Tsien & R. Saunders 177

SEYFERT GALAXIES / A. Wilson 179

Radio Continuum Observations of the Nuclei of Nearby Galaxies /
    R. Davies, A. Pedlar, & R. Booler 189
Radio Emission from the Seyfert Galaxy NGC 5548 / J. Ulvestad,
    A. Wilson, & D. Wentzel (poster) 191
Westerbork Observations of Low Luminosity Radio Sources /
    P. Parma 193
Radio Observations of Markarian 8 / D. Heeschen, J. Heidmann, &
    Q. Yin (poster) 195

SS433--OBSERVING EVOLUTION IN A PRECESSING, RELATIVISTIC JET /
    R. Hjellming & K. Johnston 197

The Compact Radio Structure of SS433 / R. Schilizzi, I. Fejes,
    J. Romney, G. Miley, R. Spencer, & K. Johnston (poster) 205
SS433: Periodic Changes in the Radio Structure of Scale
    Sizes 10^{16} cm / A. Niell, T. Lockhart, R. Preston, &
    D. Backer (poster) 207
Radiative Acceleration of Astrophysical Jets: Line-Locking in
    SS433 / P. Shapiro, M. Milgrom & M. Rees 209

MECHANISMS FOR JETS / M. Rees 211

Viscous Dissipation in Jets / M. Begelman 223
Simple Formula for Radio Jet Surface Brightness / R. Henriksen
    (poster) 227
Instabilities in Pressure Confined Beams and Morphology of
    Extended Radio Sources / A. Ferrari, S. Massaglia,
    E. Trussoni, & L. Zaninetti (poster) 229
Connections between Turbulence and Jet Morphology / G. Benford
    (poster) 231
Particle Acceleration in Radio Sources with Internal Turbulence
    / J. Eilek & R. Henriksen (poster) 233

Jets from Discs and Doughnuts / *P. Allan* (poster).................. 235
Vortex Accretion Funnel / Relativistic Beam Models of Double
  Radio Sources / *H. Scott & R. Lovelace* (poster).............. 237

INFRARED OBSERVATIONS OF RADIO GALAXIES / *G. Rieke*................ 239

THE NATURE OF THE ENERGY SOURCE IN RADIO GALAXIES AND ACTIVE
  GALACTIC NUCLEI / *F. Pacini & M. Salvati*...................... 247

BLACK HOLES AND THE ORIGIN OF RADIO SOURCES / *K. Thorne &
  R. Blandford*.. 255

Supercritical Accretion and Its Possible Relation to Quasars
  and Radio Sources / *D. Meier* (poster)....................... 263
Galactic Centers and Twin-Jets / *W. Kundt*......................... 265

X-RAY AND OPTICAL OBSERVATIONS OF QUASARS / *H. Tananbaum &
  H. Marshall*... 269

THE MILLIARCSECOND STRUCTURE OF RADIO GALAXIES AND QUASARS /
  A. Readhead & T. Pearson................................... 279

High Resolution Observations of the Quasar 3C 147 / *E. Preuss,
  W. Alef, I. Pauliny-Toth, & K. Kellermann* (poster)........... 289
Structural Evolution in the Nucleus of NGC 1275 / *J. Romney,
  W. Alef, I. Pauliny-Toth, E. Preuss, & K. Kellermann*......... 291
VLBI Observations of M87 / *M. Reid, J. Schmitt, F. Owen,
  R. Booth, P. Wilkinson, D. Shaffer, K. Johnston,
  & P. Hardee*... 293
Compact Radio Sources: Their Use and Size / *K. Johnston*........... 295
Spectral Shapes of Compact Extragalactic Radio Sources /
  S. Spangler.. 297
Polarization of the Compact Radio Structure of 3C 454.3 /
  W. Cotton, B. Geldzahler, & I. Shapiro..................... 301
A Millimetre/Submillimetre Study of Optically Selected Quasars
  / *W. Sherwood, G. Schultz, E. Kreysa, & H. Gemünd*........... 305
Detection of a Broad HI Absorption Feature at 5300 km sec^{-1}
  Associated with NGC 1275 (3C 84) / *P. Crane,
  J. van der Hulst, & A. Haschick*............................. 307
A Search for HI in Elliptical Galaxies with Nuclear Radio
  Sources / *L. Dressel, T. Bania, & R. O'Connell*.............. 309
Changes in the HI Absorption Line Spectrum of AO 0235+164 /
  M. Davis & A. Wolfe.. 311
Theoretical Models to Explain the Variable 21 cm Absorption
  Spectrum in AO 0235+164 / *A. Wolfe*......................... 313

VARIABLE RADIO SOURCES / *R. Fanti, L. Padrielli, & M. Salvati*...... 317

Interstellar Scintillations as A Tool for Investigations of
  Hyperfine Structure in Extragalactic Radio Sources /
  L. Ozernoy & V. Shishov (poster)........................... 325

TABLE OF CONTENTS

Radio Flux Flicker of Extragalactic Sources / *D. Heeschen*
(poster).. 327
Broadband Studies of Compact Sources / *T. Jones & L. Rudnick*
(poster).. 329
Polarization Variability of Some Compact Radio Sources /
M. Komesaroff, D. Milne, P. Rayner, J. Roberts, & D. Cooke
(poster).. 331
Cm-Wavelength Fluxes and Polarizations of Compact Extra-Galactic
X-Ray Sources / *M. Aller, H. Aller, & P. Hodge* (poster)....... 335
Rotating Structures in Extragalactic Variable Radio Sources /
H. Aller, P. Hodge, & M. Aller (poster)........................ 337
Depolarization of Extragalactic Radio Sources / *M. Inoue &*
H. Tabara (poster)... 339
The Optical Polarization of QSOs / *R. Moore*...................... 341

SUPERLUMINAL RADIO SOURCES / *M. Cohen & S. Unwin*................. 345

Superluminal Expansion of 3C 273 / *T. Pearson, S. Unwin,*
M. Cohen, R. Linfield, A. Readhead, G. Seielstad, R. Simon,
& R. Walker (poster)... 355
Superluminal Expansion of the Quasar 3C 345 / *S. Unwin* (poster).... 357
Superluminal Motion in NRAO 140 and a Possible Future Method
for Constraining H_o and q_o / *A. Marscher & J. Broderick*........ 359
Superluminal Expansion in 3C 179 / *R. Porcas*...................... 361
Relativistic Jets as Radio and X-Ray Sources / *A. Königl*......... 363
Compton Rockets: Radiative Acceleration of a Relativistic
Fluid / *S. O'Dell*.. 365
VLA Observations of the Palomar Bright Quasar Survey /
D. Shaffer, R. Green, & M. Schmidt............................... 367
Optical Spectra of Radio-Loud and Radio-Quiet Active Galactic
Nuclei / *D. Osterbrock*... 369
Optical Spectra and Radio Properties of Quasars / *B. Wills*....... 373
Characteristics of Nebulosity Associated with Parkes Quasars /
P. Wehinger, S. Wyckoff, & T. Gehren............................. 375

BL Lac OBJECTS AND THEIR ASSOCIATED GALAXIES / *D. Weistrop*....... 377

X-Ray Emission from BL Lac Objects: Comparison to the
Synchrotron Self-Compton Models / *D. Schwartz,*
G. Madejski, & W. Ku... 383
Evidence for Relativistic Motion in the Millisecond Structure
of BL LAC / *R. Mutel & R. Phillips*............................... 385
Mark III VLBI Observations of the Nucleus of M81 at 2.3 and
8.3 GHz / *N. Bartel, B. Corey, I. Shapiro, A. Rogers,*
A. Whitney, D. Graham, J. Romney, & R. Preston (poster)....... 387
Radio Observations of the Galactic Center / *D. Backer* (poster)..... 389
Extragalactic Radio Supernovae in NGC 4321 and NGC 6946 /
R. Sramek, K. Weiler, & J. van der Hulst (poster)............... 391

THE ANGULAR SIZE DISTRIBUTION OF RADIO SOURCES AT LOW FLUX
DENSITIES / *A. Downes*.. 393

THE EVOLUTION OF LINEAR SIZES / V. Kapahi & C. Subrahmanya......... 401

Hot-Spots and Radio Lobes of Quasars / G. Swarup, R. Sinha, &
    C. Salter... 411

THE OPTICAL AND INFRARED PROPERTIES OF 3CR RADIO GALAXIES /
    S. Lilly & M. Longair 413

Redshift Estimates for Distant Radio Galaxies Based on Broad-
    band Photometry / J. Puschell, F. Owen, & R. Laing............ 423
Radio Evolution in High Redshift Clusters / W. Jaffe............... 425
Colors of Radio Galaxies at High Redshifts / R. Windhorst,
    R. Kron, D. Koo, & P. Katgert................................ 427
A Study of Small Angular Size Ooty Sources / T. Menon.............. 433
A Comparison of the Structures of 3CR Quasars and Blank Field
    Radio Sources / F. Owen, J. Puschell, & R. Laing.............. 435

SPACE DISTRIBUTION OF QUASARS BASED ON OPTICALLY SELECTED
    SAMPLES / M. Schmidt & R. Green............................... 437

COSMOLOGICAL EVOLUTION OF QSOs AND RADIO GALAXIES FROM
    RADIO-SELECTED SAMPLES / J. Wall & C. Benn.................... 441

Gravitational Lenses and Cosmological Evolution / J. Peacock....... 451

THE INTERGALACTIC MEDIUM / A. Fabian & A. Kembhavi................. 453

Modelling the Gravitational Lens of the Double Quasar /
    P. Moore & S. Harding... 461
Superluminal Velocities of Compact Radio Sources: A
    Gravitational Lens Effect / J. Barnothy....................... 463

SYMMETRY IN RADIO GALAXIES / R. Ekers............................. 465

A CONSEQUENCE OF THE ASYMMETRY OF JETS IN QUASARS AND ACTIVE
    NUCLEI OF GALAXIES / I. Shklovsky............................. 475

OTHER PAPERS PRESENTED AT THE SYMPOSIUM........................... 483

SUBJECT INDEX... 485

OBJECT INDEX.. 487

PREFACE

IAU Symposium 97, Extragalactic Radio Sources, was held at Albuquerque, New Mexico August 3-7, 1981. It was co-sponsored by IAU Commissions 28, 40, 47 and 48 and by URSI Commission J. Financial and organizational support were provided by the National Radio Astronomy Observatory, the University of New Mexico, and the National Science Foundation.

A wide variety of interesting objects and phenomena can be covered under the heading "Extragalactic Radio Sources", and a diverse set of topics was in fact discussed at the symposium. Radio galaxies, quasars, Seyfert galaxies and BL Lacertids received the most attention, but normal galaxies, the galactic center, and even SS433 were also discussed. While the unifying theme of the symposium was radio emission, studies at all wavelengths--X-ray, UV, optical, IR, and radio--were included. In general, the emphasis was on individual objects and the physical processes associated with them, but there were also papers on statistical studies and cosmology.

The symposium was attended by 209 scientists from 18 countries. Twenty-seven invited papers and 121 short contributions were presented during the 5-day conference. Of the latter, 57 were poster presentations. Most of the invited and contributed papers are included in this volume. A full-day break midway through the scientific sessions was used by most participants to visit the VLA, 100 miles southwest of Albuquerque. That excursion concluded with a popular public lecture by Prof. Philip Morrison of MIT. His talk, Cosmic Waterfalls, Whirlpools and Fountains, was well attended and well received by the local community and by symposium participants.

It was left to the initiative (whim?) of discussion participants to write up and submit their discussion contributions. Few chose to do so. In addition, some of the discussion following contributed papers, though submitted, was deleted to conserve space. As a result, the discussions published herein are only a small, rather arbitrarily selected sample of the extensive and lively discussions that actually occurred.

We thank authors for the timely submission of their camera-ready manuscripts and for the fact that most really were "camera-ready", requiring little effort from the editors. Finally, we gratefully acknowledge the expert and dedicated assistance of Mrs. Phyllis Jackson in the preparation of this volume.

D. S. Heeschen
C. M. Wade

SCIENTIFIC ORGANIZING COMMITTEE

- R. Giacconi
- R. Ekers
- D. Jauncey
- K. Kellermann (Chairman)
- Yu. Parijskij
- I. Pauliny-Toth
- G. Setti
- A. Stockton
- G. Swarup
- J. Wall
- L. Woltjer

LOCAL ORGANIZING COMMITTEE

- J. Burns
- R. Hjellming
- R. Perley
- R. Price
- E. Rigby
- R. Sramek
- D. Swann
- A. Thompson (Chairman)
- C. Wade
- S. Weeke

LIST OF PARTICIPANTS

Acronyms for institutions are given at the end of this listing.

Abell, G. O.	Univ. of Calif., Los Angeles, *USA*
Allan, P. M.	Sterrewacht, Leiden, *NL*
Aller, H. D.	Univ. of Michigan, Ann Arbor, *USA*
Aller, M.	Univ. of Michigan, Ann Arbor, *USA*
Andernach, H.	MPIfR, Bonn, *FRG*
Baath, L. B.	Onsala Space Observatory, Onsala, *Sweden*
Backer, D. C.	Univ. of Calif., Berkeley, *USA*
Baldwin, J. E.	Cavendish Laboratory, Cambridge, *UK*
Balonek, T. J.	Univ. of Mass., Amherst, *USA*
Barnothy, J. M.	Evanston, *USA*
Bartel, N. H.	MIT, Cambridge, *USA*
Barthel, P. D.	Sterrewacht, Leiden, *NL*
Basart, J. P.	NRAO, Socorro, *USA*
Begelman, M.	Univ. of Calif., Berkeley, *USA*
Benford, G. A.	Univ. of Calif., Irvine, *USA*
Bignell, R. C.	NRAO, Socorro, *USA*
Bijleveld, W.	Sterrewacht, Leiden, *NL*
Birkinshaw, M.	Univ. of Calif., Berkeley, *USA*
Bjornsson, C. I.	Lick Observatory, Santa Cruz, *USA*
Booth, R. S.	NRAL, Jodrell Bank, *UK*
Bridle, A. H.	NRAO, Socorro, & Univ. of N. M., Albuquerque, *USA*
Browne, I.	NRAL, Jodrell Bank, *UK*
Burbidge, G. R.	KPNO, Tucson, *USA*
Burke, B. F.	MIT, Cambridge, *USA*
Burns, J. O.	Univ. of N.M., Albuquerque, *USA*
Callahan, P. S.	JPL, Pasadena, *USA*
Cavaliere, A. G.	Inst. Astron., Rome, *Italy*
Chu, Y.	Univ. of Sci. & Tech., Anhwei, *PRC*
Clark, B. G.	NRAO, Socorro, *USA*
Cohen, M. H.	Caltech, Pasadena, *USA*
Condon, J. J.	NRAO, Charlottesville, *USA*
Condon, M. A.	Charlottesville, *USA*
Conway, R. G.	NRAL, Jodrell Bank, *UK*
Cornwell, T. J.	NRAO, Socorro, *USA*
Costain, C.	DRAO, Penticton, *Canada*
Cotton, W. D. Jr.	NRAO, Charlottesville, *USA*
Crane, P. C.	NRAO, Green Bank, *USA*
Crane, P.	ESO, *FRG*
Daishido, T.	Waseda Univ., Tokyo, *Japan*

LIST OF PARTICIPANTS

Davies, J. G.	NRAL, Jodrell Bank, *UK*
Davies, R. D.	NRAL, Jodrell Bank, *UK*
Davis, M. M.	Arecibo Observatory, Arecibo, *USA*
De Young, D. S.	KPNO, Tucson, *USA*
Dent, W. A.	Univ. of Mass., Amherst, *USA*
Douglas, J. N.	Univ. of Texas, Austin, *USA*
Downes, A.J.B.	Cavendish Laboratory, Cambridge, *UK*
Dreher, J. W.	NRAO, Socorro, *USA*
Dressel, L. L.	NASA/GSFC, Greenbelt, *USA*
Duffett-Smith, P.J.	Cavendish Laboratory, Cambridge, *UK*
Eilek, J. A.	N.M. Inst. of Mining & Technology, Socorro, *USA*
Ekers, R. D.	NRAO, Socorro, *USA*
Epstein, E. E.	Aerospace Corp., Los Angeles, *USA*
Erickson, W. C.	Univ. of Maryland, College Park, *USA*
Fabian, A. C.	Inst. of Astronomy, Cambridge, *UK*
Fanti, C. G.	Instituto di Radioastronomia, Bologna, *Italy*
Fanti, R.	Instituto di Radioastronomia, Bologna, *Italy*
Feigelson, E. D.	MIT, Cambridge, *USA*
Feretti, L. G.	Instituto di Radioastronomia, Bologna, *Italy*
Ferrari, A. M.	Instituto di Cosmo-geofisica del C.N.R., Torino, *Italy*
Fomalont, E. B.	NRAO, Charlottesville, *USA*
Fosbury, R. A.	Royal Greenwich Observatory, Herstmonceux, *UK*
Gavazzi, G.	Lab. Fisica Cosmica, Milan, *Italy*
Gibson, D. M.	N.M. Inst. of Mining and Technology, Socorro, *USA*
Giovannini, G.	Instituto di Radioastronomia, Bologna, *Italy*
Goss, W. M.	Kapteyn Astronomical Institute, Groningen, *NL*
Gower, A. C.	Univ. of Victoria, Victoria, *Canada*
Gregorini, L.	Instituto di Radioastronomia, Bologna, *Italy*
Gull, S. F.	Cavendish Laboratory, Cambridge, *UK*
Harris, D. E.	CFA, Cambridge, *USA*
Heckman, T. M.	Steward Observatory, Tucson, *USA*
Heeschen, D. S.	NRAO, Charlottesville, *USA*
Henriksen, R. N.	Stanford Univ., Palo Alto, *USA*
Hjellming, R. M.	NRAO, Socorro, *USA*
Howard, W. E. III	NSF, Washington, D. C., *USA*
Hummel, E.	Univ. of N.M., Albuquerque, *USA*
Hutchings, J. B.	DAO, Victoria, *Canada*
Inoue, M.	Tokyo Astronomical Observatory, Tokyo, *Japan*
Jaffe, W. J.	NRAO, Charlottesville, *USA*
Johnston, J. K.	NRL, Washington, D. C., *USA*
Jones, C.	CFA, Cambridge, *USA*
Jones, T. W.	Univ. of Minn., Minneapolis, *USA*
Kapahi, V. K.	Tata Institute, Bangalore, *India*
Kardashev, I. S.	Institute for Space Research, Moscow, *USSR*
Kellermann, K. I.	NRAO, Green Bank, *USA*
Königl, A.	Univ. of Calif., Berkeley, *USA*
Kronberg, P. P.	DAO, Univ. of Toronto, Toronto, *Canada*
Kuhr, H.	Univ. of Ariz., Tucson, *USA*
Kundt, W. H.	Univ. of Bonn, Bonn, *FRG*
Laing, R. A.	NRAO, Charlottesville, *USA*

LIST OF PARTICIPANTS

Lilley, S. J.	Royal Observatory, Edinburgh, *UK*
Linsley, J.	Univ. of N. M., Albuquerque, *USA*
Little, A. G.	Univ. of Sydney, Sydney, *Australia*
Longair, M.	Royal Observatory, Edinburgh, *UK*
Lovelace, R.V.E.	Cornell Univ., Ithaca, *USA*
Machalski, J.	Jagiellonian Univ., Krakow, *Poland*
Mantovani, F.	Instituto di Radioastronomia, Bologna, *Italy*
Marcaide, J. M.	MIT, Cambridge, *USA*
Marom, A.	Haifa, *Israel*
Marscher, A. P.	Univ. of Calif., San Diego, *USA*
Meier, D. L.	JPL, Pasadena, *USA*
Menon, T. K.	Univ. of B.C., Vancouver, *Canada*
Mills, B. Y.	Univ. of Sydney, Sydney, *Australia*
Moffet, A. T.	Caltech, Pasadena, *USA*
Moore, P. K.	NRAL, Jodrell Bank, *UK*
Moore, R. L.	Caltech, Pasadena, *USA*
Moran, J. M.	CFA, Cambridge, *USA*
Morison, I.	NRAL, Jodrell Bank, *UK*
Morrison, P.	MIT, Cambridge, *USA*
Mutel, R.	Univ. of Iowa, Ames, *USA*
Neff, S.	NRAO, Charlottesville, *USA*
Nicolson, G. D.	Radio Astronomy Observatory, Johannesburg, *S. Africa*
Niell, A. E.	JPL, Pasadena, *USA*
Okeke, P. N.	Univ. of Nigeria, Nsukka, *Nigeria*
Okoye, S. E.	Univ. of Nigeria, Nsukka, *Nigeria*
Oort, J. H.	Sterrewacht, Leiden, *NL*
Osterbrock, D. E.	Lick Observatory, Santa Cruz, *USA*
Owen, F. N.	NRAO, Socorro, *USA*
O'Dea, C. P.	Univ. of Mass., Amherst, *USA*
Pacholczyk, A. G.	Steward Observatory, Tucson, *USA*
Pacini, F.	Osservatori Astrofisico, Florence, *Italy*
Parijskij, Y. N.	Special Astrophysical Observatory, Leningrad, *USSR*
Parma, P.	Instituto di Radioastronomia, Bologna, *Italy*
Peacock, J. A.	Royal Observatory, Edinburgh, *UK*
Pearson, T. J.	Caltech, Pasadena, *USA*
Pelletier, G. J.	Univ. de Grenoble, Grenoble, *France*
Perley, R. A.	NRAO, Socorro, *USA*
Pooley, G. G.	Cavendish Laboratory, Cambridge, *UK*
Porcas, R. W.	MPIFR, Bonn, *FRG*
Preston, R. A.	JPL, Pasadena, *USA*
Preuss, E.	MPIFR, Bonn, *FRG*
Price, R. M.	Univ. of N.M., Albuquerque, *USA*
Puschell, J. J.	Univ. of Calif., San Diego, *USA*
Radhakrishnan, V.	Raman Research Inst., Bangalore, *India*
Raimond, E.	NFRA, Dwingeloo, *NL*
Readhead, A. C.	Caltech, Pasadena, *USA*
Rees, M. J.	Inst. of Astronomy, Cambridge, *UK*
Reid, M. J.	CFA, Cambridge, *USA*
Rieke, G.	Steward Observatory, *USA*

Robertson, J. G.	Anglo-Australian Obs., Epping, *Australia*
Roberts, J. A.	CSIRO, Sydney, *Australia*
Roberts, M. S.	NRAO, Charlottesville, *USA*
Robinson, B. J.	CSIRO, Sydney, *Australia*
Romney, J. D.	MPIfR, Bonn, *FRG*
Ronnang, B. O.	Onsala Space Obs., Onsala, *Sweden*
Rots, A. H.	NRAO, Socorro, *USA*
Rudnick, L.	Univ. of Minn., Minneapolis, *USA*
Sanders, R. H.	Kapteyn Laboratory, Groningen, *NL*
Scheuer, P.A.G.	Cavendish Laboratory, Cambridge, *UK*
Schilizzi, R. T.	NFRA, Dwingeloo, *NL*
Schmidt, M.	Caltech, Pasadena, *USA*
Schreier, E. J.	CFA, Cambridge, *USA*
Schwartz, D. A.	CFA, Cambridge, *USA*
Scott, H. A.	Cornell Univ., Ithaca, *USA*
Scott, P. F.	Cavendish Laboratory, Cambridge, *UK*
Seaquist, E. R.	Caltech, Pasadena, *USA*
Setti, G.	Instituto di Radioiastronomia, Bologna, *Italy*
Shaffer, D. B.	NASA/GSFC, Greenbelt, *USA*
Shapiro, P. R.	Univ. of Texas, Austin, *USA*
Shaver, P. A.	ESO, Garching, *FRG*
Sherwood, W. A.	MPIfR, Bonn, *FRG*
Shklovsky, I. S.	Institute for Space Research, Moscow, *USSR*
Simon, R. S.	Caltech, Pasadena, *USA*
Slee, O. B.	CSIRO, Sydney, *Australia*
Smarr, L. L.	Univ. of Ill., Urbana, *USA*
Smith, M. D.	Univ. of Ill., Urbana, *USA*
Smith, R. M.	Mt. Stromlo Observatory, Canberra, *Australia*
Spangler, S. R.	NRAO, Socorro, *USA*
Sparke, L. S.	Univ. of Calif., Berkeley, *USA*
Sramek, R. A.	NRAO, Socorro, *USA*
Stannard, D.	NRAL, Jodrell Bank, *UK*
Stockton, A. N.	Institute of Astronomy, Honolulu, *USA*
Swarup, G.	Tata Institute, Bangalore, *India*
Tananbaum, H. D.	CFA, Cambridge, *USA*
Terrell, J.	Los Alamos Natl. Lab., Los Alamos, *USA*
Thompson, A. R.	NRAO, Socorro, *USA*
Thorne, K. S.	Caltech, Pasadena, *USA*
Trimble, V. L.	Univ. of Maryland, College Park, *USA*
Tsien, S. C.	Mullard Radio Astronomy Observatory, Cambridge, *UK*
Turtle, A. J.	Univ. of Sydney, Sydney, *Australia*
Ulvestad, J. S.	NRAO, Charlottesville, *USA*
Unwin, S. C.	Caltech, Pasadena, *USA*
Valentijn, E.	ESO, Garching, *FRG*
van Breugel, W.	KPNO, Tucson, *USA*
van Gorkom, J.	NRAO, Socorro, *USA*
Veron, M.-P.	ESO, Garching, *FRG*
Vestrand, W. T.	NRAO, Charlottesville, *USA*
Wade, C. M.	NRAO, Socorro, *USA*

LIST OF PARTICIPANTS

Walker, R. C.	NRAO, Charlottesville, *USA*
Wall, J. V.	Royal Greenwich Observatory, Herstmonceux, *UK*
Wardle, J.F.C.	Brandeis Univ., Waltham, *USA*
Wehinger, P. A.	Ariz. State Univ., Tempe, *USA*
Weistrop, D. E.	NASA/GSFC, Greenbelt, *USA*
White, R. A.	NASA/GSFC, Greenbelt, *USA*
Wielebinski, R.	MPIfR, Bonn, *FRG*
Wilkinson, P. N.	NRAL, Jodrell Bank, *UK*
Willis, A. G.	NFRA, Westerbork, *NL*
Wills, B. J.	Univ. of Texas, Austin, *USA*
Wills, D.	Univ. of Texas, Austin, *USA*
Wilson, A. S.	Univ. of Maryland, College Park, *USA*
Windhorst, R. A.	Sterrewacht, Leiden, *NL*
Wolfe, A. M.	Univ. of Pittsburgh, Pittsburgh, *USA*
Woltjer, L.	ESO, Garching, *FRG*
Wright, J. P.	NSF, Washington, D. C., *USA*
Wrobel, J. M.	NRAO, Charlottesville, *USA*
Wyckoff, S.	Ariz. State Univ., Tempe, *USA*
Yin, Q. F.	Beijing Univ., Beijing *PRC*
Zaninetti, L.	Instituto di Fisica Generale, Torino, *Italy*

CFA	Harvard-Smithsonian Center for Astrophysics
CSIRO	Commonwealth Scientific & Industrial Research Organization
DAO	Dominion Astrophysical Observatory
DRAO	Dominion Radio Astrophysical Observatory
ESO	European Southern Observatory
GSFC	Goddard Space Flight Center
KPNO	Kitt Peak National Observatory
MIT	Massachusetts Institute of Technology
MPIfR	Max-Planck-Institut fur Radioastronomie
NASA	National Aeronautics and Space Administration
NFRA	Netherlands Foundation for Radio Astronomy
NRAL	Nuffield Radio Astronomy Laboratory
NRAO	National Radio Astronomy Observatory
NRL	Naval Research Laboratory
NSF	National Science Foundation
JPL	Jet Propulsion Laboratory

INTRODUCTORY LECTURE

J.H. Oort
Sterrewacht, Leiden.

On the subject of radiogalaxies my most vivid reminiscence is of an afternoon in Santa Barbara Street with a marvellous meeting with Baade and Minkowski where they told me that they had discovered that Cygnus A had been identified with a faint galaxy.

What a perspective did this open for penetration to distances far beyond our reach up to that date, distances where effects of the structure of the Universe must become wonderfully important!

The deep impression made by this discovery was only to be rivaled by that felt when a few years later Ryle and his co-workers actually penetrated through observations of radio sources into a past when the Universe was *actually* different from the present.

New discoveries thereafter succeeded each other in rapid succession: The discovery by Moffet at Owens Valley of the large *double* structure of radio galaxies; the discovery of the polarization and the synchrotron nature of their radio emission; and next the discovery of the quasars by Schmidt and Greenstein. The quasars gave great expectations for cosmology. But they did more: they also opened our eyes to the essential role which extremely small and massive nuclei might play in the formation of large radio sources.

That these cores must be very *massive* is demonstrated by the often tremendous radio as well as X-ray powers of the radio galaxies. They must also be *small*, as is shown by the frequent occurrence of flat spectra and by the large variations in brightness and polarization on time scales of days (and sometimes even *hours*, as in the X-ray emission from NGC 5128). Perhaps the most conspicuous property of the core engines that produce the energy for the radio galaxies is their directionality as evidenced by the mostly two-sided radio lobes and the jets leading to them. This has led to the idea that the central engines would be rotating black holes surrounded by accretion disks. However, no massive black holes have ever been observed, and it remains possible that the actual source is something else.

Evidently it would be important to find *direct dynamical evidence* for nuclear black holes, and to determine their masses. So far the attempts have been unsuccessful. A few years ago it looked as if an increase in the velocity dispersion near the centre of M87 indicated the presence of a non-luminous central mass of roughly 3×10^9 $M_\odot$; but later observations did not confirm this. They only indicated an upper limit of about 1×10^9 $M_\odot$. Apparently, the Space Telescope will be needed to determine the mass of M87's central engine. That will be an exciting programme, which can doubtlessly be extended to other radio galaxies.

The gradual loss by radio galaxies as well as quasars of their status as an entirely separate class of objects has been an intriguing development. More and more evidence has shown their *resemblance to other types* of active galaxies, in the first place the Seyfert systems, which may have the same sort of central machine, but of smaller power. Double and triple structures closely resembling that of radio galaxies but on two orders of magnitude smaller scale have been discovered in several Seyfert spirals (Wilson & Willis 1980); this shows that they have collimated ejection.

Shklovskii (1978) has suggested that a mass measurement might be possible for the nucleus of NGC 1275. At 1.35 cm the nucleus is double, with a separation of $0\overset{''}{.}0014$, or 0.5 pc. The optical spectrum suggests that the N_1, N_2 and H_β emission lines might consist of two components whose velocities differ by about 800 km s^{-1}. Combination of these data led Shklovskii to a, clearly very hypothetical, mass estimate for the nucleus, of about 3×10^8 $M_\odot$, and to the suggestion that this may represent the mass of a central black hole.

A binary nucleus might also offer an explanation of the unusual further milliarcsec structure (Matveyenko *et al* 1980) and the equally unusual radio features in the 1-7 kpc range found by Noordam & de Bruyn (1981) after they had succeeded to subtract the bright central point source (Figure 1).

As it might well be that all large galaxies pass intermittently through stages of Seyfert activity it seemed possible that they all possess the same potency for a high energy production in their nuclei. It becomes meaningful then to extend the quest for massive black holes also to ordinary galaxies. Again, the Space Telescope may provide the means. In this connection it is intriguing that Seyfert-like activity has recently been found in the ultra-compact nucleus of M81 (cf. Peimbert & Peimbert 1981).

In connection with what I have just said I want to turn your attention for a few moments to the *centre of our Galaxy*. Here we do not need the extreme resolution of the Space Telescope to penetrate to the core.

In principle the existence of a dark mass in the centre can be discovered if from the motions and density distribution of objects very close to the centre the total mass within, say, 0.5 pc can be determined.

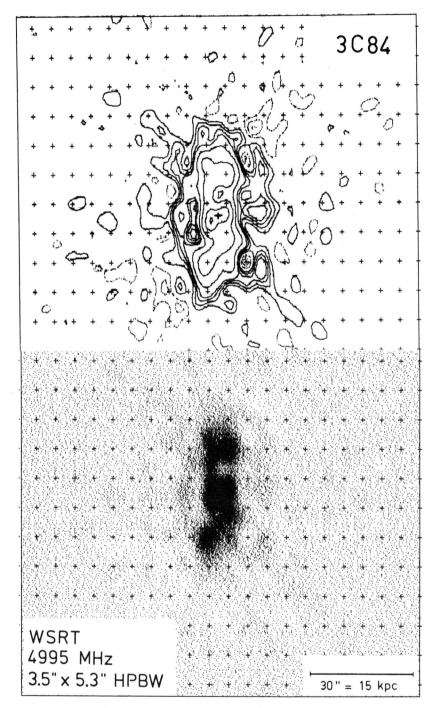

Figure 1. 3C 84 = NGC 1275 at 6 cm after "removal" of 59600 mJy core at + sign. Lowest contours about 1/36000th of original peak. North is on top, East to the left. Below: radio picture (Noordam & de Bruyn 1981).

Members of Townes' group have attempted to do this from the dozen fast moving clumps of ionized gas which they discovered in the 12.8-μ NeII fine structure line. Their observations indicated a total mass of 5.4×10^6 $M_\odot$ within 0.5 pc (cf. Lacy et al 1979, 1980). The total *stellar* mass within this radius may be inferred from the 2.2-μ radiation as measured by Becklin & Neugebauer (1968). This gives 2.0×10^6 $M_\odot$ (Sanders & Lowinger 1972; Oort 1977). The comparison between the two numbers shows that there might be a central black hole of 3×10^6 $M_\odot$. Clearly the reasoning leading to this result involves some quite uncertain suppositions; the outcome is therefore no more than a marginal indication that a black hole with a mass of at most $\sim 3 \times 10^6$ $M_\odot$ might be present.

The fast moving clumps present several interesting problems. What is their origin? What causes their rather peculiar state of ionization, and what happens with the large quantity of gas produced by their decay? As Lacy et al (1980) point out a black hole may help to solve these problems, but alternative solutions also exist.

A further interesting feature at the galactic centre is the radio source Sgr A West, and the extremely compact source contained in it (diameter $\sim$ 80 light-minutes). This source lies within 1 or 2" from the dynamical centre indicated by the NeII observations; it is still uncertain whether the two positions co-incide exactly.

The galactic centre appears also to be a remarkably strong source of radiation in the positron-electron annihilation line at 511 keV. It is unknown how the large quantity of positrons is produced; but the NeII clumps provide suitable regions for their annihilation, and give a natural explanation of the unexpected narrowness (half width 2 keV) of the observed line. It has been claimed that the γ-radiation at 511 keV is variable on a time scale of the order of half a year. Confirmation of this variability would evidently be of utmost importance. It is a desideratum deserving highest priority (Riegler et al 1982 and ref. therein).

In the further surroundings of the centre there are a number of large expanding features, partly in molecular form, partly in HI. The masses involved are or the order of 10^8 $M_\odot$. The molecular clouds lie in an inclined layer making a considerable angle with the galactic plane.

The cause of the anomalous motions and positions is still uncertain. One possibility is that they are due to expulsion from the surroundings of a rotating black hole whose axis makes a large angle with the galactic axis. Such a possibility is suggested by what we observe on a much grander scale in radio galaxies, where ejection of large masses in directions making large angles with the symmetry axis of the central galaxy is not uncommon. It is still more strongly suggested by the anomalous arms observed in NGC 4258 (see below) which are believed to be due to expulsion from the nucleus in two opposite directions in the equatorial plane of this spiral or at small angles with it. In this case the anomalous arms are ionized, and they emit synchrotron radiation, which is not observed in our Galaxy.

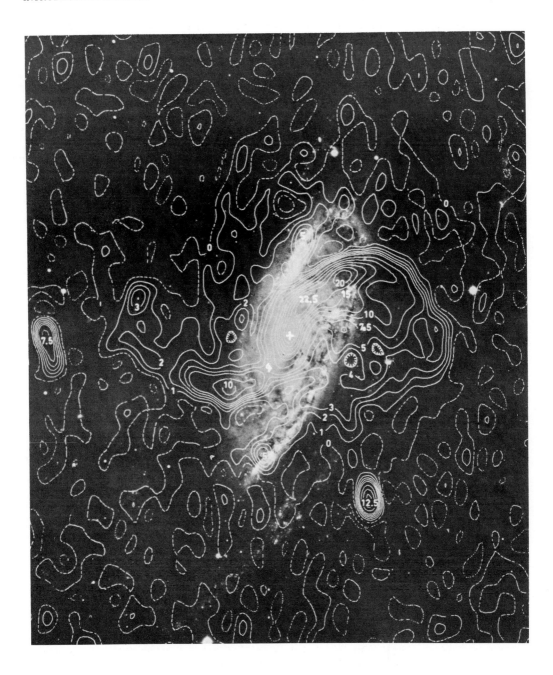

Figure 2. NGC 4258, with radio contours at 1415 MHz. Contour units 0.9 °K brightness temperature. Scale 1 cm = 48". The position of the optical nucleus is indicated by a white cross (van der Kruit et al 1972).

In the case of our Galaxy alternative interpretations may be possible, such as that suggested by Burton & Liszt, who interpret the observations by a kinematic model involving strongly excentric orbits (Liszt & Burton 1980).

That large masses of gas are sometimes expelled from the nuclei of "common" galaxies is amply shown by the Seyfert phenomena, but in particular also by NGC 4258. This is a bright spiral, at an estimated distance of 6.6 Mpc, which in optical appearance does not differ in any way from a normal massive Sbc system, with an indication of a central bar. However, when observed in the radio continuum it shows a most remarkable spiral-like structure of synchrotron emission which cuts almost perpendicularly through the optical spiral arms, and extends to the outermost limit of the gaseous disk (Figure 2). In the inner part the anomalous arms continue as thin filamentary HII arms right into the core (cf. Courtès & Cruvellier 1961; van der Kruit et al 1972; van Albada 1978, 1980). The anomalous arms are probably produced by two opposite streams of gas ejected from the nucleus which interact with the disk gas of the spiral, the synchrotron emission being due either to the ejected plasma itself, or to compression of the disk magnetic field, or to re-acceleration of relativistic particles in the shocks caused by the ejected gas. From an extremely crude model of the ejection it was found that between 10^7 and 10^8 solar masses of gas should have been ejected 20 - 40 million years ago at velocities ranging from 800 - 1600 km s^{-1}. Van Albada (1978) has pointed out that there is evidence of intermittent ejection, beginning with a large eruption about 40 million years ago and continuing in the same direction almost up to the present. As he mentions, there is some similarity with the ejections that give rise to head-tail radio galaxies, as well as to the common collimated double radio sources. In the most extensively studied head-tail galaxy, NGC 1265, the total mass in the tail has been estimated at 5×10^8 $M_\odot$ (Miley et al 1975), roughly an order of magnitude larger than the ejected mass in NGC 4258; for the ejection velocities values of 9000 km s^{-1} have been suggested in a model discussed by Owen et al (1978), again considerably higher than those which seem to have occurred in NGC 4258.

One reason why the peculiar phenomena that are so striking in NGC 4258 do not occur more often in spirals may be that they can occur only when the axis of ejection (presumably the rotation axis of the central black hole's accretion disk) is almost perpendicular to the rotation axis of the galaxy.

As there is no sign of any present activity in the nucleus of NGC 4258 the ejection has now probably ceased.

A closer comparison with radio galaxies will have to await better models, for which calculations are presently being made by van Albada (1981).

NGC 1275 may, in a way, be similar to NGC 4258 in the queer shaped anomalous synchrotron emission pattern shown in Figure 1.

Most interesting evidence showing that nuclei of some *normal* E and So galaxies eject matter – or at least relativistic particles – in narrow cones extending from ~ 1 pc to kpcs has recently been given by Jones, Sramek & Terzian (1981). The observations strongly suggest that these galaxies contain something like black holes with rotating accretion disks. They also suggest that these might be of common occurrence.

I must now turn to the radio galaxies proper. Having never done any direct research in this field I feel very little qualified to introduce the subject to an audience of experts. The only justification I can claim is that my lack of specialist knowledge makes me free from preoccupation with any particular aspect.

Let me therefore give a list of some of the principal problems (see next page) as they present themselves to a greatly interested outsider, who hopes to get a little understanding of many of these questions in the present conference.

For the problems of cores and jets reference should be made to a most illuminating review prepared by Rees *et al* (1981) for the Tenth Texas Symposium on Relativistic Astrophysics (December 1980).

It seems that the best information on problem 14 will come from the tailed radio galaxies, where the approximately known motion of the galaxy relative to the cluster gas gives us a hold on the travel times through the tails. Combination with density estimates from Faraday rotation yields a rough idea of the flow. The, still quite uncertain, present data indicate flows of the order of 1 $M_\odot$ per year.

Finally, the problem of *evolution*. This is perhaps the most important subject connected with quasars and radio galaxies: their own evolution, and the evolution of the galaxies from which they emerge (on this latter very interesting question (cf. Katgert *et al* 1979) I want to refer to a communication to be presented by Windhorst, Kron & Koo).

The enormous increase in number density of quasars with increasing redshift must reflect the birth rates of galaxies. It also indicates that the quasar phenomenon develops during the galaxy's youth, and lasts only a small fraction of its lifetime. From the steepness of the drop in proper density between $z \sim 2$ and $z \sim 1$ we may conclude that the average duration of the quasar cannot be longer than 10^8–10^9 years.

The very large number density of quasars in the past may indicate that a large fraction of large ellipticals – if not all – have gone through a quasar stage and therefore contain a massive "black hole".

Recent observations indicate that *non-quasar radio galaxies* show similar strong evolution effects. They may well be descendants of the early quasars. The fact that old elliptical galaxies are sometimes fairly strong radio emitters shows that radio emission is either recurrent, or that it can last during the whole life of an elliptical up to the present time (possibly due to situation in the centre of a cluster?).

A LIST OF QUESTIONS

Problem	Sub-problems	Suggested answers
THE CENTRAL ENGINE		
1. total energy	energy of relativistic protrons? equipartition? source life times? γ-ray power of 3C273	accretion by massive compact objects; how massive? (up to 10^9–10^{10} $M_\odot$?)
2. formation of "black holes"	prevention of fragmentation; disposal of excess angular momentum? restriction in mass? does b.h. form galaxy or does galaxy form b.h.?	Rees, I.A.U. Symp. No. 77, p.239. or: primordial black holes
3. expulsion in narrow cones		"nozzle"; but how to explain milliarcsec jets in same direction?
4. origin of accreting matter	infalling gas must get rid of angular momentum; must its spin axis coincide with that of black hole? how are "loss-cone" orbits refilled?	infall of cooling gas from surrounding cluster; merging galaxies; dynamical instability of galaxy.
5. why do strong radio sources always originate in ellipticals and never in spirals?		difference in gaseous environment?
6. long-term constancy of beam direction	divergences of radio axes from gal. minor axes, and probably also from *rotation* axes; how much matter must flow into nucleus during life of collimated structure?	rotation of very massive "black hole"

INTRODUCTORY LECTURE

7. Z-shape and cross-like sources — precession of rot. axis; caused by binary black hole? (which must then be fairly common); in some cases vel. gradient in intergal. medium

THE NUCLEAR REGION ($\lesssim 1$ pc)

8. variability on time scales of hours to years (optical, radio, X-ray; polarization) — longer-term (10-20 y) decrease in brightness; variation at low frequency — outbursts; possibly, absorbing nuclear clouds (but how can these exist?)

9. superluminal expansion — is it restricted to cases of beams near line of sight? — ultra-relativistic ejection; or beam sweeping over screen

10. milliarcsec (pc) jets — are they always one-sided? — ultra-relativistic

JETS

11. why do we see so many one-sided jets (M87, 3C273, NGC 6215, NGC 315, etc.?) — intrinsic one-sided ejection; visibility of jet by interaction with intergalactic medium — ultra-relativistic velocity (M87?); cloud structure in core; asymmetry of intergalactic medium

12. why are $\sim 1/5$ of radio quasars one-sided? — — merger?

13. anomalous optical jet in DA 240 — large angle with radio axis; 3000 km s^{-1} vel.?(is jet stellar?)

EXPULSION AND ACCRETION

14. rate and velocity of expelled matter — — from tailed sources: vel.1000-10000 km s^{-1}; ~ 1 M$_\odot$ per year?

RECURRENCE OF NUCLEAR ACTIVITY

15. multiple lobes in radio galaxies — what causes recurrence? — dynamical instability of galaxy; mergers

16. what evidence do Seyfert galaxies give? — percentage Seyferts

17. what is time scale? — $10^6 - 10^8$ years? Activity in NGC 4258 may indicate similar time scale

18. is quasar stage recurrent? — how long does it last? — steep increase in number density suggests occurrence in youth of galaxies, and life times < 0.01 Hubble time

NORMAL GALAXIES

19. do all massive galaxies go through Seyfert stage(s)?

20. have they all gone through quasar stage? — quite possibly, in view of past frequency of quasars and present numbers of giant ellipticals

21. do most, or all major galaxies have central "black holes" ($10^6 - 10^8 M_\odot$)? — Space Telescope (M87, M31, M81, ..)

In this connection my last question is: Do all massive ellipticals go through one or more quasar or radio galaxy stages? If not, and radio emission is therefore a characteristic of only a fraction of the massive E galaxies, what gives them this distinction?

I am much obliged to Colin Norman, George Miley, Ger de Bruyn, Harry van der Laan and several other members of the Leiden staff for enlightening discussions.

REFERENCES

Becklin,E.E.,and Neugebauer,G.:1968,Ap.J. 151,p.145.
Courtès,G.,and Cruvellier,P.,1961:C.R. Acad. Sci. Paris 253,p.218.
Jones,D.L.,Sramek,R.A.,and Terzian,Y.:1981,Ap.J.(Letters) 247,p.L57.
Katgert,P.,de Ruiter,H.R.,and van der Laan,H.:1979,Nature 280,p.20.
Lacy,J.H.,Baas,F.,Townes,C.H.,and Geballe,T.R.:1979,Ap.J.(Letters)
    227,p.L17.
Lacy,J.H.,Townes,C.H.,Geballe,T.R.,and Hollebach,D.J.:1980,Ap.J.241,p.132.
Liszt,H.S.,and Burton,W.B.:1980,Ap.J. 236,p.779.
Matveyenko,L.I.,Kellermann,K.I.,Pauliny-Toth,I.I.K.,Kostenko,V.I.,
    Moiseev,I.G.,Kogan,L.R.,Witzel,A.,Preuss,E.,Ronnang,B.O., and
    Shaffer,D.B.:1980,Pis'ma Astron.Zh. 6,p.77.
Miley,G.K.,Wellington,K.J.,and van der Laan,H.:1975,Astron.Astrophys.
    38,p.381
Noordam,E.,and de Bruyn,A.G.:1981,submitted to Nature.
Oort,J.H.:1977,Ann.Rev.Astron.Astrophys. 15,p.295.
Owen,F.N.,Burns,J.O.,and Rudnick,L.:1978,Ap.J.(Letters) 226,p.L119.
Peimbert,M.,and Torres-Peimbert,Silvia:1981,Ap.J. 245,p.845.
Rees,M.J.,Begelman,M.C.,and Blandford,R.D.:1981,Tenth Texas Symposium
    on Relativistic Astrophysics,Ann.New York Ac.Sc.,in preparation.
Riegler,G.R.,Ling,J.C.,Mahoney,W.A.,Wheaton,W.A.,Willett,J.B.,Jacobson,
    A.S.,and Prince,T.A.:1981,Ap.J.(Letters) 248,p.L13.
Sanders,R.H.,and Lowinger,T.:1972,Astron.J. 77,p.292.
Shklovskii,I.S.:1978,Pis'ma Astron. Zh. 4,p.493;
             Sov. Astr. Lett. 4,p.266 (1979).
van Albada,G.D.:1978,Doctor's Thesis, Leiden .
van Albada,G.D.:1980,Astron. Astrophys. Suppl. 39,p.283.
van Albada,G.D.:1981,in preparation.
van der Kruit,P.C.,Oort,J.H.,and Mathewson,D.S.:1972,Astron. Astrophys.
    21,p.169.
Wilson,A.S.,and Willis,A.G.:1980,Ap.J. 240,p.429.

LARGE SCALE EMISSION FROM RADIO GALAXIES

H. Andernach and R. Wielebinski
Max-Planck-Institut für Radioastronomie, Bonn, F.R.G.

Radio galaxies are known to exhibit a variety of scales in their structure. First we have the nuclear sources, which so far have not been completely resolved even on the scale of 1/10 milliarcsecond with VLBI observing methods (e.g. Preuss, 1981). Then we have the 'jets' (which at some stage break up into 'blobs') which are considered to transfer energy from the 'nuclear engine' to the outer heads. The latter appear to be the sites of transfer of the collimated jet energy into a diffuse emission region. Despite their usually low brightness these diffuse emission regions dominate the internal energy content in particles and fields, even for the collimated doubles. Note that only 1% of the total energy in Cyg A is in the hot spots (Perola, 1981).

The diffuse emission regions of the nearest radio galaxy, Cen A, subtend an angle of nearly 10° in the sky. There are a number of radio galaxies (e.g. 3C 236, NGC 6251, NGC 315) which subtend $\sim 1°$. An unfilled aperture instrument with a pencil beam, like the Mills Cross at a lower radio frequency, or a single dish telescope at high frequencies are well suited to study the extended emission. Synthesis telescopes in a very compact array (e.g. the 150 MHz Cambridge array, 610 MHz Westerbork system or 1420 MHz D-configuration of the VLA) can also be used to study the diffuse emission. One limit on the largest size of the object that can be studied is given by the field of view of the individual synthesis array elements, since the combination of separate fields has turned out to be rather difficult in practice. Another precaution which must be taken, especially when frequency comparisons for the determination of the spectral index are made, is the inclusion of the missing spacing information. Synthesis maps made at two frequencies with the same array, without the restoration of all the missing spatial components, cannot be used for spectral index studies. For both the pencil beam instruments and synthesis arrays the dynamic range imposes a crucial limit for studies of extended emitting regions. Bearing all the instrumental problems in mind, which lead to considerable selection effects, we can turn to the discussion of the accumulated data on diffuse emission in radio galaxies.

The steps in the accumulation of knowledge about radio galaxies are intimately connected with the increase of both angular resolution and sensitivity of radio telescopes. Cen A was the first object which could be studied (e.g. Sheridan, 1958) when the best angular resolution available was $\sim 1°$. The development of single dishes, like the Caltech 90-ft or Parkes 240-ft antennas, which operated at decimetre wavelengths giving resolutions of tens of arcminutes, resulted in the next step forward in the study of Cen A. The basic structure of a radio galaxy was then established: a nuclear source coincident with the E0 galaxy NGC 5128, the guided emission from the nucleus, a suggestion of tumbling motion in the outer diffuse regions. In the work of Cooper et al. (1965) observations of Cen A made at numerous frequencies from 19.7 MHz up to 4.8 GHz were used to discuss the spectral characteristics. A polarisation study showed very high degrees of linear polarisation. Cen A is shown in Figure 1, observed at 408 MHz by Haslam et al. (1981) with the Parkes telescope. In this map we see the foreground galactic radiation and in particular the peculiar 'spur' which was noted by Bolton and Clark (1960). We now know that this is not a bridge connecting Centaurus A to our Galaxy but a rather peculiar coincidence.

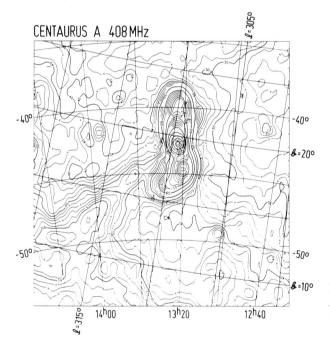

Figure 1:
A 408 MHz map of the Cen A region (from Haslam et al., 1981). (HPBW = 51')

The major advance of our knowledge was connected with the successful application of the synthesis technique to the mapping of radio galaxies. In a series of papers from Cambridge (e.g. Macdonald et al., 1968; Mackay, 1969) a large sample of extended sources from the 3C catalogue were mapped, revealing an astounding wealth of structural detail.

It must be noted here that the next steps of increasing angular resolution by the increase of baselines and frequencies resulted in a loss of capability of mapping large and diffuse emission regions. The Molonglo Cross could add new information on extended sources (e.g. Cameron, 1971; Schilizzi and McAdam, 1975). An important finder survey was the study of a complete sample of intense radio sources using the NRAO 300-ft telescope by Bridle et al. (1972). The Westerbork synthesis telescope made its impact in this field, particularly due to a considerable increase in the dynamic range, which resulted in the detailed study of numerous large radio galaxies (e.g. Miley and van der Laan, 1973; Miley, 1973; Willis et al., 1974). At the low frequency end of the spectrum the 150 MHz array in Cambridge (e.g. Waggett et al., 1977) and the 160 MHz Culgoora telescope (Slee, 1977) added necessary information. High frequency maps of radio galaxies have been made in Effelsberg (Baker et al., 1975; Hachenberg et al., 1976). In the southern sky only the Fleurs synthesis telescope (e.g. Christiansen et al., 1977) is available for mapping with < 1' resolution. Finally the VLA (e.g. Perley et al., 1979; Burns and Christiansen, 1980) has come on line to produce the beautiful maps of both the diffuse regions and the inner structures of numerous radio galaxies. All these efforts during the last two decades have shown that observations of each single radio galaxy at various angular resolutions were necessary to arrive at today's understanding of the structure of radio galaxies, which we will outline in the following paragraphs.

First it was the classical double sources (as example see Figure 2) with their strong outer heads aligned with the core source that instigated the twin-jet picture of their origin. In these sources one suspects lossless beams of relativistic plasma ejected along a line on both sides of the core which at a certain distance (a few hundred kpc) are randomized in a bow shock front. Magnetic field amplification due to compression of the plasma combined with particle acceleration via turbulent processes then leads to the enhanced emission of the hot spots. The diffuse tails extending back towards the nucleus can be viewed as the embers of former activity. This picture is confirmed by observation, since the heads are invariably observed with a flatter spectrum,

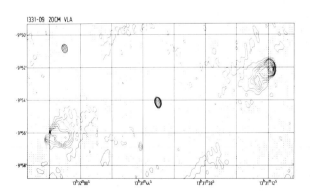

Figure 2:
A 1446 MHz VLA map of the large radio galaxy 1331-09. Contours are given at 4, 8, 12, 16, 20, 30, 50, 70, 90, 120 mJy/beam (HPBW = 22" x 13" in p.a. 13°). Polarisation bars show the E-vector and are proportional to polarised intensity. The overall extent is 1.7 Mpc (H_0 = 50 km s^{-1} Mpc^{-1}).
Courtesy of D.A. Graham.

while there is a continuous spectral steepening along the extended ridges towards the nuclei (e.g. Högbom, 1979). Most of the polarisation observations give additional support to this view (Laing, 1981; Dreher, 1981). At the outer boundaries a net field orientation perpendicular to the ejection axes is often observed showing a high degree of disorder as well (i.e. low percentage polarisation), whereas in the diffuse inward extensions the fields are mostly found to be well ordered and aligned with the main emission ridge. Thus it seems as if an originally random field were compressed at the shock front and, in the fainter regions, continuously stretched by shear forces, caused by the flow of matter relative to a surrounding medium (Laing, 1980).

Apart from the classical doubles we have complex non-aligned sources, which do not exhibit hot spots, and which have lower radio luminosities. Brightness peaks occur near the cores, while streamers of emission often extend far beyond these maxima fading away continuously. High resolution observations with present synthesis telescopes nevertheless suggest a common origin of classical doubles and complex sources. In what follows we like to summarise the effects which are thought to determine the large scale appearance of radio galaxies (cf. Miley, 1980).

First we must remember the omnipresent projection effects (e.g. Reynolds, 1980). Then a number of kinematic effects may play a role. One of these effects is the *precession* of the ejection axis, which can be inferred from the rotational symmetry of the ridges extending from some of the hot spots in classical doubles (cf. Figure 2). The effect becomes more pronounced with increasing opening angle of the precession cone like in NGC 326 (Ekers et al., 1978) or 3C 315 (Högbom, 1979). The ridges of the radio emission are then thought to delineate the path traced out by the end of the jet in the past. There are also sources which show a *reflection symmetry* about an axis through the nucleus. Instead of precession one invokes here an encounter of the parent galaxy and a neighbouring galaxy, as Blandford and Icke (1978) illustrated for the case of 3C 31. Depending on the time scale of the encounter compared to that of the ejection of matter, there is again a continuous variety of morphologies with reflection symmetry, starting with HB 13 (Masson, 1979) or IC 708 (Vallée et al., 1981) in which probably a single encounter is observed. Other sources (e.g. 3C 449, Lupton and Gott, 1981; or 3C 40, Andernach, 1981) show evidence for the parent galaxy orbiting around a massive neighbour. While such a model gave reasonable fits to the first few bends of 3C 31 and 3C 449, the outer lobes usually deviate from the expected trajectories. This effect may be due to the most frequent factor shaping extragalactic radio sources — the *motion* of the parent galaxy *relative to an ambient medium*. Again there is a continuous sequence of deviation from collinearity of the two-sided ejecta that starts with the wide-angle tails and ends with the comet-like head-tail radio galaxies. The preferred occurrence of these sources in rich clusters of galaxies is in fact independent evidence for the existence of a hot dense intracluster medium (ICM). This medium is also believed to confine the extended and less dense blobs of relativistic plasma via its static thermal pressure and thus prevent these plasmons from diffusion at the relativistic sound speed of $c/\sqrt{3}$.

Less evident possibilities affecting cluster source morphology are either *buoyant* bubbles of plasma rising against the density gradient of a denser ambient medium or, vice versa, heavy bubbles falling in a gravitational field through a less dense medium. Some of the distortions in radio structure could also arise from a *large scale shear* in some intergalactic winds, for which we do not have as yet any other evidence.

Finally we must mention the *core-halo sources* which at first glance do not fit into the outlined scenario. The number of halo sources known today is much less than previously claimed (e.g. Sramek, 1970; Fomalont, 1971), since most of them have been reclassified. When seen at higher resolution, the halo-like extensions either resolve into classical doubles (3C 103, 3C 105, 3C 236) or wide-angle tails (3C 40) or head-tails (3C 264, 2247+11). We are essentially left with the two famous halo sources Per A (Gisler and Miley, 1979; Reich et al., 1980) and Vir A (Andernach et al., 1979; Kotanyi, 1980). Since the mapping of just these halos is largely limited by dynamic range problems due to their strong cores, we might argue that either too little structural detail is as yet seen, or that due to projection we simply see the emission of one of their extended lobes in front of the core. High resolution observations of Vir A and Per A show them to have the same twin-jet origin as the aligned double sources.

With this picture of morphological origin of the sources we still have to explain what makes the particles radiate over Mpc scales. From synchrotron theory we expect the spectrum to steepen with increasing distance from the sites of particle injection (either nuclei or hot spots). This is only in qualitative agreement with most of the two-frequency spectral index maps produced so far. In most cases the observed spectral steepening is less than expected from synchrotron aging calculations, i.e. the bare existence of high frequency emission far away from nuclei or hot spots implies that in situ particle acceleration must take place over large parts of the extended lobes.

Theory predicts a spectral break at a frequency that depends on magnetic field, particle age and rate of injection of fresh particles. Investigations of the spectral behaviour across the sources on the basis of maps at various frequencies have only recently localized source regions with such a spectral break (Winter et al., 1980; Andernach, 1981). In the complex sources these regions are the faint outer streamers, as e.g. in the case of the wide-angle tailed radio galaxy 3C 465 (Figure 3). The spectral distributions (Figure 3c) show a steepening of $\Delta\alpha \sim 1.3 \pm 0.6$ between 0.6 and 10 GHz to occur in the components A, F, G and H. From the equipartition field strength and the break frequency we may derive an age of the emitting particle ensemble. Again this age, together with the projected core distance, gives a lower limit to the mean transfer speed of particles in the plasmon. Here and in many other cases this speed is found to be highly super-Alfvénic, which implies that particle diffusion cannot be the mechanism of energy

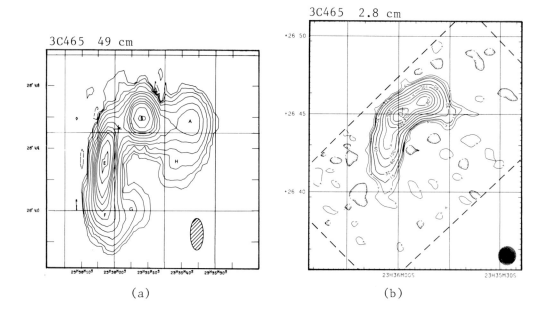

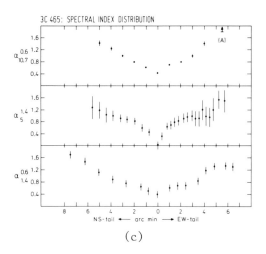

Figure 3: (a) 3C 465 as seen with the WSRT at 610 MHz (van Breugel, 1980); (b) 3C 465 as seen with the 100-m telescope at 10.7 GHz. Contours are at -5 (ticked), +5, 10, 15, 30 ... mJy per beam area (= hatched circle). The cross marks NGC 7720. (c) Spectral index distributions along the ridge line of the source defined in van Breugel (1980). The top figure results from a convolution of the 0.6 and 10.7 GHz data to 1.′2 × 2′ resolution, while the figures below are from van Breugel (1980) with a resolution of 23″ × 51″ and 0.′9 × 2′ (bottom).

transfer. The time scale derived from the spectral break must then be viewed as the time elapsed since injection of fresh particles has ceased in these components. If particle reacceleration is triggered by the plasmon's motion through the ambient medium, then the derived age may also be the time since ram pressure has stopped their motion. We see that even the existence of a spectral break in a source component does not imply the absence of particle injection throughout the source.

Finally we stress the need for in situ particle acceleration in the extended lobes of the so-called *relaxed doubles* like 3C 219 (Perley et al., 1980) or 3C 388 (Burns and Christianson, 1980). Here the older particles had time to diffuse out of the beams and hot spots to form extended envelopes around them, which do not seem to interact with the ICM. Recently Strom et al. (1981) have carried out a careful spectral index comparison across DA 240 using a 610 MHz Westerbork map and a 5 GHz Effelsberg map. They argue that neither departure from equipartition nor super-Alfvénic diffusion from the hot spots is likely to explain the lack of sufficient spectral steepening in the extended envelopes surrounding the hot spots. Stimulated by the fine scale structure observed in the lobes of the relaxed doubles 3C 310 and 3C 326 Smith and Norman (1980) have proposed a model in which shock waves are generated by supersonically moving knots accelerate particles. At higher resolution the extended lobes of DA 240 (Strom and Willis, 1981) also exhibit fine scale structure suggesting the plasmons to be in a turbulent state. Today we have more and more evidence for such fine structure from VLA maps of extended lobes (compare e.g. the maps of Cen A by Christiansen et al. (1977) and Schreier et al. (1981)). We may well ask whether in these lobes we start to resolve the shock waves proposed by the models. Thus with further multifrequency VLA observations of at least the brighter lobes of relaxed doubles there seems to be hope that the working mechanism for reacceleration can be tied down in the near future.

REFERENCES

Andernach, H.: 1981, Dissertation, University of Bochum
Andernach, H., Baker, J.R., von Kap-herr, A., Wielebinski, R.: 1979, Astron. Astrophys. 74, 93
Baker, J.R., Green, A.J., Landecker, T.L.: 1975, Astron. Astrophys. 44, 173
Blandford, R.D., Icke, V.: 1978, Monthly Notices Roy. Astron. Soc. 185, 527
Bolton, J.G., Clark, B.G.: 1960, Publ. Astron. Soc. Pacific 72, 29
Breugel, W.J.M. van: 1980, Astron. Astrophys. 88, 248
Bridle, A.H., Davis, M.M., Fomalont, E.B., Lequeux, J.: 1972, Astron. J. 77, 405
Burns, J.O., Christianson, W.A.: 1980, Nature 287, 208
Cameron, M.J.: 1971, Monthly Notices Roy. Astron. Soc. 152, 439
Christiansen, W.N., Frater, R.H., Watkinson, A., O'Sullivan, J.D., Lockhart, I.A., Goss, W.M.: 1977, Monthly Notices Roy. Astron. Soc. 181, 183

Cooper, B.F.C., Price, R.M., Cole, D.J.: 1965, Australian J. Phys. 18, 589
Dreher, J.W.: 1981, Astron. J. 86, 833
Ekers, R.D., Fanti, R., Lari, C., Parma, P.: 1978, Nature 276, 588
Fomalont, E.B.: 1971, Astron. J. 76, 513
Gisler, G.R., Miley, G.K.: 1979, Astron. Astrophys. 76, 109
Hachenberg, O., Fürst, E., Harth, W., Steffen, P., Wilson, W., Hirth, W.: 1976, Astrophys. J. 206, L19
Haslam, C.G.T., Klein, U., Salter, C.J., Stoffel, H., Wilson, W.E., Cleary, M.N., Cooke, D.J., Thomasson, P.: 1981, Astron. Astrophys. 100, 209
Högbom, J.A.: 1979, Astron. Astrophys. Suppl. 36, 173
Kotanyi, C.: 1980, Astron. Astrophys. 83, 245
Laing, R.A.: 1980, Monthly Notices Roy. Astron. Soc. 193, 439
Laing, R.A.: 1981, Monthly Notices Roy. Astron. Soc. 195, 261
Lupton, R.H., Gott III, J.R.: 1981, Astrophys. J. (in press)
Macdonald, G.H., Kenderdine, S., Neville, A.C.: 1968, Monthly Notices Roy. Astron. Soc. 138, 259
Mackay, C.D.: 1969, Monthly Notices Roy. Astron. Soc. 145, 31
Masson, C.R.: 1979, Monthly Notices Roy. Astron. Soc. 187, 253
Miley, G.K.: 1973, Astron. Astrophys. 26, 413
Miley, G.K.: 1980, Ann. Rev. Astron. Astrophys. 18, 165
Miley, G.K., van der Laan, H.: 1973, Astron. Astrophys. 28, 359
Perley, R.A., Willis, A.G., Scott, J.S.: 1979, Nature 281, 437
Perley, R.A., Bridle, A.H., Willis, A.G., Fomalont, E.B.: 1980, Astron. J. 85, 499
Perola, G.C.: 1981, Fund. Cosmic Phys. 7, 59
Preuss, E.: 1981, Proc. 2nd ESO/ESA workshop "Optical Jets in Galaxies", ESA SP-162, p. 97
Reich, W., Stute, U., Wielebinski, R.: 1980, Astron. Astrophys. 84, 204
Reynolds, J.E.: 1980, Proc. Astron. Soc. Australia 4, 74
Schilizzi, R.T., McAdam, W.B.: 1975, Mem. Roy. Astron. Soc. 79, 1
Schreier, E.J., Burns, J.O., Feigelson, E.D.: 1981 (in press)
Sheridan, K.V.: 1958, Australian J. Phys. 11, 400
Slee, O.B.: 1977, Australian J. Phys. Astrophys. Suppl. 43, 1
Smith, M.D., Norman, C.A.: 1980, Astron. Astrophys. 81, 282
Sramek, R.A.: 1970, Ph.D. Thesis, California Institute of Technology
Strom, R.G., Baker, J.R., Willis, A.G.: 1981, Astron. Astrophys. 100, 220
Strom, R.G., Willis, A.G.: 1981, Proc. 2nd ESO/ESA workshop "Optical Jets in Galaxies", ESA SP-162, p. 83
Vallée, J.P., Bridle, A.H., Wilson, A.S.: 1981, Astrophys. J. (in press)
Waggett, P.C., Warner, P.J., Baldwin, J.E.: 1977, Monthly Notices Roy. Astron. Soc. 181, 465
Willis, A.G., Strom, R.G., Wilson, A.S.: 1974, Nature 250, 625
Winter, A.J.B., Wilson, D.M.A., Warner, P.J., Waldram, E.B., Routledge, D., Nicol, A.T., Boysen, R.C., Bly, D.W.J., Baldwin, J.E.: 1980, Monthly Notices Roy. Astron. Soc. 192, 931

EVOLUTIONARY TRACKS OF EXTENDED RADIO SOURCES

J.E. Baldwin
Mullard Radio Astronomy Observatory, Cavendish Laboratory,
Cambridge, U.K.

We know almost nothing about the evolutionary tracks of extragalactic radio sources but those tracks are, however, strongly constrained by the distribution of sources in the radio luminosity, P, overall physical size, D, diagram. This is the radioastronomer's H-R diagram, an analogy which two lines of algebra shows is exact. Fig. 1 is the P-D diagram for the 3CR 166 source sample of Jenkins et al. (1977) with later additions. Most of the sources are identified and have known redshifts. It is a flux density limited sample so that the numbers at any P are weighted relative to the true space density by $P^{3/2}$ because of the differing volumes of space sampled. The important feature of the diagram is the lack of sources greater than 1 Mpc in size. Because of doubts about the completeness of the sample in this region, we have made searches in the 6C 151MHz survey for sources having surface brightnesses lying between the two lines of slope 2 on the right of Fig.1. The numbers found to a limiting flux density of 1-2 Jy suggest that there is no serious underestimate of the numbers in the 166 source sample.

Interpretation of the diagram is helped by the reasonable assumption that sources expand and their lifetimes are $<< H_o^{-1}$, which I think is true for all except the lowest P sources. The P-D diagram then reflects a steady state in which the density of points is, among other factors, $\propto$(velocity of expansion, V)$^{-1}$. If we adopt the simple view that the evolutionary tracks of all sources have the same shape, some tracks can be excluded. For example, P and V both constant leads to numbers of sources in successive decades in D in the ratio 1:10:100:1000, which is evidently untrue. To proceed further we need a physical model of radio sources and the only quantitative ones we have are those of Scheuer (1974). In his model A, two beams of constant cone angle impact on a uniform intergalactic medium giving hotspots in which particles are accelerated and then escape to fill a cigar-shaped bridge. The velocity of outward motion of the hotspots, $V \propto D^{-1}$ whilst the radio luminosity of the bridge, $P \propto D^{7/8}$. This model is a very bad fit to the points in Fig.1 because it gives a strong piling up of powerful, large sources. A cut-off in numbers at large sizes requires the sources to die rapidly so that we rarely see them dying; only two sources with $P > 10^{26}$ show

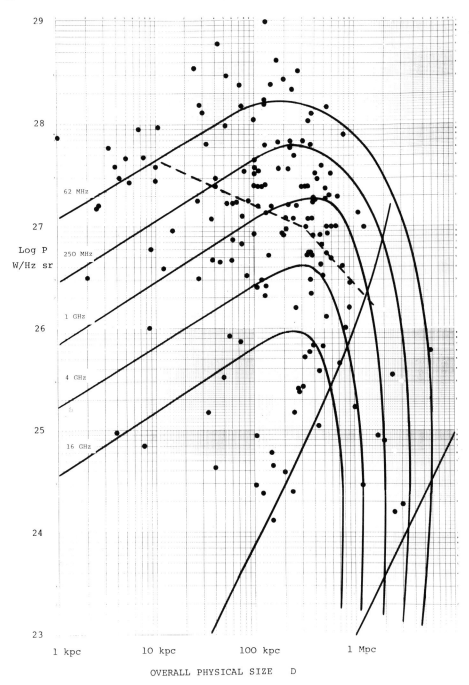

Fig 1. The P-D diagram for the 3CR 166+ sample.
Dashed line: Evolutionary track of a typical source.
Full lines : Surface brightness limits in 6C search.
Full curves: Synchrotron cut-off frequencies half way to hotspots.

no hotspots. It turns out to be difficult to kill them quickly enough; switching off the beams and allowing adiabatic expansion to take its toll is too slow.

So Scheuer's models need modification and there is one change which we know we ought to make. He took a uniform external medium purely for algebraic reasons and in the absence of other information. We now know from Xray observations that galaxies may have their own gaseous haloes and more extended atmospheres associated with a cluster. All models, and those Xray observations which have been reduced, show the density, ρ, falling increasingly steeply with radius. Scheuer's model is easily calculable for a power law variation of ρ and the corresponding dependence on D of P, V and the number, N, of sources in the P-D diagram per unit interval of log D is:

$$\rho \propto R^{\alpha} \; ; \; P \propto D^{(14+11\alpha)/16} \; ; \; V \propto D^{-(1+\alpha/2)} \; ; \; N \propto D^{(106+49\alpha)/32} \; .$$

In Fig.1, N increases slowly with D up to sizes of $\sim$ 300 kpc and thereafter falls rapidly. Using the density of points to deduce the best-fitting power law variation of ρ with R gives $\alpha = -1.9$ (10 kpc < D < 300 kpc) and $\alpha = -2.9$ (D > 300 kpc), values which are a good approximation to the density variation in an isothermal galactic atmosphere, with a radial scale similar to that of the Xray observations of M87 (see Fig.2). The shape of the corresponding evolutionary track is shown as the dashed line in Fig.1. The main conclusion from this analysis is that Scheuer's models still give a good account of the behaviour of radio sources and that the speed of expansion will be roughly constant over the range 10 < D < 500 kpc but is likely to be greater at both smaller and larger values of D.

In Fig.1 there is a dearth of sources with $P > 10^{26}$, D > 1 Mpc which would be required as the later stages of the higher P sources. Are they really there but disguised? To answer this we need to determine how fast the evolutionary tracks are traversed. The only good evidence is that of synchrotron ageing which several authors have used to deduce speeds, V, of $\sim c/20$. If we accept the models above with V almost constant, then the cut-off frequency at a specified point between the hotspot and the nucleus, is determined uniquely by P and D provided the geometry of all sources is the same and we ignore the contribution to P from the hotspots. Thus contours of constant ν_c on the P-D diagram can be plotted for any given value of V; c/20 was chosen in Fig.1. Note the very large range in ν_c which implies that small, low P sources should show no ageing effects whereas in high P sources the effects are so large as to remove the bridge completely except at very low frequencies. This gives a good account of the structures seen in doubles at high frequencies; all double sources with $P < 1.5 \times 10^{27}$ have bridges whilst for larger P there are no continuous bridges and at very high P even the short tails to the hotspots disappear. Such objects with large values of D would be hard to pick out and may have been missed. We have one or two candidates under suspicion.

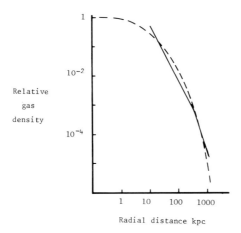

Fig 2. Relative gas density with radius in a galaxy.
Full line : Density deduced from P-D diagram and Scheuer's model.
Dashed line: Model atmosphere for galaxy with r_e = 30 kpc.

Overall, my conclusion is an optimistic one, that studies of complete samples of radio sources do provide a powerful method of investigating their evolutionary development.

REFERENCES

Jenkins, C.J., Pooley, G.G. & Riley, J.M.: 1977, Mem. R. astr. Soc., 84, pp 61-99.
Scheuer, P.A.G.: 1974, Mon. Not. R. astr. Soc., 166, pp 513-528.

DISCUSSION

LAING: Do you believe that the model of twin hot-spots moving the intergalactic medium applies to the weak sources (through jets and tails)?

BALDWIN: The right model must be different for those sources. Indeed, they grow bigger the harder one looks at them! My suspicion is that some of the jet sources could be in a steady-state and have an indefinitely large age.

SWARUP: What is the predicted effect of increasing distance on the size of radio lobes on hot spots in your model in which the density around the galaxy decreases as R^{-1} or R^{-2}?

BALDWIN: In Scheuer's original model with a uniform external medium, the bridge between the hot spots is cigar-shaped. For a density falling as R^{-2}, the bridge has a similar shape in the outer parts but is pinched to zero width near the central galaxy.

ROTATING GAS AND THE SHAPES OF RADIO SOURCES

L.S. Sparke,
Astronomy Department, University of California,
Berkeley, CA 94720.

The power for a strong extragalactic radio source comes from deep within the nucleus, but the extended radio structure is clearly related to the larger-scale properties of the galaxy in which it lives. Very large sources are found in elliptical rather than spiral galaxies, and big galaxies have stronger radio sources than small ones. The narrow jets mapped in weaker radio galaxies do not expand with a constant opening angle, but become better focussed along their length, suggesting that they are confined by an external pressure. This paper discusses how the rotation of a radio galaxy affects the distribution of gas within it, and consequently the radio structure in elliptical and Seyfert galaxies. A model is proposed which leads to a specific prediction, relating the width of radio jets to the rotation speed of the galaxy in which they lie.

Gas lost from aging galactic stars flows back into the interstellar medium, and is heated by collisions to a temperature approximately corresponding to the stellar velocity dispersion. This material can fall into the nucleus to fuel an accreting black hole; how much of it reaches the center depends on how hot it is, and how it rotates. If the galaxy rotates, the ejected gas shares that motion, and spins faster and faster as it nears the center. Centrifugal forces push the gas away from its rotation axis, leaving a partially emptied channel there. The angular width of this evacuated funnel increases with the galactic rotation speed V_R approximately as V_R/σ, where σ is the stellar velocity dispersion; this is 6-12° for an elliptical galaxy.

VLBI observations suggest that the energy for radio emission is supplied as a relativistic beam of plasma shot out from the innermost parsec of a radio galaxy; theoretical arguments indicate that this should point along the angular momentum vector of whatever material is falling onto the nucleus. In an isolated system this will be the galactic spin axis, so that the beam is injected along the centrifugally-emptied funnel. A fast-moving beam is unstable to disturbances at its interface with stationary material; the more so, the less the flow or the denser the surrounding gas. Growing instabilities may give rise to shocks, which disrupt the ordered motion and accelerate particles, producing a radio-bright 'hot spot'. If this process occurs within the emptied channel, and the ambient gas pressure is sufficient, radiating particles will be confined near the galactic spin axis. The radio-emitting plasma then drifts outwards, filling a jet, the width of which is set by the galactic rotation speed. In the more powerful radio sources, the beam must carry more energy and may be denser or faster, and so more stable against surface disturbances; the 'hot-spot' marking its disruption will occur further from the nucleus. Beams in the strongest souces will remain relativistic until they meet the intergalactic medium in a 'working surface' far outside the galaxy. In spiral galaxies, which contain interstellar gas a thousand times denser than that in ellipticals, the relativistic beam is likely to be disrupted before it has left the galactic bulge; dense galactic gas then confines the radiating particles to small double lobes similar to those observed in Seyfert galaxies (A.S. Wilson, this volume).

This model has a number of immediate consequences. Most obviously, galaxies with radio jets must contain gas extending beyond their nuclear regions, exerting sufficient pressure to balance that of the radiating plasma. Extended radio doubles should point along the spin axis of the gas confining them; this need not be the galactic rotation axis if the galaxy has recently merged or, for example, if the gas in the bulge of a disk system does not share the general galactic rotation. Radio jets and small double-lobed sources should show opening angles which increase with the rotation speed of the galaxy; this could easily be checked with presently-available optical telescopes and VLA radio maps. The jets should be focussed by the pressure of the surrounding gas, rather than expanding freely away from the nucleus.

Details of the calculations will appear in the Astrophysical Journal.

DISCUSSION

TOHLINE: It's dangerous to assume that the gas will settle to an orientation in which its spin axis aligns with the spin axis of the stars. As Tohline, Simson and Caldwell (1982, Ap. J., 251, 000) have argued, the gas--irrespective of the initial orientation of its angular momentum vector--will respond to the global shape of the galaxy's potential well and will align its angular momentum vector with the symmetry axis of a spheroidal galaxy. Cen A may be an example, and NGC 5363 more likely is, of a "tumbling prolate" galaxy in which the spin axis of the gas is orthogonal to the spin axis of the stars.

SPARKE: I agree that, in a galaxy which rotates about an axis which is not one of symmetry, the angular momentum of the gas may not be parallel to that of the stars; then, the radio jets should lie along the spin axis of the gas surrounding them. However, radio galaxies tend to be round, so that the gas and stars spin about the same axis.

APERTURE SYNTHESIS OBSERVATIONS OF CYGNUS A AT 86.2 GHz

M. Birkinshaw and M.C.H. Wright
Department of Radio Astronomy, University of California,
Berkeley, CA 94720

ABSTRACT

Recent observations of Cygnus A with the Hat Creek interferometer at 86.2 GHz limit the spectral curvature of the hotspots and show that the diffuse lobe emission has a spectral index of about 1.5. The central component is, at most, weakly variable and its spectrum shows a distinct break to a spectral index near 0.65 at about 20 GHz.

OBSERVATIONS

The improved 2-element Hat Creek interferometer was used over the period 1981 June - July at four EW antenna spacings from 30 - 120 m to map Cygnus A in order to establish the high-frequency spectral properties of the central component, the diffuse lobe emission and hotspots A and D (using the nomenclature of Hargrave & Ryle 1974). Two observations were made at each of the four antenna spacings, and almost complete 12-hr tracks of Cygnus A were obtained. The baselines were calculated from observations of a number of quasars with well-determined positions, and phase drifts were removed using frequent observations of the nearby point source V2005+403. The synthesized beam had half-power widths 4 x 6 arcsec, and the CLEANed map of Cygnus A had a noise level of about 70 mJy at its centre.

Two earlier syntheses of Cygnus A, using antenna spacings from 13 - 55 m have also been made, and have been used to measure the flux density of the central component on several dates. The 13-m data also provide good measurements of the integrated flux densities of the Np and Sf lobes of the source.

RESULTS

In the Sf component, only the hotspot D is seen clearly on the full-resolution map. D is unresolved (angular size < 2 arcsec) and has a spectral index (defined in the sense $S_\nu \propto \nu^{-\alpha}$) of 0.94 ± 0.10 from 15 to 86 GHz. The diffuse emission in the Sf lobe has a spectral index $\alpha 5^{86} = 1.4 \pm 0.1$ (the 5 and 15 GHz flux densities are taken from Hargrave & Ryle 1974 and 1976 respectively). In the Np lobe, component A is seen to be resolved with an angular size ~ 4 arcsec and a spectral index $\alpha 5^{86} = 0.82 \pm 0.08$. In this lobe, the diffuse emission has $\alpha 5^{86} = 1.7 \pm 0.2$. These hotspot and diffuse emission spectral indices agree with previous determinations at lower frequencies (Hargrave & Ryle 1976), indicating little spectral curvature over the range 5 to 86 GHz.

The central component, however, has a flux density of only 0.39 ± 0.05 Jy at 86.2 GHz (averaged over three epochs of observation), significantly below the result of about 1 Jy found at most other frequencies, and, in particular, significantly below the single-dish measurements at 35, 99 and 150 GHz (Hachenberg et al. 1976; Hobbs et al. 1978; Kafatos et al. 1980). This discrepancy is not produced by variability (on the evidence of our three independent observations of the flux density of the central component over an interval of 20 months) and is probably a result of the severe problems of measuring the flux density of the central component in the presence of the bright lobes of Cygnus A. Discarding the single dish results, the spectrum of the central component appears almost flat from 1.7 to 15 GHz, then breaks to a spectral index of 0.65 ± 0.16 from 23 to 86 GHz. This spectral shape resembles that produced by a single-component, isotropic source with a magnetic field ~ 2×10^{-7} T, a size ~ 0.001 arcsec (as found by VLBI; Kellermann et al. 1981) and an electron energy spectrum cut off below about 0.5 GeV.

REFERENCES

Hachenberg, O., Furst, E., Harth, W., Steffen, P., Wilson, W. & Hirth, W., 1976. Astrophys. J., 206, L19.
Hargrave, P.J. & Ryle, M., 1974. Mon. Not. R. astr. Soc., 166, 305.
Hargrave, P.J. & Ryle, M., 1976. Mon. Not. R. astr. Soc., 175, 481.
Hobbs, R.W., Maran, S.P., Kafatos, M. & Brown, L.W., 1978. Astrophys. J., 220. L77.
Kafatos, M., Hobbs, R.W., Maran, S.P. & Brown, L.W., 1980. Astrophys. J., 235, 18.
Kellermann, K.I., Downes, A.J.B., Pauliny-Toth, I.I.K., Preuss, E., Shaffer, D.B. & Witzel, A., 1981. Preprint.

THE SPECTRAL INDEX DISTRIBUTION OF CYGNUS A

P.F. Scott
Mullard Radio Astronomy Observatory, Cavendish Laboratory,
Madingley Road, Cambridge, England

Abstract. High dynamic range maps of Cygnus A at 2.7 and 5 GHz have been used to investigate the variation of spectral index over the extended parts of the source. Although both components show a steepening of spectral index away from the hotspots there is a marked asymmetry between the two components. This is interpreted as being due to a higher value of magnetic field in the S_f component.

We have recently reobserved Cygnus A with the Cambridge 5-km telescope at 5 GHz and 2.7 GHz in order to investigate the variation of spectral index over the extended parts of the source. Previous measurements (Hargrave & Ryle 1974; De Young, Hogg & Wilkes 1979; Dreher 1979) have indicated a steepening of the spectral index away from the hotspots but were limited by dynamic range or signal/noise to areas in the vicinity of the hotspots. Winter et al. (1980) were able to map the emission at 151.5 MHz across the whole source, but due to the absence of quadrature phase information, had to assume source symmetry. These earlier measurements have been generally consistent with synchrotron ageing of electrons left behind by the outward motion of the hotspots.

The new data consist of a 64-spacing map at 2.7 GHz and a 128-spacing map at 5 GHz. Separate measurements were made of I-Q and Q-U, providing both total intensity and polarisation distributions. The oversampling of the U-V plane (by a factor of $\sim$ 2) results in a virtually constant offset of the zero level over the area of the source, facilitating spectral comparisons. Standard point-source calibrators were used to determine system phases and the total source flux-density at each frequency has been normalised to the values given by Baars et al. (1977).

In addition to the standard Fourier inversion the data have

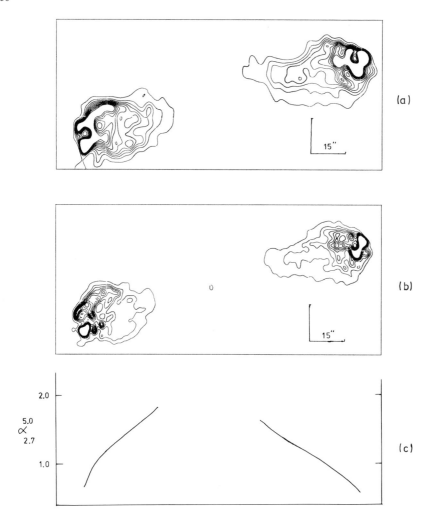

Figure 1. (a) and (b). Total intensity maps of Cygnus A at 2.7 GHz and 5 GHz. The lowest contour is at 44 mJy arcsec^{-2} and the contour interval 88 mJy arcsec^{-2}. (c) The variation of spectral index along the source axis.

been analysed by a two-stage maximum-entropy (ME) technique. At first ME map was used to re-determine values for the mean phase and x-y coordinates of the 8 telescopes for each 12-hour observation. A new ME map was then made incorporating these corrections. Tests on artificially perturbed data showed that this process rapidly converges to the original (i.e. 'correct') values. Two iterations were made on the 5 GHz data and one at 2.7 GHz. The resulting maps of total intensity are shown in

Fig. 1 (a) and (b), the first contour being at 44 mJy arcsec^{-2} and the contour interval 88 mJy arcsec^{-2} on each map. The central component has flux densities of 0.82±0.15 Jy and 0.89±0.10 Jy at 2.7 and 5 GHz respectively. These values do not differ significantly from those obtained in previous observations (at 2.7 GHz in 1976 and at 5 GHz in 1973) and do not confirm the variation found using the RATAN 600 telescope over a similar period (Soboleva 1981).

For spectral index comparison the ME maps have each been convolved to a resolution of 4 x 6 arcsec. The variation of spectral index along the major axis of the source, on the same scale as the contour maps, is shown in Fig. 1 (c). Both components show the trend of spectral index found in previous observations but there is a pronounced difference between the rate of variation in the two components, the S_f component having a more rapid steepening away from the hotspot. In addition the N_p component has a marked curvature away from the source axis which is even more obvious in the lower contours of the 2.7 GHz map, the emission near the central component lying at an angle of $\sim 45°$ to the line of the hotspots. The source asymmetries could be the result of a higher value of magnetic field in the S_f component which may be related to the much higher rotation measure found for this component (Mitton 1973). This would explain the more rapid steepening of spectral index and, because of the lower energy density in the N_p component, make this component more susceptible to distortion. Since the energy density in the extended region of this component is comparable with that in the gas which gives rise to the extended X-ray emission (Fabbioni et al. 1979) this distortion could be due to bouyancy effects.

REFERENCES

Baars, J.W.M., Genzel, R., Pauliny-Toth, I.I.K. & Witzel, A., 1977. Astr. Astrophys., 61, pp. 99-106.
De Young, D.S., Hogg, D.E. & Wilkes, C.T., 1979. Astrophys. J., 228, pp. 43-63.
Dreher, J.W., 1979. Astrophys. J., 230, pp. 687-698.
Fabbiano, G., Doxsey, R.E., Johnston, M., Schwarz, D.A. & Schwarz, J., 1979. Astrophys. J., 230, pp. L67-L72.
Hargrave, P.J. & Ryle, M., 1974. Mon. Not. R. astr. Soc., 166, pp. 305-327.
Mitton, S., 1973. Astrophys. Lett., 13, pp. 19-22.
Soboleva, N.S., 1981. Astrofiz. Issled., 14, pp. 15-71.
Winter, A.J.B., Wilson, D.M.A, Warner, P.J., Waldram, E.M., Routledge, D., Nicol, A.T., Boysen, R.C., Bly, D.W.J. & Baldwin, J.E., 1980. Mon. Not. R. astr. Soc., 192, pp. 931-944.

OBSERVATIONS OF RADIO GALAXIES AND QSR WITH RATAN-600

N. S. Soboleva and Y. N. Parijskij
Special Astrophysical Observatory of the
Academy of Science, Leningrad, USSR

We shall tell you briefly about the main observational program connected with radio galaxies. Different theories of radio galaxies predict different types of spectral index variations across the main body of the source. One would expect that the best solution of the problem is the construction of two-dimensional maps at a number of frequencies. However, we suggest that in some cases (i.e., for standard well-aligned structures) one-dimensional images with filled aperture may be much more accurate in determination of the variations of the spectral index along the major axes of radio galaxies. We now have 47 one-dimensional multifrequency images of all sources brighter than 1 Jy at centimeter wavelengths in the declination range $-43°$ – $+53°$ resolvable with our beam. Up to 7 frequencies were used (1.35, 2.08, 3.9, 6.5, 8.2, 13, and 31 cm). Cyg A is the best example showing structures of different scale: nuclear sources, bridges, main bubbles, and hot spots.

The main results are:

1. We have found no well established cases of any variation of the spectral index over the one-dimensional images, excluding the nuclear source. If they exist they are below the sensitivity of our measurements. The better the signal-to-noise ratio the smaller the upper limit on the spectral index variations.

The mean rms deviations of spectral index across the sources in our sample are below 0.04 (with dispersion of the integrated spectral indices 0.14) and are fully explained by the sensitivity of our maps.

2. For the classical doubles the flux density ratio of the main components is independent of frequency over the whole observable range to very high accuracy, even for the case of curved integrated spectrum. accuracy. We interpret this fact as direct evidence of small speed of separations of the components. For Cyg A, we estimated $v < 0.03$ c. In comparing our results with existing theories, we think that the best agreement may be found with in situ acceleration theories.

3. We have found "Cyg A features" around two superlight-velocity sources, 3C 273 and 3C 120. In both cases the position of the blobs coincide well with the jet-like features. The energy content in these blobs is much greater than in the bright portion of the sources.

We have now observed a much weaker sample of ~2000 RG and QSR in our deep sky survey program at a level ~1 mJy at 7.6 cm.

A POSSIBLE EVOLUTIONARY FEATURE IN THE SPECTRA OF RADIO GALAXIES

O.B. Slee
Division of Radiophysics, CSIRO, Sydney, Australia

1. INTRODUCTION

A recent analysis (Slee, 1981) of the spectra of extragalactic sources derived by Slee et al. (1981) has revealed a relationship between spectral index and redshift in the sense that the spectra of identified radio galaxies steepen with increasing redshift; this behaviour is consistent with an effect first reported by Kellermann et al. (1969) that the average spectral index of identified radio galaxies is lower than that of sources in empty optical fields, which are now generally regarded as distant radio galaxies (rather than QSOs). However, because of the selection effects present in the sample, Slee was not able to decide whether the basic correlation is between spectral index and redshift (implying an evolutionary origin), between spectral index and radio power, or between spectral index and linear dimension.

This paper draws attention to another characteristic of the radio galaxy spectra evident in the analysis of Slee et al. (1981), namely that third-degree polynomials fitted to the log s-log ν data of identified radio galaxies possess in many cases a minimum in slope at ~500 MHz; the effect is absent (or much less evident) in the spectra of the un-identified (blank field) sources. If, as seems likely, the unidentified sources are predominantly radio galaxies which formed at approximately the same epoch as the identified radio galaxies, the spectral differences between the two classes probably arise because of evolutionary processes. It seems probable that the correlation between spectral index (slope of the best-fitting first-degree polynomial) and redshift identified by Slee (1981) may be partly due to the presence of redshift-dependent structure in the spectra of radio galaxies.

2. THE AVERAGED BEST-FITTING THIRD-DEGREE POLYNOMIALS

Figure 1 shows the result of fitting third-degree polynomials to the log s-log ν data of Slee et al. (1981), to which the reader is referred for a full discussion of the properties of the radio source

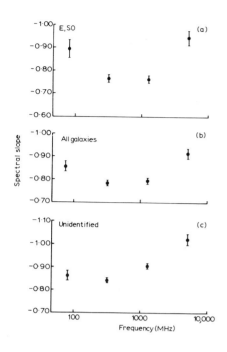

Figure 1. Average slopes of the best-fitting third-degree polynomials to the log s-log ν data of Slee et al. (1981). The slopes have been averaged at frequencies of 80, 320, 1280 and 5120 MHz; the vertical line about each mean has a half-length equal to its standard error. (a) refers to the results from 169 E, S0 radio galaxies, (b) to 404 sources identified with N, D, cD, E, S0, db and G galaxies, and (c) to 850 unidentified sources in blank optical fields.

sample, origins of the flux densities used in constructing the spectra and corrections to flux scales used by various observers. The points plotted are spectral slopes at four frequencies averaged over all sources in the identification class, together with the standard error of each average. It is clear that a minimum in the spectral slope occurs at ~500 MHz for the identified radio galaxies. The minimum is especially clear for the E, S0 class but is also very significant when all identified radio galaxies (N, E, S0, D, cD, db and G) are grouped. On the other hand, the Class III sources (blank optical fields), show no clear minimum in spectral slope; the average spectrum of an unidentified source tends to be straight at low frequencies but steepens progressively as frequency increases.

In order to investigate in more detail this apparent redshift-dependent trend in spectral shape, the identified radio galaxies have been subdivided into four intervals of redshift with approximately equal numbers of sources (~22) in each interval. Figure 2 shows the averaged spectral slopes at four frequencies in each redshift range resulting from fitting third-degree polynomials to E, S0 galaxies; a similar set of figures is obtained for the 'all galaxies' classification. It is clear that the averaged spectrum of the E, S0 radio galaxies in the two lower intervals of redshift show deep minima in spectral slope at ~500 MHz. The third redshift interval (Fig. 2c) shows a much less pronounced minimum while the highest redshift interval (Fig. 2d) yields an averaged spectrum with no evidence for a minimum; Figure 2(d) is in fact similar to 1(c), which applies to the unidentified sources.

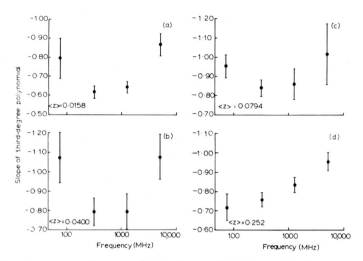

Figure 2. Average slopes of the best-fitting third-degree polynomials to the subset of 87 sources with measured redshifts drawn from the E, S0 identification class of Slee et al. (1981). (a) to (d) refer to sources grouped in successively higher ranges of redshift, whose average values are given in the lower left-hand corners. Each redshift interval contains ~22 sources.

It can be shown (see e.g. Slee et al., 1981) that the spectral minima in Figures 1(a) and 1(b) and Figures 2(a) and 2(b) are statistically significant. The effect is unlikely to be a spurious one, introduced by systematic differences in flux scales used by various observers; if this were the case then Figures 1(c) and 2(d), which are derived from the same sources of flux data, should also show significant minima.

3. DISCUSSION

The presence of minima in the spectral slopes of many of the closer radio galaxies is probably responsible for the correlation between spectral index (first-degree fit) and redshift (Slee, 1981). Slee pointed out that as redshift increases existing samples preferentially select increasingly more powerful radio galaxies; in addition, the subsets of resolved radio galaxies preferentially select those with increasingly larger linear dimensions as redshift increases. If the minima of Figure 2 are in fact properties of the less powerful and/or smaller radio galaxies and are not due to general evolutionary effects, then perhaps such sources have components with different spectra which combine to give a complex spectrum in the total flux density. Alternatively, evolutionary effects such as synchrotron losses, inverse Compton losses, bremsstrahlung losses, expansion losses and the injection, escape and acceleration of relativistic electrons may be more or less effective in the weaker/smaller sources than they are in the powerful/larger sources.

It is simpler (possibly simplistic) to interpret Figure 2 in terms of the general evolution of a radio galaxy irrespective of its power output or linear dimensions. If we can assume that all these radio galaxies were first formed within say 10^7 years, so that evolutionary processes within them can be considered to have started almost simultaneously, then significant changes have occurred in the time interval of 4.1×10^9 years (H_0 = 50 km s^{-1} Mpc^{-1}) represented by the averaged redshifts of the galaxies in Figures 2(d) and 2(b).

It is difficult to propose an explicit evolutionary model that could reproduce the observed spectral features; the theory of source evolution is still in rather a primitive state and treats only the simplest source geometries (Melrose, 1980) incorporating only one or two of the several possible evolutionary processes. According to the idealized model of Kellermann (1966) with recurring injections of flat-spectrum particles into a magnetic field and the subsequent influence on the spectrum of synchrotron losses (ignoring other evolutionary influences), the low-frequency spectrum reflects the flatter initial electron energy distribution in the particle injections. The intermediate frequency spectrum is steeper and reflects the presence of an electron energy range in which the synchrotron losses are balanced by the injection of fresh electrons; at higher frequencies the radio spectrum steepens further owing to the full effects of synchrotron losses.

The only spectra which approximate the predictions of Kellermann's (1966) model are those of Figures 1(c) and 2(d). If, as seems probable, this model applies to the most distant detectable radio galaxies, then there have been repeated injections of fresh electrons during the time interval of $\sim 5 \times 10^9$ years following the formation of these galaxies. A further $\sim 4 \times 10^9$ years corresponding to the spectra of Figures 2(a) and 2(b) is sufficient to show the effects of other evolutionary influences which have yet to be evaluated in a quantitative manner.

References

Kellermann, K.I.: 1966, Astrophys. J. 146, p. 621.
Kellermann, K.I., Pauliny-Toth, I.I.K. and Williams, P.J.S.: 1969, Astrophys. J. 157, p. 1.
Melrose, D.B.: 1980, In 'Plasma Astrophysics', Vol. 2, p. 106, Gordon & Breach.
Slee, O.B.: 1981, 'Some features in the spectra of extragalactic radio sources', Proc. Astron. Soc. Aust. (in press).
Slee, O.B., Siegman, Betty C. and Mulhall, P.S.: 1981, 'Spectra, dimensions and luminosities of radio sources in the Culgoora-3 list', Proc. Astron. Soc. Aust. (in press).

HIGH RESOLUTION VLA OBSERVATIONS OF QUASARS WITH DISTORTED RADIO STUCTURE

J. Stocke
Steward Observatory, University of Arizona

W. Christiansen
Morehead Observatory, University of North Carolina

J. Burns
University of New Mexico

ABSTRACT: Most quasars with extended radio structures have the classical linear double structure (Class II) which is also characteristic of the most luminous radio galaxies. A very small fraction of quasars do, however, show deviations from the classical double structure which are similar to the distortions observed in some low luminosity radio galaxies, e.g., bent double sources. Among radio galaxies, structural distortions seem to be associated with membership in a cluster of galaxies. Indeed, Scott and Hintzen, 1978, have suggested that such distortions might be indicators of cluster membership for quasars as well. To test this hypothesis and examine the physics of distorted quasars, we have mapped the quasars 3C 270.1 and 3C 275.1 in both I and P using the VLA A-Array at 20 cm and 6 cm. Our high resolution (0.1 arc sec at 6 cm) and high dynamic range (200:1) observations demonstrate that the displaced components reported previously for these sources are indeed connected with the quasars themselves and, therefore, the distortions are not simply projection effects. We also report the discovery of a very narrow northwesterly pointing jet in the quasar 3C 275.1.

REFERENCE

Scott, John S. and Hintzen, P. B., 1978, Ap. J. Lett., 224, L47.

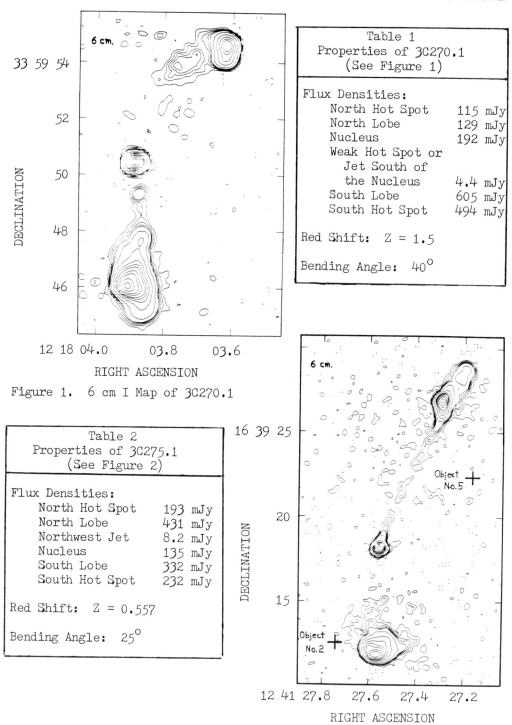

Figure 1. 6 cm I Map of 3C270.1

Table 1
Properties of 3C270.1
(See Figure 1)

Flux Densities:
  North Hot Spot   115 mJy
  North Lobe   129 mJy
  Nucleus   192 mJy
  Weak Hot Spot or
    Jet South of
    the Nucleus   4.4 mJy
  South Lobe   605 mJy
  South Hot Spot   494 mJy

Red Shift: Z = 1.5

Bending Angle: 40°

Table 2
Properties of 3C275.1
(See Figure 2)

Flux Densities:
  North Hot Spot   193 mJy
  North Lobe   431 mJy
  Northwest Jet   8.2 mJy
  Nucleus   135 mJy
  South Lobe   332 mJy
  South Hot Spot   232 mJy

Red Shift: Z = 0.557

Bending Angle: 25°

Figure 2. 6 cm I Map of 3C275.1

THE RADIO SPECTRUM ACROSS THREE TAILED RADIO GALAXIES

H. Andernach
Max-Planck-Institut für Radioastronomie, Bonn, FRG

Multifrequency observations with large single dishes are the ideal tool to examine the variation of the shape of the radio spectrum across extended extragalactic radio sources. Three complex low luminosity sources with angular extent $20' < \theta < 30'$ have been mapped with the 100-m telescope of the MPIfR at frequencies 2.7, 4.9 and 10.7 GHz (HPBW = 4!4, 2!6 and 1!2 resp.). To extend the frequency range we used published low frequency maps for the spectral comparison, too. Thus at least four maps with angular resolution $\leq 4!4$ were available for each source. All maps were (if necessary) corrected for sidelobes, then cleaned from obvious background sources and finally smoothed to the same beam of 4!4 HPBW. To look for spectral curvature a spectrum of the form $\ln S = a_1 + a_2 \ln \nu + a_3 (\ln \nu)^2$ was fitted to the brightness data for each sampling point of the map. We chose two parameters to characterize the spectral shape. The first is the *mean spectral index*, $\bar{\alpha}$, defined as the slope of the fit curve at the geometric mean of the lowest and highest observing frequency. As a measure of the spectral curvature we derived the *change of spectral index*, $\Delta\alpha$, along the fit curve between the lowest and highest observing frequency, i.e. $\Delta\alpha$ positive for a concave (= flattening) spectrum and vice versa. In the following the results are briefly summarized (see Andernach, 1981, for details).

Figure 1a shows our 10.7 GHz map of *3C40* in Abell 194. In this map the main part of the source appears as an asymmetric wide-angle-tailed radio galaxy with N547 as the parent galaxy. A separate component is due to emission from N541 and "Minkowski's object". The 2.7 and 4.9 GHz data suggest a highly polarized, steep spectrum radio bridge connecting N541 and the N545/7-system. The far northern tail is the only place in our sample of sources, where we could detect a spectral break in the radio spectrum as expected from synchrotron losses. The age derived from the break frequency and equipartition arguments corresponds to a particle transport speed of 2000 km s^{-1}, not too far from bulk velocities in other radio jets inferred from different arguments. The need for in situ particle acceleration is evident from Figure 1b, since the spectrum *flattens* with frequency (i.e. $\Delta\alpha > 0$) over a large extent of the source. This feature is also present in the two other sources, the

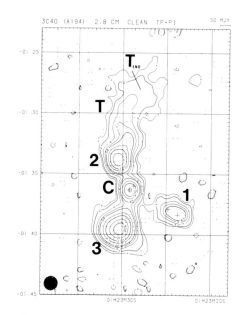 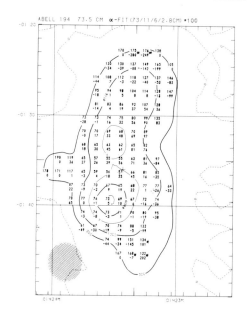

Figure 1: (a) 3C40 at 10.7 GHz with sidelobes removed. Contours in mJy/beam and E-vectors of polarized intensity are given. The crosses mark (in incr. R.A.) N541, Minkowski's object, N545 and N547. (b) Radio spectrum across 3C40 as derived from maps at 408 MHz (Schilizzi et al., 1972), 2.7, 4.9 and 10.7 GHz. $\bar{\alpha}$ is given above $\Delta\alpha$ (see text, each multiplied by 100). For a few points at the edge only two ($\Delta\alpha = 0$) or three frequencies (asterisk!) could be used.

twin-jet source *HB13* and the head-tail radio galaxy *2247+11 (N7385)*. For these sources we compared our data with low frequency maps of Masson (1979) and Schilizzi and Ekers (1975). In all three sources the observed flattening of the spectrum in the inner jets cannot be explained by the influence of the flat spectrum cores alone. Instead the data suggest the inner jets to have a rather flat high frequency spectrum ($\alpha \approx 0.4$). In the case of 2247+11 we observe this flattening even at the far end of its tail. Since our maps also indicate a bifurcation of the tail, we propose the plasmon with the flattening spectrum to be in turbulent motion (either falling or rising) through the ambient medium, leaving behind a secondary "detached" tail.

I thank the Deutsche Forschungsgemeinschaft for financial support under grants Re 304/7 and Re 304/9.

Andernach, H.: 1981, Dissertation, University of Bochum
Masson, C.R.: 1979, Monthly Notices Roy. Astron. Soc. 187, 253
Schilizzi, R.T., Lockhart, I.A., Wall, J.V.: 1972, Australian J. Phys. 25, 545
Schilizzi, R.T., Ekers, R.D.: 1975, Astron. Astrophys. 40, 221

VLA AND OPTICAL MAPPING OF THE QUASAR PKS 0812+020

F.D. Ghigo and L. Rudnick, University of Minnesota
K.J. Johnston, Naval Research Laboratory
P.A. Wehinger, Arizona State Univ. and Northern Arizona Univ.
S. Wyckoff, Arizona State University

Observations are reported of the remarkable object PKS 0812+020, the first quasar found to have optical emission in one of its radio lobes. A distinct radio jet is also seen, and there is radio and optical evidence that the quasar is near the center of a galaxy cluster.

Because the quasar PKS 0812+020 (z = 0.402) had a faint associated nebulosity (Wyckoff, Wehinger and Gehren 1981) a number of studies at radio and optical wavelengths were undertaken (Wyckoff et al. 1981). The number of diffuse images in the optical field suggests that the quasar may be in a galaxy cluster. New VLA observations at $\lambda\lambda 20$, 6, and 2 cm amplify the previous radio studies. Also, new optical astrometry on the deep ESO plates has allowed accurate relative positioning of the radio and optical maps, which has strengthened the case for a faint optical object ($m_R \approx 22$) being associated with the north lobe radio hot spots.

The north radio lobe has two hot spots, about 1" (5 kpc) apart, aligned roughly perpendicular to the main source axis. At high resolution (2 cm), emission is seen between the spots. A straight radio jet points from the quasar toward the stronger of the two north hot spots.

The polarization position angle differs by 90° between the spots. The spectral index in the stronger (west) spot is $\alpha \sim 1.1$ and in the weaker (east) spot is $\alpha \sim 1.5$. Faraday rotation is seen in the east spot, for which $\log N_e (cm^{-3}) \approx -5$ is estimated, but not in the west. The west spot thus appears to be a younger, higher energy place than the east, making it tempting to suggest that the west spot is where the beam is stopped by the intergalactic medium.

Assuming equipartition, the synchrotron lifetime for the extended optical emission is about 300 years, considerably less than the light travel time between the two hot spots. Acceleration must occur at many places in the north lobe. The north lobe optical spectrum shows a nonthermal continuum.

To study the extended radio halo, we removed the high surface brightness flux from the low resolution (5".2) 20 cm map by cleaning the 1".5 resolution 20 cm map of 250 components at 20% loop gain, restoring it with a 5".2 beam, and subtracting the result from the low resolution map (see Fig. 1). Intensity profiles through the resulting difference map appear consistent with two spherical to slightly shell-like distributions of emission centered 6" south and 7".5 north of the quasar. These profiles suggest largely static confinement by an intergalactic medium.

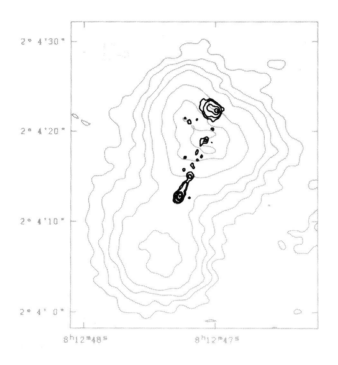

Fig. 1: Deconvolved 20 cm low surface brightness halo (5".2 beam) superimposed on 6 cm (0".5 beam; bold lines) map. Contour levels are λ20 cm: 8, 16, 25, 33, 49, 66, 78 mJy; λ6 cm: 2, 5, 30, 75, 180 mJy.

Work at U of Minnesota is supported in part by NSF grant 79-000304. The VLA (NRAO) is operated by Associated Universities, Inc., under contract with NSF. ESO and the Max Planck Gesellschaft have supported the optical observations.

REFERENCES

Wyckoff, S., Wehinger, P.A. and Gehren, T.: 1981, Ap.J. 247, 750-761.
Wyckoff, S., Johnston, K., Ghigo, F., Rudnick, L., Wehinger, P.A. and
    Boksenberg, A.: 1981, Ap.J. (in press).

WHAT BENDS WIDE-ANGLE TAILED RADIO SOURCES?

Jack O. Burns
University of New Mexico, Albuquerque, NM, USA
Jean A. Eilek
New Mexico Tech, Socorro, NM, USA
Frazer N. Owen
National Radio Astronomy Observatory, Socorro, NM, USA

It has been generally assumed that wide-angle tailed (WAT) sources like 3C465 are formed in a manner similar to that of the more strongly bent U-shaped sources such as NGC 1265, i.e., by ram pressure arising from galaxy motion through a dense intracluster medium (ICM). The WAT sources were thought to be less strongly bent because of the smaller ratio of tail plasma flow momentum flux to galaxy velocity. However, as noted recently by Burns (1981), there is a serious discrepancy between the ram pressure model requirements for bending WATs and the dynamics of the associated radio galaxy. To bend the tails, we calculate that the galaxy must typically move at velocities of $0.7-1 \times 10^3$ km s^{-1} for distances comparable to the length of the radio tails ($\sim$200 kpc for 3C465). This implied galaxy motion is inconsistent with the nature of the massive cD galaxies generally associated with WATs. Cluster galaxy velocity data, X-ray observations, and recent models suggest that these giant galaxies are nearly at rest at the bottoms of cluster potential wells, at most moving $\sim$200 km s^{-1} in an oscillatory motion of small amplitude (<0.3 of a core radius, Malumuth, 1981, private communication). Thus it appears that some other mechanism is responsible for bending WAT sources.

The actual bending of the radio tails results from an interaction between the ICM and the extended radio plasma. Pressure gradients within the ICM will distort the plasma flow from linearity. Such pressure gradients could be seen as asymmetries in the X-ray emission produced by the hot cluster gas. Therefore, by observing the distribution of X-rays from the ICM, one may be able to gain insight into the mechanism which bends the WAT sources. We undertook such X-ray observations of the A2634/3C465 field with the IPC on the Einstein Observatory. A comparison between the X-ray and VLA emissions (Eilek et al. 1981) for the inner core of A2634 is presented in the Figure. Unlike the large-scale structure, the inner X-ray emission has an anisotropic, egg-shape near the cD with the excess between the radio tails. The difference in X-ray emissivities at distances of 1'-2' on opposite sides of the cD (p.a.$\sim$+45°) is $\sim$20-25%. This corresponds to a density difference of $\Delta n \sim 5$% if the two sides of the galaxy have about the same temperature.

What is the origin of this gas anisotropy and what does it

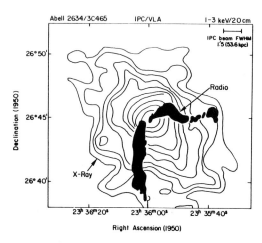

imply about the bending of the 3C465 tails? We consider briefly four models:

(1) <u>Buoyancy</u>--if the ICM is not spherical and the radio tails are less dense than the ICM, a buoyancy force will bend the radio streams away from regions of largest ρ_{ICM}. This is unlikely in the case of 3C465 since the radio tails appear to bend toward a region of enhanced X-ray emission.

(2) <u>Gravitational Bending</u>--"Heavy" tails will fall toward higher density regions in an anisotropic ICM. However, an excess mass of $\sim 10^{13}$ $M_\odot$ to the SW of 3C465 would be needed to bend the tails by gravity. No such mass is present in gas or in galaxies.

(3) <u>Motion through a Supercluster</u>--A2634/A2666 may form a supercluster, both at $z \sim .03$. If an intrasupercluster medium exists and A2634 is moving through that medium, it is conceivable that a ram pressure gradient transmitted through the A2634 medium might bend the radio tails. In this case, however, one would not expect to see distortions in just the inner parts of the cluster ICM, but possibly throughout.

The above models do not require motion by the cD, but none seem to be in agreement with the X-ray data. We consider below a model involving slow galaxy motion (~ 200 km s^{-1}) as a potential explanation of the X-rays.

(4) <u>Accretion with Slow Galaxy Motion</u>--There is evidence of radiatively induced accretion flows into the cores of rich clusters near cDs. If the galaxy associated with 3C465 is moving subsonically and accreting gas, might this produce the observed egg-shaped asymmetry? According to De Young, <u>et al</u>. (1980) who have studied this process for M87, $\sim 5\%$ density enhancement is expected on the downstream side of the galaxy. This is the same excess which is implied by our X-ray data for 3C465; however, this model must be tested for a massive galaxy with high accretion to see if the results scale from M87 to a cD.

If accretion with slow motion is the answer to the X-ray anisotropy, it still does not explain the radio tails since the pressure is only $\rho_{ICM} v_g^2$ which is small compared to that needed to bend the tails. The question remains: What bends wide-angle tailed radio sources?

References

Burns, J. O. 1981, <u>M.N.R.A.S.</u>, <u>195</u>, 523.
De Young, D. S., Condon, J. J., and Butcher, H. 1980, <u>Ap.J.</u>, <u>242</u>, 511.
Eilek, J. A., Owen, F. N., Burns, J. O., and O'Dea, C. 1981, in preparation.

NUCLEAR EJECTION - ONE SIDE AT A TIME

Lawrence Rudnick
Department of Astronomy
School of Physics and Astronomy
University of Minnesota

ABSTRACT

Examination of the structures of extragalactic radio sources shows a distinct asymmetry in addition to the more general symmetries which are well known. The most likely explanation for the observed asymmetries is that ejection from the active nucleus occurs in only one direction at a time. This direction then switches back and forth to form the large scale double structures. Implications of this picture for the nuclear engine and the radio source environment are discussed.

The well-known symmetry of extragalactic radio sources has become better appreciated in recent years (see review by R. Ekers, this volume). However, when examined in detail, most double sources are very different on the two sides. I will argue that these differences are of a very specific nature, illustrated in the figure below. Namely, when emission is found on one side of a source, there is generally *not* emission at the same distance on the other side; where one side turns on, the other side turns off. This on/off property may well apply to all extragalactic sources, but is seen in its extreme in one-sided jets (see review by A. Bridle, this volume) and in the D2 sources.

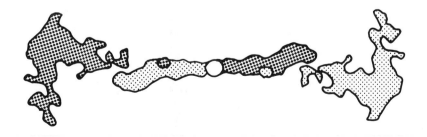

Fig. 1: 4CT 74.17.1, 6cm map (dark, adapted from van Breugel and Willis 1980) overlaid by a copy of itself (light), rotated 180° around core.

To examine the asymmetries in an unbiased manner, a sample of QSOs was selected from the interferometer surveys of Miley and Hartsuijker (1978) and Fanti et al. (1977), at the WSRT, and Owen, Porcas and Neff (1978) and Potash and Wardle (1979), at Green Bank. All sources were included which had a compact central component, and more than four synthesized beam lengths along the source. The angular distances from the peak on each side to the central component (Θ_1 and Θ_2) were calculated from models of the brightness distributions, without reference to the maps. The values of Θ_1/Θ_2 and Θ_2/Θ_1 were then histogrammed, as in Fig. 2, so that each of the 47 sources has been plotted twice.

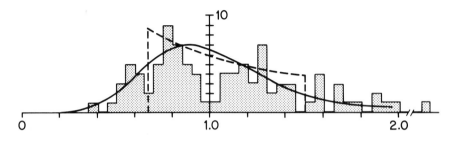

Fig. 2: Histograms of (Θ_1/Θ_2) and (Θ_2/Θ_1) for 47 QSOs, as described in the text. The dashed curve shows the expected distribution for intrinsically symmetric sources which undergo the time delay effect due to a fixed ejection velocity of $v = 0.2c$. The solid curve is the distribution expected if the length of each side of a double is independently drawn from a gaussian population with mean/rms = 3.

The observed distribution was also compared with those expected from random (i.e. mean free path) stopping of ejections, and projection effects on intrinsically bent sources. None of the above effects reproduce the dip observed near (Θ_1/Θ_2) = (Θ_2/Θ_1) = 1. Formally, one can rule out a flat or rising distribution near (Θ_1/Θ_2) = 1 with 80-90% confidence. As caveats, note that the statistics are still small, some pollution by bent sources exists and the resolution cutoff was chosen somewhat arbitrarily. In support of this result, see Ingham and Morrison (1976), who deduced an intrinsic arm length difference, and Longair and Riley (1979).

Assuming that the dip near 1 in Fig. 2 is real, what are its implications? First, there must be some coordination or physical relation between events on each side of a source, so that they can avoid occuring at the same distance from the nucleus. One possibility is the slingshot mechanism (Valtonen 1979), which preferentially ejects objects with different masses (and velocities, to conserve momentum). Alternatively, a drag force due to motion through the intergalactic medium could cause both bending and a length difference between the two sides. However, both of these mechanisms assume that the shorter side is essentially a delayed version of the longer one. This is difficult to reconcile with many sources which show complex, differing structures on the two sides, yet mesh in jigsaw fashion when displayed as in Fig. 1.

If the dip in Fig. 2 results, instead, from alternating one-sided ejection, a number of consequences follow. Relativistic particles cannot, by definition, be continuously re-energized by the nuclear source. Close to the central engine, some assymetry must be allowed in any otherwise symmetric models (e.g. Wiita and Siah 1981). Subsequent ejections must be made at the same velocity (within 5-10%), so the dip near $(\Theta_1/\Theta_2) = 1$ is not wiped out. Similarly, there can be only little slowing down of the ejected material, or stopping by intergalactic clouds. Finally, ejection velocities must, in general, be $<.05$-$0.1c$, so that time delay effects (see, e.g. Longair and Riley 1979), do not fill in the dip.

As a caveat, it is clear that much of the low surface brightness emission in double sources does not follow this one-sided picture. Although trailing and diffusion offer natural explanations for this failure, they complicate further work. Also, the one-sided ejection picture is incomplete in that it does not predict the overall symmetries of double radio sources (see below).

Thanks to B.K. Edgar for work on the analysis, and A. Bridle, R. Ekers, V. Icke, and T. Jones for stimulating discussions. This project is supported in part by NSF grant AST 79-000304.

REFERENCES

Fanti, C. et al.: 1977, Astr. Ap. Suppl. Ser., 28, pp. 351-362.
Ingham, W. and Morrison, P.: 1975, M.N.R.A.S., 173, pp. 569-577.
Longair, M.S. and Riley, J.M.: 1979, M.N.R.A.S., 188, pp. 625-635.
Miley, G.K., Hartsuijker, A.P.: 1978, Astr. Ap. Suppl., 34, pp. 129-163.
Owen, F.N., Porcas, R.W., Neff, S.G.: 1978, Astr. J., 83, pp. 1009-1020.
Potash, R.W. and Wardle, J.F.C.: 1979, Astr. J., 84, pp. 707-717.
Valtonen, M.J.: 1979, Ap.J., 231, pp. 312-319.
van Breugel, W.J.M. and Willis, A.G.: 1981, Astr. Ap., 96, pp. 332-344.
Wiita, P.J. and Siah, M.J.: 1981, Ap.J., 243, pp. 710-715.

DISCUSSION

M. Rees: Is your evidence for asymmetry in component separations consistent with what Dr. Ekers told us about symmetry along some jets, etc?
L. Rudnick: These symmetries are consistent with, but not implicit in the one-sided picture. Producing symmetric large scale brightness distributions in twin jets, or accounting for the overall correlation in length between the two sides of a double may call for a (too?) careful arrangement of time scales for ejection, switching, and particle life.
R.M. Hjellming: One thing that may be reasonable is to have the "engine" orbiting around a dynamic center, with ejection characteristics which are strongly affected by gradients in the surrounding galactic atmosphere.
L. Rudnick: Yes. Some current work by V. Icke suggests that orbits do give alternating jets, but the periods are too long, e.g., for NGC 6251. Icke finds an instability in the equatorial plane of a jet which may produce single jets alternating on the required timescales.

TIME DEPENDENT ENERGY SUPPLY IN RADIO SOURCES AND MORPHOLOGY OF RADIO LOBES

W. A. Christiansen
Moorehead Observatory, University of North Carolina

A. G. Pacholczyk
Pachart Corporation, Tucson, Arizona

John S. Scott
Steward Observatory, University of Arizona

The fact that radio jets, which are often one-sided, are nevertheless associated with extended lobes of nearly equal luminosity indicates when analyzed in detail (1) that the energy supply by the nuclear engine to the lobes is strongly time-dependent. Time dependent ejection from the nucleus of a parent galaxy produces low density channels which are ploughed in the background medium by the passage of intermittently ejected radio emitting plasmons (2), (3), (4) and (5). An analysis of the dynamics of radio emitting plasmons in such channels leads to unique morphological features consistent with observations of both narrow jets (resulting from splashback or reflections from the channel ends) and conical lobes (resulting from slower hydrodynamic deceleration).

An outgoing plasmon upon striking a dense relic of previous ejecta at the end of a channel flows around the perifery and splashes back into the trailing channel. The plasmon's energy is thermalized causing its expansion to the sides and rapid expansion back down the evacuated channel. The reflected shock front is strongly collimated both by the two-dimensional Sedov radial density distribution in the channel (2) and by exponential atmosphere behind the plasmon (5). The blowout time (i.e. the lifetime of the reflection jet) is approximately 10^6 years. See figure 1 and compare it with maps of e.g. 3C 388 and NGC 6251.

After the plasmon - relic contact, quasistatic ram deceleration phase begins and secular nature of deceleration establishes pressure balance between dense compressed head material and hot tenuous channel gas, resulting in a formation of a conical lobe behind the head of a radio source. See figure 2 and compare it with a map of e.g. Cyg A.

REFERENCES

(1) Christiansen, W.A. and Scott, J.S. 1981 (preprint)
(2) Christiansen, W.A., Pacholczyk, A.G. and Scott, J.S. 1981 (preprint A)

(3) Christiansen, W.A. 1973 M.N.R.A.S. 164, 211
(4) Christiansen, W.A., Pacholczyk, A.G. and Scott, J.S. 1979 Nature 266, 593
(5) Christiansen, W.A., Pacholczyk, A.G. and Scott, J.S. 1981 (preprint B)

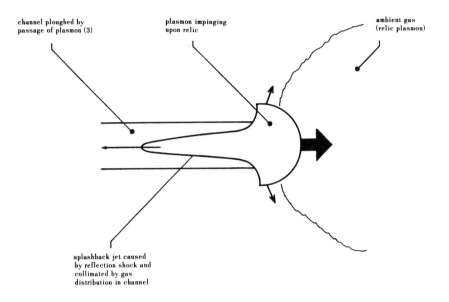

Figure 1. Transient splashbacks (first phase)

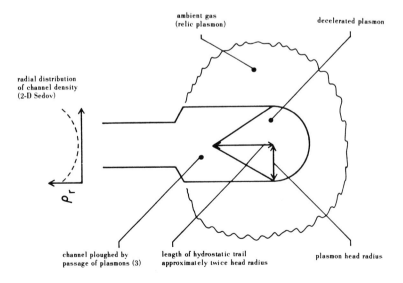

Figure 2. Hydrostatic trails (second phase)

MULTI-FREQUENCY POLARIZATION STUDIES OF RADIO GALAXIES

G.G. Pooley
Mullard Radio Astronomy Observatory
Cavendish Laboratory
Cambridge, England

This is a brief report on some of the work in progress at Cambridge, with emphasis on studies of extragalactic radio sources using the One-mile and 5-km telescopes; together with the 6C survey and a new instrument now being commisioned at 151 MHz, we have a frequency range of 100:1 available for high-resolution mapping. The 6C survey has resulted in the discovery of a number of giant radio galaxies, most notably NGC 6251. At 151 MHz, the beautiful jet is not prominent; we have mapped it with the One-mile and 5-km instruments and a paper (by Saunders *et al.*) will shortly appear in *Monthly Notices*. Dr Willis will speak later on the structure of this source.

Many of the galaxies being studied are old favourites - low-luminosity sources (FR class I) like 3C465 and 3C272.1, and 3C66B which is the subject of a joint study with the Leiden group. Using 3C192 as an example of a 'typical' FR class II source (Fig. 1), we see the main features of the emission which have to be explained. In the bridge region, the projected magnetic field is perpendicular to the axis; in the tails (near the hot-spots) it tends to be parallel, and around the hot-spots themselves it is circumferential. If the field were initially tangled, we can understand the field in the hot-spots by using Robert Laing's model of compression into a shell; in the bridge, we need to expand the field transversely.

When we consider the morphology of the sources, we find that many of the bridges are distorted. Those like 3C430 are distorted to one side; plausible explanations include motion relative to the IGM, or a gradient in its density. For bridge like that of 3C192 (or more extreme cases like 3C315), do we need precessing beams or could rotation of gas in the galaxy itself cause the distortion? The hot-spot structure of some sources does suggest precessing beams (for example, 3C20), but others lack the rotational symmetry and might be better explained in terms of shocks in an IGM which has strong density gradients, leading to plausible explanations for the orientations of the hot-spots and their magnetic fields.

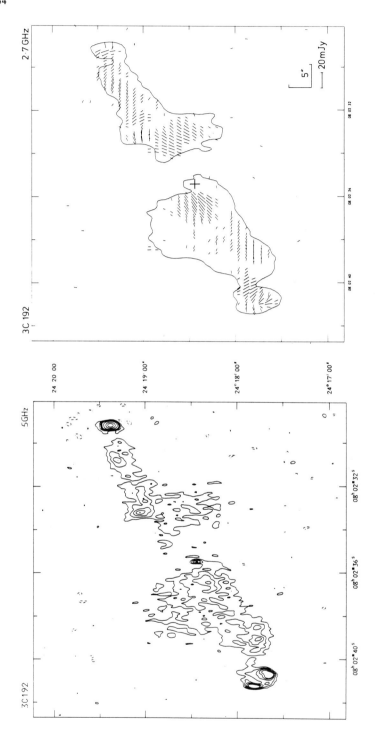

Fig. 1. 3C192, observed with the 5-km telescope.
Left: 5 GHz, resolution 2" x 5"
Right: 2.7 GHz linear polarization (E- vectors). The Faraday rotation is nearly uniform; to obtain the projected magnetic-field direction, rotate the lines counter-clockwise by 70°. Resolution 3".6 x 9".

A COMPLETE SAMPLE OF RADIO GALAXIES

P.A. Shaver[1], I.J. Danziger[1], R.D. Ekers[2], R.A.E. Fosbury[3],
W.M. Goss[4], D. Malin[5], A.F.M. Moorwood[1], J.V. Wall[3]

[1] European Southern Observatory, Garching, W. Germany
[2] N.R.A.O., Socorro, New Mexico, U.S.A.
[3] Royal Greenwich Observatory, Herstmonceux, England
[4] Kapteyn Astronomical Institute, Groningen, The Netherlands
[5] Anglo-Australian Observatory, Epping, N.S.W., Australia

We report here some preliminary results of a multi-wavelength study of a complete sample of radio galaxies. The sample is comprised of 93 radio sources from the Parkes 11 cm catalog which are identified with galaxies of 17th magnitude or brighter in the declination zone $-17°$ to $-40°$. Our objective is to cross-correlate the radio, infrared, optical, and other properties of a properly defined sample of radio galaxies.

By going to relatively low flux densities and using an optical magnitude limit we are approximating a volume-limited sample and consequently have a class of objects which are more representative of radio galaxies as a whole than the 3C sample. Since the 3C radio galaxies are generally of high luminosity, many are also very distant and therefore difficult to study optically. Another common difficulty in the study of radio galaxies is that radio astronomers usually only have access to the heterogeneous optical data which happen to be published on the relevant galaxies and vice-versa. It is the objective of the present work to obtain uniform and comprehensive data at radio, infrared, and optical wavelengths of a complete sample of relatively nearby radio galaxies.

We have mapped all the galaxies in the sample using the VLA, with supplementary data for the larger sources from Molonglo, Fleurs, and Parkes. Near-infrared photometry is being done using the 3.6m telescope at La Silla, and UKIRT. Spectrophotometry of all the radio galaxies and many of their companions has been completed using telescopes at La Silla and Las Campanas, and the AAT. In addition we have UV and X-ray data for several of the galaxies.

This sample contains a smaller fraction of classical double radio sources than the 3C sample; these galaxies exhibit more distorted radio structures and are of lower radio and optical luminosity. 90% of them are within $z = 0.1$. Radio cores, mostly of the flat-spectrum type, have been detected in 80% of the galaxies. Only 6% have obvious radio jets, although there may be some indication of possible jet structure in as many as a third of them. Half of these radio galaxies are smaller than 75 kpc, as

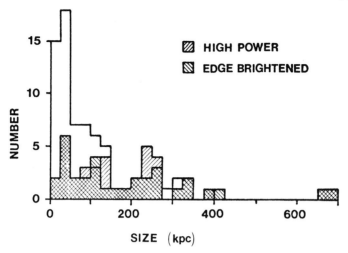

Figure 1. Linear size distribution. High power means $\geq 10^{24}$ WHz^{-1} at 5000 MHz; edge brightened corresponds to Fanaroff and Riley (1974) class II.

shown in fig. 1, and the high-power, edge-brightened sources are the biggest. In the infrared about 20% of the galaxies are abnormal in H-K, and 30-40% in K-L (fig. 2). 40% have so far been found to have optical emission lines, and almost all of these are narrow; those galaxies with the strongest radio cores all have emission lines.

Fig. 3 shows that there is a preference for the radio major axis to be parallel to the optical minor axis, as Guthrie (1979) and Palimaka et al. (1979) also found. This preference may be somewhat stronger for isolated and/or more flattened galaxies.

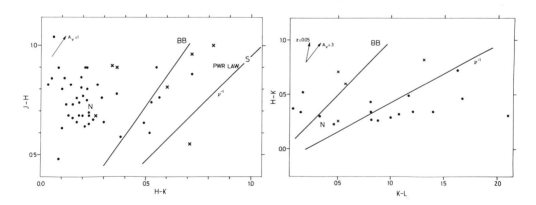

Figure 2. Near-infrared photometry for a sub-sample of the radio galaxies. "N" and "S" indicate where normal and Seyfert (i.e. active) galaxies occur in such plots. Crosses denote spirals and S0s, and filled circles denote ellipticals.

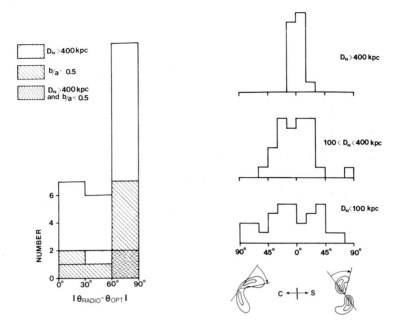

Figure 3 (left). Comparison of the position angles of the radio and optical major axes for 30 galaxies with optical axis ratio b/a ≤ 0.9.
Figure 4 (right). Histograms of the C and S distortions for different values of D_N, the (projected) distance to the nearest galaxy.

In fig. 4 it can be seen that both the "C" and "S" distortions of the extended radio structure (mirror symmetry and inversion symmetry respectively - cf. Ekers, this volume) occur only when other galaxies are nearby (within a few hundred kiloparsec). The isolated radio galaxies are much less distorted, more symmetrical, more edge-brightened, and exhibit higher contrast in their radio brightness distribution. It is well known that the C-distortion is related to a high density of galaxies; radio galaxies in Abell clusters almost all exhibit C-type distortions (Ekers et al., 1982). The S-distortion, on the other hand, is more often associated with the presence of a single nearby galaxy than with a high density of galaxies. The close passage of another galaxy could perhaps cause the S-distortion by indirectly influencing events in the nucleus, although these galaxies are not exceptional in their radio power, infrared excess, or optical line emission; alternatively, the S-distortion may be produced in the outer regions by tidal effects and re-collimation of the beam.

REFERENCES

Ekers,R.D., Fanti,R.,Lari,C.,Parma,P. 1982, Astr. Ap. (in press)
Fanaroff,B.L., Riley,J.M. 1974, Mon.Not.Roy.astr.Soc. 167, 31p
Guthrie,B.N.G. 1979, Mon.Not.Roy.astr.Soc. 187, 581
Palimaka,J.J.,Bridle,A.H.,Fomalont,E.B.,Brandie,G.W. 1979, Ap.J. 231, L7

EXTENDED STRUCTURE IN HIGH-REDSHIFT RADIO SOURCES

P.J. Duffett-Smith & A. Purvis
Cavendish Laboratory,
Cambridge, England.

We have compared measurements of several hundred 3C and 4C radio sources at large redshifts to investigate how radio-source structure changes over a factor of 5-10 in luminosity. Our results show that for $z \gtrsim 0.6$:
(i) most sources (both 3C and 4C) have hotspots about 3.5 kpc in size ($H_o = 50$ km s^{-1} Mpc^{-1}, $\Omega = 1$);
(ii) <u>lower-luminosity</u> sources (bottom of 4C) have <u>less-extended</u> outer lobes.

Our observations were made at 81.5 MHz with the Cambridge 3.6-hectare Array using the method of interplanetary scintillation (IPS: Readhead, Kemp & Hewish 1978). The 3C sources were all measured individually to determine the angular diameter, θ, in the range 0.2 to 2 arcsec, and 'compactness', R, which is the fraction of the total flux density originating in the compact feature. The fainter sources in the range 2-3 Jy at 81.5 MHz were too weak to show up individually on the records but contributed to an IPS background. We applied the method of background-deflection analysis (P(D)) to determine weighted average values of θ and R for these sources (Duffett-Smith, Purvis & Hewish 1980). Wall, Pearson & Longair (1980) have shown that sources in this range of flux density occur mainly at redshifts between 0.6 and 2. We were thus able to compare the mean IPS properties of these sources with the properties of the high-redshift members of the 3C167-sample to investigate how radio source structure varies with luminosity.

The results show clearly that the fainter sample has larger average values of both θ and R implying that the less-luminous sources are more compact on average than their 3C counterparts but which yet have a larger average angular diameter. We can typify the structure of the high-redshift members of the 167-sample by an idealised equal-double source model (Fig. 1a) in which the lobes are at least 10 arcsec apart and contain hotspots with $\theta = 0\farcs4$ contributing 60% of the total flux density. Taking this model as the starting point, how must we modify it to reproduce the IPS background measurements (which correspond to a factor of 5-10 lower luminosity)? The most likely possibility is that

a) high luminosity

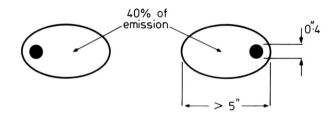

b) lower luminosity

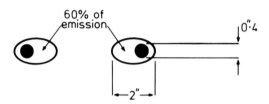

Figure 1. Schematic models of radio sources with $z \gtrsim 0.6$.

the hotspots have about the same average linear sizes as their 3C counterparts but that the extended structure has a much smaller angular extent so that it now contributes significantly to the scintillations. (This requires its angular size to be of the order of 2 arcsec; Fig. 1b). The IPS measurements then reflect a blend between the hotspot and its surroundings giving larger values of both θ and R (Duffett-Smith 1980).

These are of course only schematic representations of a whole range of brightness distributions undoubtedly present in both samples. However, it is hard to escape the general result that outer lobes in high-redshift radio sources are smaller when the luminosity is lower. We should expect such a result if the extended structure is in pressure balance with its surrounding medium. Sources at the same redshift should have similar environments. Hence the more powerful is the source, the further its extended structure can expand until its energy density equals that of the medium around it.

References

Duffett-Smith, P.J.: 1980, Mon. Not. R. astr. Soc., 192, pp 33-39.
Duffett-Smith, P.J., Purvis, A. & Hewish, A.: 1980, Mon. Not. R. astr. Soc., 190, pp 891-901.
Readhead, A.C.S., Kemp, M.C. & Hewish, A.: 1978, Mon. Not. R. astr. Soc., 185, pp 207-225.
Wall, J.V., Pearson, T.J. & Longair, M.S.: 1980, Mon. Not. R. astr. Soc., 193, pp 683-706.

EXTENDED OPTICAL LINE EMISSION ASSOCIATED WITH RADIO GALAXIES

Wil van Breugel
Kitt Peak National Observatory
Tucson, Arizona

Tim Heckman
Steward Observatory
Tucson, Arizona

I. INTRODUCTION

Using the Video Camera (e.g. Butcher et al. 1980) and the High Gain Video Spectrometer (e.g. Heckman et al. 1981) we are carrying out a program at Kitt Peak to search for optical line emission associated with the jets and lobes of radio galaxies. Several sources have been found in which extended optical line emission is clearly related to the non-thermal radio emission. Some general information on these and a few other sources is summarized in Table 1.

TABLE 1

Source	Redshift	Radio Luminosity (10^{42} erg s^{-1})	Absolute Optical Magnitude	Galaxy Morphology	References
3C 277.3	.0857	7.3	-21.8	E	Miley et al. 1981
3C 293	.0450	2.3	-22.6	peculiar	---
3C 305	.0410	1.3	-22.4	peculiar	Heckman et al. 1981
M87	.0043	1.1	-21.6	E	Ford & Butcher 1979
4C 29.30	.0650	0.9	-22.4	peculiar	van Breugel et al. 1981
4C 26.42	.0630	0.7	-23.2	CD	" "
NGC 7385	.0259	0.5	-20.5	E	Simkin & Ekers 1979
Cen A	.0016	0.1	-21.5	peculiar	Graham & Price 1981

Our optical imaging and spectroscopic data combined with accurate VLA maps of comparable resolution ($\sim 1''$) allow some preliminary, general conclusions to be made. We will briefly discuss these and illustrate several points using 4C 26.42 (Figure 1) and 4C 29.30 (Figure 2).

II. OBSERVATIONAL RESULTS

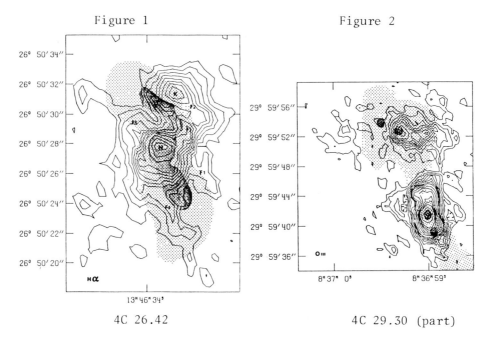

Contours represent optical line emission (Hα and OIII respectively). Shaded areas outline the radio sources, dark regions indicate radio brightness enhancements.

1. The optical line emission is predominantly found along the radio source boundaries. This particularly clear for 4C 26.42.
2. The optical line emission is usually brightest near, but slightly offset from, bright radio knots or hotspots. At these locations the radio source (jet) often deflects, the optical emission line 'knot' being near the outer bend.
3. Generally the radio source is nearly entirely depolarized at 6 cm at locations where the optical line emission overlaps with that of the radio source.
4. In the two cases where we have adequate spectroscopic data (3C 305, 3C 277.3) the pressures in the line regions are comparable to those in the radio source, granted the usual assumptions.
5. The bulk velocities of the gas in the emission line regions are typically 200-300 km s^{-1} and the velocity widths 300-500 km s^{-1}.
6. Several of the galaxies are peculiar i.e. they have spiral arm like structure, disks, dust lanes. Others however appear to be 'normal' ellipticals. All radio galaxies have relatively low radio luminosities but their associated galaxies are relatively bright.
7. Although in general the line emitting regions are located within the parent galaxies, there are cases such as 3C 277.3 and Cen A where line emission is detected as far out as 40 kpc from the galaxy nucleus.

III. PRELIMINARY DEDUCTIONS

Origin of the line emitting gas.

One can envisage the following possible origins for the line emitting gas: a) it exists 'in situ', b) it is transported outwards or c) it is falling in. Although it is difficult to observationally discriminate between these possibilities, there is circumstantial evidence that all of these can occur.

a) In 4C 29.30 the northern emission line region is located in a region of a spiral arm like structure in the galaxy. At the same position the radio source flares up and deflects. This is suggestive of a jet interacting with a locally dense environment. Similarly, our optical data indicate the presence of rapidly rotating gaseous disks in the core regions (a few kpc) of several of the other galaxies in Table 1. In all these cases, and presumably also in the core of 4C 29.30, there are strong indications based on the morphologies of the radio and optical line emission that jets are interacting with locally dense gaseous environments.

b) There is indirect evidence in some cases (Cen A, 3C 305) for outflow of the emission line gas consistent with the outflow which seems to occur on a much smaller scale in other active galaxies (Heckman et al. 1981 (II)).

c) In M87 as well as in other giant galaxies with dense gaseous (X-ray) halo's such as NGC 1275, it is generally argued that emission line filaments are formed as a result of radiatively regulated accretion of gas onto the massive central object (e.g. Cowie and Binney, 1977; Fabian and Nulsen, 1977). This might also be the case in 4C 26.42 which is identified with a cD galaxy in A 1795, a bright X-ray cluster (e.g. Perrenod and Henry, 1981).

Exitation mechanisms.

In only a few cases is sufficient spectroscopic data available to discriminate between various possible exitation mechanisms such as for example: a) photo-ionization, b) shockheating. As with the origin of the gas, it appears that different mechanisms may occur in different sources. For example in 3C 277.3 photo-ionization seems to be the most plausible process responsible for the optical line emission associated with the jet. In 3C 305 the situation is rather complex and both photo-ionization as well as shock-heating may occur. The probable occurrence of radiatively regulated accretion onto massive galaxies has already been mentioned above. In this picture the emission-line gas is excited by re-pressurizing shocks driven by the hot surrounding gas.

Depolarization mechanisms.

The preferential occurence of line emitting gas at the boundaries of radio galaxies may have profound implications for the interpretation of radio (de-)polarization measurements. For example the enhanced emission line brightness alongside 4C 26.42 is most readily interpreted as being due to a line of sight effect of a thin layer of ionized gas surrounding the radio source. It is very likely that this gas is clumpy on a scale much smaller than the observing beam (<<1" or 1 kpc). The very low percentage polarization at 6 cm and i.e. the absence of highly polarized boundaries is in clear contrast to what is usually observed (see for example van Breugel, 1980). Thus it seems that 4C 26.42 is being depolarized by a 'screen' of clumpy, magneto-ionic thermal gas which is mixed in with the outer layers of the radio source. In this case the radio emission depolarizes more quickly and at shorter wavelengths than in the (usually assumed) case of the ionized gas being mixed throughout the radio-emitting regions (see Burn, 1966).

Powering the radio sources.

Within the context of the usual assumptions (such as minimum energy etc), the relatively low bulk velocities of the emission line gas pose problems for radio source models in which it is assumed that a) the radio source is powered by a jet of large bulk kinetic energy and that b) the observed velocities are representative of the velocities in such a jet. To supply the minimum required energy, with 100% conversion efficiency, to power a source like 3C 277.3, implausibly high gas densities (>10 cm^{-3}) and mass loss rates (>10 $M_\odot$ yr^{-1}) would be required. Clearly however, a velocity gradient across a jet may exist and in cases where outflow of the emission line gas is favoured, one may argue that the observed velocities are merely those existing in the outer (entrainment ?) layers of the jet.

ACKNOWLEDGEMENTS

We thank M.H. Ulrich and H. Butcher for allowing us to use their imaging data on 3C 293.

REFERENCES

van Breugel, 1980; Thesis Leiden Observatory.
van Breugel, Heckman, Butcher, Miley, 1981; in preparation.
Burn, 1966; M.N.R.A.S. 133, 67.
Butcher, Miley, van Breugel, 1980; Ap.J. 235, 749.
Cowie and Binney, 1977; Ap.J. 215, 723.
Fabian and Nulsen, 1977; M.N.R.A.S. 180, 479.
Ford and Butcher, 1979; Ap.J. Suppl. 41, 147.
Graham and Price, 1981; preprint.
Heckman, Miley, van Breugel, Butcher, 1981; in press.
Heckman, Miley, Balick, van Breugel, Butcher, 1981; in preparation.
Miley, Heckman, Butcher, van Breugel, 1981; Ap.J. 247, L5.
Perrenod and Henry, 1981; preprint.
Simkin and Ekers, 1979; A.J. 84, 56.

EXTENDED EMISSION LINES IN RADIO GALAXIES

        R A E Fosbury
        Royal Greenwich Observatory, Herstmonceux Castle,
        Hailsham, GB-BN27 1RP

In whatever physical state the gas is found, observations of the interstellar medium in elliptical galaxies are of considerable interest. This is particularly true in the case of radio galaxies where we believe that the gas is an indespensable part of the cause of nuclear activity and plays a role in the origin and the evolution of the radio galaxy phenomenon. In a few cases we are fortunate to find some of the gas to be ionized with a temperature of about $10^4 K$ where optical spectroscopy allows us to deduce something about the excitation/ionization mechanism, about its chemical composition and about its state of motion. Here I wish to summarize observations of three Southern radio galaxies which show optical emission lines from regions tens of kiloparsecs in extent.

The 14th magnitude elliptical ($z = 0.033$) identified with *PKS 2158-380* is the brightest member of a small group of galaxies containing about ten members. The radio morphology is double but is unusually asymmetric with the components having a flux ratio of *5*. The remarkable feature of the object is a high ionization emission line region with a largest dimension of *30 kpc* ($H_o = 50$ km s^{-1} Mpc^{-1}). Both direct photography and long-slit spectroscopy show the emission to be distributed in an *S*-shaped structure centred on the nucleus. The *HeII* $\lambda 4686$/H$_\beta$ intensity ratio is about *0.3* throughout the galaxy and the emission is clearly not the result of photoionization by hot stars. Neither are the line-ratios consistent with a shock-heating mechanism and it appears likely that the gas is everywhere photoionized by an active nucleus emitting a non-thermal ultraviolet spectrum. Observations with *IUE* satellite reveal a nuclear point source with a spectral the index of about *-1.4*. The requirement to ionize gas at large distances from the nucleus does however impose a constraint on the geometrical distribution of the gas: a thin planar disk will not suffice. I believe that the observed gas velocities are the result of rotation in the gravitational potential of the elliptical galaxy but that the present configuration is not one of equilibrium. The hypothesis of a severely warped gaseous disk is consistent with all the data and in particular allows naturally for ultraviolet radiation from the nucleus to travel large distances without significant absorption before intercepting the gaseous sheet. Since dissipation processes would restrict the dynamical lifetime of the configuration to at most a few times 10^8 yrs, it appears likely that the elliptical galaxy has recently captured and disrupted a small companion whose gas content may now be providing the fuel for the nuclear activity.

The 16th magnitude galaxy ($z = 0.066$) identified with PKS 0349-27 shows even more direct evidence of a recent dynamical interaction for here we can see what appears to be the stripped nucleus of a passing galaxy which has left its gas covering a region over 70 kpc in extent. Again the ionization state is high with He II $\lambda 4686/H\beta \simeq 0.5$ throughout. Although in this case the galaxy is too faint for IUE observations, the line ratios are most straightforwardly explained as photoionization by a non-thermal source. The highly non-equilibrium dynamical state is confirmed by the long-slit spectroscopy which shows approximatley linear velocity radius relations along the three position angles observed. As well as the stripped companion 20 arcsec to the East, there is another bright elliptical at a projected distance of 140 kpc to the West and it is likely that the gas giving rise to the emission lines is subject to complex tidal forces. Notwithstanding these complexities, I believe that detailed 2D mapping of the velocity field of the gas around these ellipticals will provide valuable information about the mass distribution in such galaxies to much greater radii than are possible to reach using other techniques like absorption line spectroscopy.

The emission associated with the BL Lac object PKS 0521-36 has a different character. The emission line redshift of 0.055 was confirmed and diluted stellar absorption features at the same redshift were discovered by Danziger et al. (1979) who also reported the existence of an optical jet in position angle 305° with an extent of 10 arcsec. New observations with the VLA at 5 and 1.4 GHz show the flat spectrum core and a steep spectrum, edgebrightened structure extending SE (pa 123°), diametrically opposite the optical jet. An optical spectrum, taken with the slit aligned along the optical/radio jet axis shows extended [OII], [OIII] and Hα emission peaking at the nucleus and at the position of the steep spectrum radio component to the SE. No line emission is seen on the side of the jet seen in the B-band photograph which suggests that this is a continuum structure. If the optical jet is indeed a non-thermal continuum then PKS 0521-36 provides us with an example of a double jet which is spatially, though not energetically, symmetric.

This summary paper is the result of work carried out with many collaborators and has included observations made with a variety of different instruments. I wish to acknowledge all of these people but in particular Alec Boksenberg, John Danziger, Ron Ekers, Miller Goss, David Malin and Peter Shaver. Detailed reports of this work are being published elsewhere.

Danziger, I.J., Fosbury, R.A.E., Goss, W.M. & Ekers, R.D., 1979.
    Mon. Not. R. astr. Soc., 188, 415.

EXTENDED EMISSION LINES IN RADIO GALAXIES

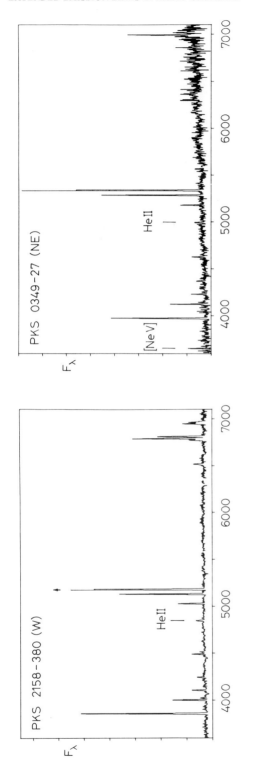

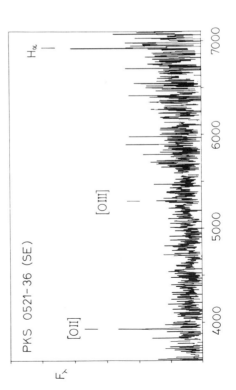

Representative optical spectra of the extended emission in the three radio galaxies discussed in the text. The spectra are from the University College London IPCS on the ESO 3.6 m telescope in Chile. The ordinate is a relative F_λ scale.

EMISSION LINES: SIGN OF A NEW ENERGY SOURCE?

David S. De Young
Kitt Peak National Observatory
Tucson, Arizona

In addition to the occurrence of emission line regions coexistent with extended radio sources which have been discussed at this Symposium, this phenomenon has been observed earlier in 3C277.3 (Miley et al. 1981) and in Centaurus A (Graham and Price 1981). This gas has been detected only in the Fanaroff and Riley "Class I" radio sources. Data concerning this class of object suggest that outflow from the nucleus is proceeding at transonic or subsonic speeds and this correlation has led to the suggestion (De Young 1981) that the origin of the emission line gas arises from entrainment of the interstellar medium into the ejected material.

Discussion of the entrainment process and a calculation of the entrainment rate are found in De Young (1981), and good agreement is obtained with the observations of Cen A and 3C277.3. What is discussed here are further observations and developments concerning this idea. It is encouraging that the new observations of 4C26.42 reported here (Heckman and van Breugel, this volume) not only find emission lines coincident with an extended radio source, but also that there appears to be an enhancement of the line emission around the periphery of the radio source. This is an expected result of boundary layer entrainment.

Entrained interstellar gas will have first been heated by the shock front produced by passage of the initial stream of material ejected from the nucleus. A jet velocity of $\sim 10^3$ km s^{-1} will heat this gas to $\sim 10^7$ K, and it will cool to emission line temperatures in $\sim 2 \times 10^6/n$ yr, where n is the number density. Recent x-ray data may give some evidence for this cooling process. Miley, Norman, and Silk (1981) have detected an x-ray flux of $\sim 1.2 \times 10^{43}$ erg s^{-1} from 3C277.3 (Coma A). In addition an extended x-ray source has been observed coincident with the emission line radio galaxy 3C305, and these data provide a limit on the mass of the x-ray emitting gas of $n_e M_x \leq 3 \times 10^8 M_\odot$ cm^{-3} for $T \leq 10^{7.5}$ K (Heckman 1981, private communication). This mass limit is consistent with the results of the earlier entrainment calculations which yield $6 \times 10^7 - 6 \times 10^6 M_\odot$ entrained over 10^8 yr for ambient number densities in the range 0.1-0.01 cm^{-3}.

As the line emitting gas cools further, star formation can occur, and in fact Graham and Price (1981) have observed at least 10^2 young massive ($\sim 10 M_\odot$) stars in the northeast radio lobe of Cen A. For Cen A and Coma A the Jeans masses and lengths at 10^2 K vary from $\sim(10^4-10^2)M_\odot$ and $\sim(10^2-1)$pc, using the observed limits on the number density. Collapse times vary from 10^5 to 10^7 years. The mass function for star formation in such an environment is unknown, but if a Salpeter function is used with cutoffs of 30 and 0.1 $M_\odot$ then 10^4-10^5 stars of mass greater than $10 M_\odot$ could be formed over 10^8 yr, assuming 10% efficiency.

The key point is that these stars are made of gas entrained by the outflow and thus also move outward. At the end of their lifetime ($\sim 10^7$ yr) these massive stars will explode and inject $10^{51}-10^{52}$ erg per star into the surrounding gas. In addition an outward moving remnant is left which can provide $\sim 10^{38}$ erg s^{-1}. It has been known for some time that the Class I radio sources very often require energy replenishment before the outflow reaches the extremities of the radio source as well as in the more distant regions themselves. The supernovae and supernova remnants due to the evolution of massive outwardly moving stars may be a significant source of the required energy. After the first 10^7 years the remnants alone can inject 10^{41} ergs s^{-1}, and ten times this value at the presumed current age of $\sim 10^8$ yr. Also possible is the formation of more massive objects whose evolutionary end point is a black hole, thus providing a more efficient source of energy within the radio source.

A supernova will not disrupt the outflow, since the radius of the remnant is ~ 50 pc for $E_{sn} = 10^{51}$ ergs and n = 10^{-2} (Chevalier 1974) which is much smaller than the size of the beam or bridge. For n $\sim 10^{-2}$ radiative losses are less important than the kinetic energy which will "stir" the outflow as well as provide a source of stochastic reacceleration of the electrons. Bright knots are commonly observed along the bridges of Class I radio sources, and the location of these knots close to the parent galaxy suggests that they might arise from the local stirring of the outflow due to one or more supernovae. Entrainment occurs principally near the onset of the turbulent boundary layer, well inside the galaxy. In a "mature" radio source the ISM near the outflow will have cooled from its original post-shock temperature. Entrainment of this gas can lead to early star formation, especially if any inhomogeneities are present, and with a lifetime of $\sim 10^7$ yr a massive star will have traveled only a few tens of kiloparsecs before a supernova occurs.

References

Chevalier, R.A. 1974, Ap.J., 188, 501.
De Young, D.S. 1981, Nature, 293, 43.
Graham, J.A., and Price, R.M. 1981, Ap.J., 274, 813.
Miley, G.K., Heckman, T.M., Butcher, H.R. and van Breugel, W.J. 1981, Ap.J. Lett, 247, L5.
Miley, G.K., Norman, C., and Silk, J. 1981, in preparation.

OPTICAL INVERSE COMPTON EMISSION IN EXTRAGALACTIC RADIO SOURCES

S.E. Okoye and O. Obinabo
Department of Physics
University of Nigeria
Nsukka, Nigeria

In this contribution the reported detection by Saslaw, Tyson and Crane (1978) of weak optical emission in the lobes of three 3C sources - 3C 265, 3C 285 and 3C 390.3 - is reappraised in the framework of three optical emission mechanisms - synchrotron (SYN), synchrotron inverse Compton (SIC) and blackbody inverse Compton (BIC). This effort has been motivated partly by the knowledge that the contribution to the synchrotron inverse Compton emission in a radio source component is likely to become significant for very compact and bright radio components (see e.g. Okoye 1972), and partly by the recent availability of high resolution radio frequency structural data on the sources in question. Another incentive arose from the demands of a separate investigation into high energy particles interactions in radio sources (reported by Okoye and Okeke in this volume) involving very high energy protons thus making it necessary to ascertain whether such highly energetic particles exist in radio source components.

Saslaw et al (1978) had already considered the feasibility of accounting for the reported weak optical emission in the radio lobes of 3C 265, 3C 285 and 3C 390.3 through BIC. According to them, the optical emission could most likely be accounted for by the inverse Compton scattering of the '3K' microwave photons by fast radio electrons, although synchrotron emission could not be ruled out. Here we estimate the relevant contributions of each of the three emission mechanisms indicated above. The expressions for the spectral power generated in a radio source by the three emission echanisms have already been derived by a number of authors (see e.g. Felton and Morrison 1966; Okoye 1972, 1973) and will not be repeated here. Using these expressions, the optical luminosities for the radio components based on the Cambridge 5-km telescope structural data (Pooley - private communication) have been estimated and the results are given in table 1. The following inferences can be drawn about the relative importance of the optical emission mechanisms considered.
(a) The calculated integrated optical synchrotron luminosities for each of the three sources dominate over inverse Compton contributions.
(b) The calculated synchrotron luminosity is about the same as the observed optical luminosity in all the three sources observed.

table 1
Source component optical luminosities based on various emission mechanisms.

Source components	$\log_{10} L_{SIC}$	$\log_{10} L_{BIC}$	$\log_{10} L_{SYN}$	$\log_{10} L_{OBS}$
(luminosity in erg s-1)				
3C265				
Np A	37.7931	39.7931	44.9320	
B	38.8371	38.5846	44.0991	
	42.2045	39.5074	43.0405	
Sf	42.9350	39.7606	43.6050	
	42.4773	41.0706	44.3575	
Integrated	43.1210	41.1419	45.1003	45.0
3C285	36.0980	37.3724	41.4612	41.5
3C390.3				
Np	33.9046	36.5853	42.2485	
	36.0902	36.9571	41.1621	
Jet	35.9614	35.9621	39.3468	
C	31.0195	32.7755	42.1364	
	30.8520	32.4000	42.0872	
Sf	38.8698	38.6848	40.3468	
Integrated	38.8711	38.6970	42.6564	43.1

It seems tempting to explain the observed optical emission from the radio source lobes as due to the optical extension of the radio synchrotron radiation. This would accord with a straightforward extrapolation of the spectral index to optical frequencies. Assuming that the radio and optical photons from these source components originate from identical volumes of space, one would expect the electron energy spectrum in the emitting volumes to extend up to electron energy values sufficient to support optical synchrotron emission, if the reported weak optical emission arise from the synchrotron process. To interpret the claimed optical emission in the radio lobes correctly, it is therefore necessary to establish whether electrons with enough energies to produce the optical synchrotron spectrum actually exist in the radio source lobes or not.

For the sake of discussion we take the estimated value of $\sim 5 \times 10^{-5}$ G as the equipartition magnetic field strength in a radio lobe of 3C285 in which case, the lifetime of the optical electrons will then be of the order of 2000 years. Thus if the optical electrons were to move away from the acceleration region at a velocity $\sim c$, they would at most move ~ 0.7 kpc during their lifetime which appears consistent with the observed scale of the optical radiation. However, if optical electrons do exist in the lobes of the extragalactic sources under consideration, then if current ideas on proton-electron energy ratios are also correct, the associated protons will have energies $\lesssim 2000$ times those of the electrons. As indicated elsewhere (see Okeke and Okoye 1981), proton-proton interactions will lead to steady injetions of secondary electrons which will produce an optical spectrum not consistent with the extrapolated radio spectrum. It seems, therefore, that for the synchrotron mechanism to survive, the primary optical electrons should have radiation lifetimes much less than values fixed by equipartition magnetic field strengths. In other words, the source magnetic field strengths must be greater than equipartition values. Since equipartition magnetic field strengths are not sacrosant, there is no reason why the actual fields cannot be greater if a mechanism exists for replenishing the quickly degraded optical electrons. The problem, however, is that because the three sources under consideration are not atypical, there is then no reason why many more radio sources should not exhibit optical synchrotron emission in their lobes. This is presently not the case within optical detection limits.

As an alternative, we consider the inverse Compton mechanism. We note that the optical inverse Compton emission (SIC and BIC) is independent of the presence of optical electrons in the radio source components. Again if the actual source magnetic field strengths can differ from equipartition values, then provided the actual values are less than equipartition values, it will be possible to account for the claimed optical emission from the radio lobes of the sources under consideration by the inverse Compton mechanism. If on the other hand, equipartition conditions apply rigidly, then it is not possible to account for the claimed optical emission by either the synchrotron or inverse Compton mechanism. For this reason, it is desirable that efforts be made to confirm the claimed optical emission in the three 3C sources discussed. Because the synchrotron mechanism makes demands not only on the electron energy spectrum but also on the source magnetic field strength, the inverse Compton mechanism constraining only the source magnetic field strength, is to be preferred of the two.

ACKNOWLEDGEMENT

One of us (SEO) thanks the Managing Director of Jimbaz (West Africa) Ltd for the bulk of a travel grant making it possible for him to present this paper at the IAU Symposium No. 97. A supplementary travel grant from the IAU is also acknowledged with thanks.

REFERENCES

Brecher, ad Burbidge, G.R., 1972, Nature, Lond., 237, p 440.
Felton, J.E. and Morrison, P., 1966, Ap.J., 146, p 686.
Okoye, S.E., 1972. Mon.Not.R.Astr.Soc., 160, p 339.
Okoye, S.E., 1973. Mon.Not.R.Astr.Soc., 165, p 413.
Okoye, S.E. and Okeke, P.N., 1982, IAU Symposium No. 97, this volume, Eds. D. Heeschen and D.M. Wade, p. 75.
Rees, M.J., 1967, Mon.Not.R.Astr.Soc., 137, p 429.
Saslaw, W.C., Tyson, J.A., and Crane, P., 1978. Ap.J., 222, p 435.

PROTON-PROTON COLLISIONS IN EXTRAGALACTIC RADIO SOURCES

S.E. Okoye and P.N. Okeke
Department of Physics
University of Nigeria
Nsukka, Nigeria

The majority of extragalactic radio sources are known to consist of two extended components straddling an optical galaxy or quasar with each component being maintained from the nucleus of the associated optical object through a beam or jet of relativistic plasma and magnetic fields. Hitherto, the energetics of radio source components have been considered essentially from the point of view of the cooling of the relativistic electrons through their interaction with ambient magnetic fields (synchrotron radiation) and with low energy photons (inverse Compton emission). Here we consider a hitherto neglected problem involving the mutual interactions between the fast particles themselves. (The results of a detailed investigation into these interactions will be reported elsewhere -- see Okoye and Okeke, 1982.)

We suggest that the beams or jets are essentially colliding nuclear beams involving high energy protons and electrons. At high energies, it is well known that proton-proton (p-p) collisions lead to pion emission. The pions, which are short-lived, quickly decay into muons which in turn quickly decay into electrons according to the following scheme:

$$pp \longrightarrow pp + \pi^+ \pi^- + \pi^0 \tag{1a}$$

$$\pi^- \longrightarrow \mu^- + \bar{\nu}_\mu \tag{1b}$$

$$\mu^- \longrightarrow e + \bar{\nu}_e + \nu_\mu \tag{1c}$$

In the discussion which follows below, we shall attempt to show that the electrons produced in the above reaction will carry away a significant proportion of the energy of a typical colliding proton. Consequently, the p-p collision represents a plausible mechanism for high energy secondary electron injection into a radio source. It can be shown (see e.g. Rosser 1964) that the threshold energy for this reaction (i.e. p-p) to occur is 936 MeV in the rest frame of the beam or jet. Under the condition of a colliding pair of protons as would happen in a jet, hot spot or even radio lobe, the proton threshold energy (in the observer's frame) becomes $936\gamma_B$ MeV, where γ_B is the Lorentz factor associated with the beam. The total proton threshold energy is consequently $(936\gamma_B + 930)$ MeV. For non-relativistic beams, the threshold energy is ~ 1.866 GeV. It is obvious from the above that the threshold energy is quite sensitive to assumptions made about the jet or beam (bulk) velocity. For example, for a highly relativistic jet, $\gamma_B \gg 1$, a very large threshold energy results which may be too high for the reactions (1) above to take place. According to current ideas, however (see e.g. Conway 1982; Readhead and Pearson 1982), the magnitude of the jet velocity is $\sim 0.2c - 0.3c$, at least for the nearby radio galaxies and quasars. This corresponds to a proton threshold energy of ~ 2.19 GeV or a proton Lorentz factor, $\gamma_p \sim 2.4$. These values are reasonably low enough as to accommodate virtually the bulk of the energetic protons in a radio source in the p-p reactions.

We now turn to the issue of the proportion of the initial proton energy carried away by secondary electrons in the p-p interaction as described above. It is a straightforward matter to show from energy conservation considerations that the resulting new particles will carry away the following amounts of energy:

$$E_e / E_\mu \lesssim 103 \gamma_e / 105 \gamma_\mu \tag{2a}$$

$$E_\mu / E_\pi \lesssim 34 \gamma_\mu / 140 \gamma_\pi \tag{2b}$$

$$E_\pi / E_p \lesssim 2/3 \tag{2c}$$

where E_p, E_π, E_μ and E_e are the relativistic energies of the protons, pions, muons and electrons respectively. From (2a), (2b) and (2c) above we obtain

$$E_e / E_p \lesssim 0.15 \gamma_e / \gamma_\pi \sim 0.15 \quad \text{if} \quad \gamma_e \sim \gamma_\pi \tag{3}$$

where $\gamma_e, \gamma_\mu, \gamma_\pi$ are the respective Lorentz factors of the electrons, muons and pions. Available radio source data suggest that electrons in the beam ejected from the galactic nucleii will have energies in the range, $1 < E_e / m_e c^2 < 10^5$. Assuming that the ratio of the proton to electron energies is similar to that found in cosmic rays (i.e. $1 < E_p / E_e < 100$), then the energy of the secondary electrons produced in the p-p collisions discussed above will be related to the energy of the original electrons ejected from the nucleus as

$$\gamma_e^s m_e c^2 \lesssim 15 \gamma_e m_e c^2 \tag{4}$$

where γ_e^s is the Lorentz factor associated with the secondary electrons. Consequently, the p-p interactions constitute a plausible mechanism for the continuous injection of secondary electrons at various sites in an extragalactic radio source and covering roughly the same energy spectrum as the initial electron beam ejected by the galactic nucleii. If so, the need for electron reacceleration in the radio jet, hot spot and lobe, as recent observations demand, may become superfluous.

REFERENCES

Conway, R.G;., 1982, IAU Symposium No. 97, this volume, Eds. D.S. Heeschen and C.M. Wade, p. 167.
Okoye, S.E. and Okeke, P.N., 1981, To be submitted to Ap.J.
Readhead, A. and Pearson, T., 1982, IAU Symposium No. 97, this volume, Eds. D.S. Heeschen and C.M. Wade, p. 279.
Rosser, W.G.V., 1964, "An introduction to the theory of relativity", p. 235. (Butterworth and Co.).

A PRELIMINARY EXAMINATION OF THE EFFECT OF CLUSTER GAS ON TAILED
RADIO GALAXIES

D.E. Harris
Harvard/Smithsonian Center for Astrophysics

ABSTRACT

From a comparison of X-ray and radio data for 20 clusters which contain tailed radio galaxies, we find evidence for the effects of buoyant forces on the low brightness parts of radio tails. In several cases, the width of the tail increases markedly in low gas density regions, strengthening the case for thermal gas confinement of radio tails. Three examples of enhanced X-ray emission around radio galaxies which are further than 2 Mpc from their cluster centers are found.

1.   INTRODUCTION

The basic features of cluster radio sources have been worked out over the last ten years predominantly from the radio observations and supporting optical data. Now, however, we are able to study the gas distribution from Einstein Observatory (EO) images of clusters. These new data will allow us to evaluate many of the hypotheses involved in models of tailed radio galaxies (TRG). The present paper is a preliminary comparison of published radio maps of TRGs with EO observations obtained with the imaging proportional counter (one degree field with 1.5 arcmin resolution, Gorenstein et al. 1981). The X-ray observations are non-uniform in sensitivity because of different exposure times. Furthermore, there is a positional uncertainty of up to one arcmin for features on most of the IPC images which were processed prior to February 1981.

For convenience we divided the radio sources into 3 groups according to the length of the tail: short (< 100 kpc), medium (100 to 400 kpc) and long ($\gtrsim$ 400 kpc). These groups are rather coarse because the available radio data do not have a uniform sensitivity, resolution, or frequency coverage.

The present sample consists of 20 clusters (Table 1). These were chosen purely on the basis of available data and for this reason, the results noted below should be considered as "exploratory".

Table 1. A Selection of Tailed Radio Galaxies for Which IPC X-ray Observations are Available

Cluster	Radio Type	Consistent with Buoyancy?		Observer	Notes
		H	L		
A84	S	−		R	
A401	L	−	−	CAL	
A478	S	−		CFA	
3C129	S,L		+	CAL	
A629	M	−		CFA	
A754	L		+	HCDE	1,2
A1314	L,S(WAT)	−	+	HCDE	
A1367	M			CFA	
A1775	M	−	+	CFA	
A1940	M(WAT)	−		CFA	
A2022	L			CFA	
Zw 1615	L	−	+	HCDE	
A2199	S(WAT)			CFA	
A2220	M(WAT)		+	B	1
A2250	L	−	+	HCDWM	2
A2255	M	−	+	CFA	1
A2256	M			CFA	
A2306	M(WAT)	+	+	HCDE	
A2319	S			CFA	
4C47.51	L(WAT)		+	B	2

Notes: The radio types are L, long; M, medium; and S, short. "WAT" means "wide angle tail".

The columns "H" and "L" contain entries which indicate if the source morphology is (+) or is not (−) consistent with buoyancy effects. "H" is for the higher surface brightness parts of the radio tail and "L" is for the lower brightness parts.

Abbreviations for the observer are:
  CFA: Center for Astrophysics
  CAL: Columbia Astrophysics Laboratory
  R: J.G. Robertson
  B: J.O. Burns
  HCDE: Harris, Costain, Dewdney, and Ekers
  HCDWM: Harris, Costain, Dewdney, Willis, and Miley
1. This entry is, or contains, a radio galaxy which is associated with an X-ray source which is separate from the general cluster emission.
2. Apparent expansion of the radio source coincides with a region of lower gas density.

2. BUOYANCY

The effects of buoyancy on extended radio lobes have been suggested by many authors. To determine if the radio morphology could be caused by buoyant forces, we have assumed that higher surface brightness regions on the X-ray map correspond to spatial regions of higher gas density. This assumption is reasonable because the X-ray emissivity is proportional to the square of the electron density. However, moderate densities with long path lengths will also increase the surface brightness as will local heating of a cool gas where the emissivity is dropping exponentially in the EO energy band.

Examination of the present sample leads us to suggest that the high brightness parts of radio tails often bend toward high density regions but that the low brightness parts usually bend towards regions of lower density. Figure 1 is an X-ray map of Abell 2255 with a sketch of the radio sources superimposed. The bright part of the TRG, 4C64.20.1A, extends into the central, high density region. At 610 MHz (not shown), the tail is weak and bends to the north, a region

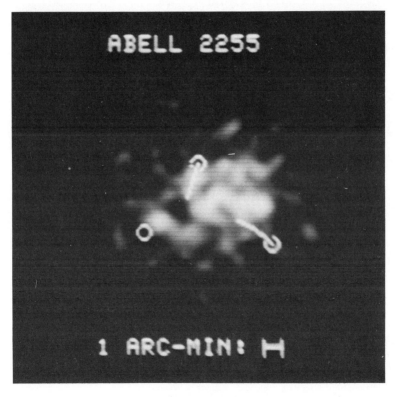

Figure 1. A2255. The X-ray map is shown by the grey scale, the radio galaxies by small circles, and the radio tails by white lines. For H_o=50 km s^{-1} Mpc^{-1}, one arcmin=138 kpc. The X-ray image has been smoothed with a Gaussian of 48".

of lower gas density. The details of the radio structure may be found in Harris et al. (1980a).

Abell 2250 is shown in Figure 2. Here, the bright part of the tail cuts across the central high density region, but the faint ends of both tails bend into low density regions (unpublished observation at 610 MHz).

An example of a wide angle tail (WAT) is shown in Figure 3 (Abell 2306). In this case, the whole structure is consistent with the effects of buoyancy.

In Table 1, we have indicated by a "+" or "-" whether or not the radio structure bends towards regions of low density. If our analysis is correct, the simplest explanation would be that the primary curvature of TRGs comes from curved trajectories of the parent galaxy, but that the older, lower surface brightness parts of the tails have had sufficient time to respond to buoyant forces.

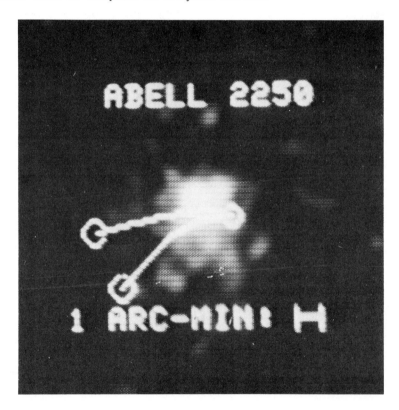

Figure 2. Abell 2250. The scale is 114 kpc for one arcmin. The two circles at the East ends of the tail indicate the positions of the expanded radio components mentioned in the text.

3. THERMAL CONFINEMENT

An approximate parity between the external gas pressure and the internal pressure in radio lobes has been found for many TRGs. The external pressure is estimated from the gas density and temperature, both derived from the X-ray observations, and the minimum internal pressure is calculated from the minimum energy density of the magnetic field and relativistic particles. Since the non-thermal pressure is often close to the external gas pressure, it follows that the internal gas pressure is probably substantially less than the external gas pressure. This is consistent with the conjecture that radio tails are less dense than the surrounding medium and are thus subjected to buoyant forces.

For three clusters in the present sample, A2250 (Fig. 2), A754, and 4C47.51, we find radio structure which shows evidence of thermal confinement. In all three cases, we see a relatively narrow tail which becomes wide and puffy in regions of low gas density.

Figure 3. A2306. The head of this wide angle tail lies close to the brightest part of the X-ray image. An extended radio source just to the north is identified with a double galaxy. For an estimated redshift of $z=0.1$, the scale would be 175 kpc/arcmin.

4. LOCAL GAS AROUND ISOLATED RADIO GALAXIES

The incidence of hot gas around radio galaxies is presently being studied by many authors. Here we note three cases of radio galaxies which lie at a projected distance of 2 to 4 Mpc from their respective cluster centers and which appear to be X-ray sources. The associated X-ray emission is quite distinct from the general cluster emission. The TRG in A2255 called "the Beaver" by Harris et al. (1980a) lies 2.6 Mpc to the south of the cluster center and is coincident with an unresolved X-ray source. The TRG in A754 (26W20, Harris et al. 1980b) is associated with an extended IPC source, and a galaxy which may be a member of A2220 (14W79, Harris et al. 1980a) also coincides with an IPC source.

While it is natural to hypothesize a higher density gas filling a local potential well, two other explanations should be investigated. For unresolved sources, the X-ray emission may arise from the nucleus of the galaxy rather than from a halo around the galaxy. Another possibility is local heating. If the outer regions of most clusters contain large amounts of gas at temperatures $\lesssim 10^6$ K, then local heating from radio galaxies could result in raising the temperature of this gas such that the 0.1 to 4 keV X-ray emission would be greatly enhanced. Evidence for gas at large distances from the center of A2255 is given by Hintzen and Scott (1980). A statistical study of normal and radio galaxies within 2 to 5 Mpc of cluster centers should help in differentiating between density enhancements and local heating.

5. THE LOCATION OF SMALL TRGs

There are a number of TRGs which are, in projection at least, physically small; on the order of 100 kpc or less. From the present sample, it appears that these "stubby" TRGs occur predominantly in regions of high gas density: 3C129.1, A478, A2199, A84, and two in A2319. Four of these six examples lie in the central regions of hot, smooth clusters and the other two are in the brighter parts of their respective clusters. The 3C129 cluster is a particularly striking example of this effect: 3C129.1 is in the center of the high brightness X-ray distribution while the long TRG, 3C129, skirts the edge of the visible gas.

Normally one assumes that short TRGs are just those that have a low velocity with respect to the surrounding gas. However, other possibilities should be considered. Is there any process which might suppress the creation of a long, well defined tail? Could high gas density and/or high temperature lead to excessive turbulence and rapid dispersion? If so, a dispersive loss of relativistic electrons might generate an extended radio halo of the type found in the Coma Cluster and in A2319.

6. SUMMARY

From a comparison of radio and X-ray observations of 20 clusters which contain tailed radio galaxies, we find:

(a) evidence for the effects of buoyancy on the old, faint parts of tails,
(b) evidence for thermal confinement of tails,
(c) the occurrence of enhanced X-ray emission from radio galaxies in the outer regions of clusters, and
(d) an indication that the formation of long, well defined tails may be suppressed in regions of high density and/or temperature.

ACKNOWLEDGEMENTS

J.G. Robertson and J.O. Burns kindly allowed me to use their guest observations and the group at the Columbia Astrophysical Observatory is acknowledged for permission to examine their EO data. Several clusters come from guest observations of Harris, Dewdney, Costain, Willis, Miley, and Ekers, and the remainder are from the CFA program on clusters which is headed by C. Jones.

REFERENCES

Gorenstein, P., Harnden, F.R., Jr., and Fabricant, D.G.: 1981, Trans. IEEE Nuc. Sci. NS-28, 869.
Harris, D.E., Kapahi, V.K., and Ekers, R.D.: 1980a, Astron. Astrophys. Suppl. 39, 215.
Harris, D.E., Costain, C.H., Strom, R.g., Pineda, F.J., Delvaille, J.P., and Schnopper, H.W.: 1980b, Astron. Astrophys. 90, 283.
Hintzen, P. and Scott, J.S.: 1980, Ap.J. 239, 765.

RADIO-OPTICAL STUDIES OF A COMPLETE SAMPLE OF ABELL CLUSTERS

B.Y. Mills and R.W. Hunstead
School of Physics
University of Sydney

Study of the properties of radio galaxies in clusters is beset with problems of selection and identification. To reduce these problems we have selected a complete sample of Abell clusters and are using optical spectra of identified galaxies to determine their cluster membership. The sample is in a volume of space defined by the limits of coverage and sensitivity of the two Sydney University radiotelescopes and the Anglo-Australian Telescope; it comprises the 42 clusters with $\delta \leq -8°$ and distance class 4 or closer. Radio maps have been prepared using archive 408 MHz data from the Molonglo Cross. Possible optical identifications have been selected from the Palomar Sky Survey and low-dispersion optical spectra have been taken of these and sometimes other galaxies or stellar objects within the cluster area. When needed, maps of the cluster areas have been produced at 1415 MHz using the Fleurs Synthesis Telescope. Some details of our early results have been published (Mills et al. 1978, 1979) and here a brief summary of the overall statistics is presented. Observations on five clusters remain to be completed but should not greatly affect the present conclusions.

The 408 MHz maps cover areas ranging from 0.25 to 4 square degrees, depending on the cluster distance. Mostly the coverage extends to the Abell radius or beyond. Results currently available for 37 clusters are as follows:

Total number of radio sources	263
Optical spectra obtained	84
Number of radio galaxies in clusters	48 (54)

The number in parenthesis includes six possible identifications with cluster galaxies which are at present uncertain. Confining attention to the more certain identifications, the distribution of radio galaxies among the clusters is as follows:

Number of radio galaxies/cluster	0	1	2	3	4	5
Number of clusters	13	11	5	6	1	1

This result is not significantly different from a Poisson distribution with a mean of 1.3 radio galaxies per cluster.

In order to examine the relationship between cluster properties and the occurence of radio galaxies, the clusters have been allocated to three distinct groups, those containing no radio galaxies (13), those containing one radio galaxy (11) and those rich radio clusters containing three or more radio galaxies (8). For these clusters we have calculated the mean richness R as defined by Abell, the mean redshift z, and the mean Bautz-Morgan classification:

Radio galaxies in cluster	0	1	≥ 3
$<R>$	$0.69 \pm .21$	$0.73 \pm .27$	$0.75 \pm .16$
$<z>$	$0.071 \pm .006$	$0.067 \pm .005$	$0.052 \pm .004$
$<BM\ class>$	$2.75 \pm .16$	$1.94 \pm .13$	$2.08 \pm .42$

Although the richest radio clusters are also the richest optically, the correlation is weak and not statistically significant. The significant negative correlation between radio richness and z is to be expected because lower luminosity radio galaxies can be detected at closer distances. The correlation with the BM classification arises because all BM types I or I-II in the sample contain at least one radio galaxy (not always the cD).

An important result is the very weak correlation between radio and optical richness, a result which has appeared in various guises in other investigations. The expected number of radio galaxies in a cluster is pN where p is the probability of a galaxy becoming a radio galaxy and N is the number of galaxies. If p is constant, a strong correlation between radio and optical richness would be expected. In the present sample N covers a range of about 6:1. The most natural explanation of the poor correlation here and elsewhere is that p independently varies over a greater range. There is some indirect evidence for variations of p in the present sample. Each of the seven clusters classified as BM types I or I-II contains at least one radio galaxy and the average is 2.3, more than twice the average of the remaining clusters. Additionally, it appears that the presence of strong X-ray emission is associated with a high value of p. Six of these clusters have been recorded as X-ray sources in published surveys. Five possess three or more radio galaxies each and the sixth, A1060, which has no clear-cut example, was detected as an X-ray source only because of its proximity (z = .010). Radio emission from three 'normal' galaxies in A1060 was detected for the same reason. Another obvious cluster property which may be related to p is the proportion of elliptical galaxies; this property is currently under examination.

To summarise, the majority of clusters in our sample contains at least one radio galaxy; radio galaxies are the rule rather than the exception in Abell clusters. They occur preferentially in clusters with above-average amounts of gas as shown by the BM and X-ray correlations.

REFERENCES

Mills,B.Y.,Hunstead,R.W.,Skellern,D.J.:1978,M.N.R.A.S.,185,pp.51P-56P.
Mills,B.Y.,Hunstead,R.W.,Shobbrook,R.R.,Skellern,D.J.:1979,
    New Zealand Jnl. of Science 22, pp.365-367.

STUDIES OF A COMPLETE SAMPLE OF ABELL CLUSTERS AT 1400 MHz

Richard A. White
NASA/Goddard Space Flight Center, NAS/NRC Research Associate

Frazer N. Owen
National Radio Astronomy Observatory

Robert J. Hanisch
University of Maryland

We present here analyses of a radio survey of Abell clusters at 1400 MHz using the NRAO 91-m telescope. Details will appear in a paper to be submitted to the Astronomical Journal where we present two lists. The first contains sources within 0.5 of an Abell radius (hereafter R_a, 3 Mpc if H=50) of the center of an Abell cluster. The second contains those clusters for which there were no sources within that limit. The flux limit is 100 mJy, the beam size ~10 1/2 arcmin, and the declination limit -19°30'. For consistency we use Corwin's (1974) m_{10} - z calibration throughout. The errors in m_{10} and therefore z and R_a, combined with a beam large compared to galaxy and cluster size (preventing identifications) preclude all but the simplest analyses which we present here.

We observed 1476 clusters and detected sources within 0.5 R_a in ~1/3 of them. The log N - log S relation of Fomalont et.al.(1974) indicates ~60% of the detections are expected to be random. In the remaining discussion we consider only sources within 0.3 R_a; all numbers are corrected for expected random detections. We define two statistical subsamples, both north of -19°. The first includes all clusters with $m_{10} \leq 16.9$ (in distance group 5). The second includes all clusters richness 3 or greater, and is 100% complete. Because of the large beam size and large average distance of these clusters, we will not discuss them further here. The first sample contains 538 clusters, 530 (98.5%) observed and 103+8 (19%+2%) detected. We present in the figures below correlations for various characteristics of this sample. Errors are 1 sigma.

In Fig. 1 we see the richer clusters are more likely to contain a radio source, possibly because they simply contain more galaxies. The data are consistent with the probability of a galaxy being a radio source being independent of cluster richness, but may show some Bautz-Morgan (BM) effect. In Fig. 2 equal normalized probabilities would mean that the radio galaxy distribution follows a King model. There may be an indication

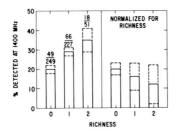
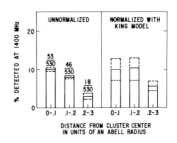

Fig. 1. Detection Rate vs. Richness   Fig. 2. Detection Rate vs. Radius

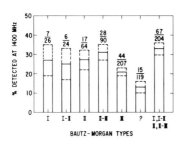
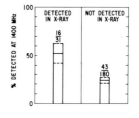

Fig. 3. Detection Rate vs. BM Type   Fig. 4. Correlation with HEAO A-2 Results of McKee et.al.(1980).

of a BM effect. In Fig. 3, comparing rates for BM III with BM I to II-III clusters, there is some evidence that clusters which contain a central galaxy with some degree of dominance are more likely to contain a radio source. This may be simply because brighter galaxies are intrinsically more likely to be radio sources; but the degree of dominance does not seem to matter. The central location may also play a role. In Fig. 4 the correlation of detection rates is most likely due to higher radio detection rates in rich clusters, which are also preferentially detected by McKee et.al.(1981). The BM effect may enter in the same way.

We are pursuing the appropriate VLA, optical, and X-ray studies to investigate further the effects of cluster environment on radio sources, particularly those associated with dominant central galaxies.

REFERENCES

Corwin, H.G.: 1974, Astron. J. 79, 1356.
Fomalont, E.B., Bridle, A.H., and Davis, M.M.: 1974, Astron. Astrophys. 36, 273.
McKee, J., Mushotzky, R., Boldt, E., Holt, S., Marshall, F., Pravdo, S., and Serlemitsos, P.: 1980, Astrophys. J. 242, 843.

TWO PECULIAR RADIO GALAXIES IN A1367

G. Gavazzi
Istituto di Fisica Cosmica, CNR, Milano, Italy

W. Jaffe
NRAO, Edgemont road, Charlottesville, Virginia

We observed the cluster of galaxies A1367 to map the structure of the cluster and that of the radio galaxy 3C264. We report on 3C264 and on the peculiar galaxy UGC6697.

3C264
Fig.1 shows the 1.4 GHz map of the radio galaxy. The large scale morphology of the source belongs to the classical head-tail type with a twin-arm structure. The head contains a very large fraction of flux. A bright unresolved core coincides in position with the nucleus of the galaxy. Like in other sources of this type the spectral index increases from the core along the two arms as expected in the 'trail' model where radiative losses steepen the electron spectrum away from their origin. What is striking about this source is the broad, amorphous structure of the head as shown in Fig.2 at much higher resolution. In fact this component does not contain any jet or sign of collimation, the latter being, on the contrary, the characteristic of the source on the large scale. The observed morphology of the head seems to imply that energy collimation from the nucleus has, in the recent past, been destroyed, while the nucleus itself remained on an active stage, as demanded by the luminosity of the head. This fact is unique among other well known head-tail sources like NGC1265, where the collimation is mantained on all scales.

UGC6697
The present observations seems to rule out the interpretation given in Gavazzi (1978, A.A., $\underline{69}$, 355) where the asymmetry of the radio source associated with this late type galaxy was interpreted as the relativistic material was swept out from the galaxy disk by ram pressure stripping (trail model). In fact the high resolution radio and optical observations shown here match very closely: both are very asymmetrical with a steep brightness gradient to one side and a long smooth tail on the other (Fig.4). The distribu-

tion of the spectral index (Fig.3) is remarkably steep and constant along the source. It does not show high frequency cut-off or ageing of the electrons along the radio tail. The evidence of optical low surface brightness features in the tail suggests that star formation is undergoing in the region where radio radiation shows up. Ram pressure stripping could provide the necessary mechanism to sweep matter from the disk of the galaxy in to the far tail.

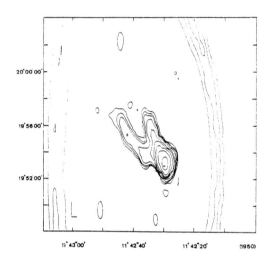

Fig.1: 1.4 GHz map of 3C264

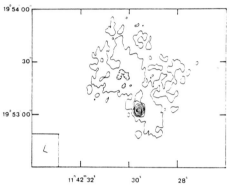

Fig.2: 4.9 GHz map of the head and core of 3C264

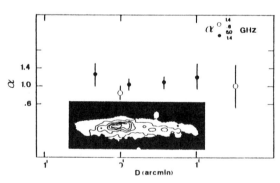

Fig.3: 1.4 GHz VLA map of UGC6697 (2".6x5".1 res.) with three frequencies spectral index distribution

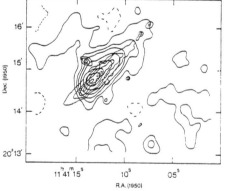

Fig.4: 1.4 GHz VLA map of UGC 6697 (17"x20" res.) superposed on blue isophotes

RADIO OBSERVATIONS AT 1.4 GHz OF ABELL CLUSTERS

C.Fanti, R.Fanti, L.Feretti, A.Ficarra, I.M.Gioia,
G.Giovannini, L.Gregorini, F.Mantovani, B.Marano*,
L.Padrielli, P.Parma, P.Tomasi, G.Vettolani
Istituto di Radioastronomia, via Irnerio 46, Bologna, Italy
*Istituto di Astronomia, via Zamboni 33, Bologna, Italy

We have observed with the Westerbork Synthesis Radio Telescope (WSRT) at 1.4 GHz the Abell clusters included in the HEAO-2 satellite observing program, for which radio information was not available. In practice we excluded the clusters south of 15°, in order to ensure an adequate angular resolution in declination; moreover we did not observe distance class (d.c.) 6 clusters for which better resolution and sensitivity would be necessary. The final list includes 61 clusters. These have been observed to a limiting peak flux density of about 5 mJy, corresponding to average minimum radio powers ranging from $\sim 5 \times 10^{21}$ to $\sim 2 \times 10^{23}$ W/Hz for d.c. 1 to 5. By adding the present data to those already available in the literature, we have radio information about all the clusters of d.c. 1 and 2, north of 15°, except A1185. The sample of d.c. 3 clusters contains 60% of the clusters of this class, but seems unbiased both for richness criteria and for radio characteristics; therefore it is useful for statistical studies. The observed clusters of d.c. 4 and 5, instead, are richer than average. The list of the observed clusters is presented in table 1: the Abell name is given in column 1, the d.c. in column 2, the richness class in column 3, the number of radio detected cluster galaxies in column 4 (in parentheses the number of galaxies for which the membership of the cluster is doubtful or the radio identification is not certain is given). The identification of the radio sources with the cluster galaxies of d.c. 5 is still in progress. While a more complete discussion of the properties of radio sources in clusters will be performed later, using the data about all the clusters of the sample, here we summarize the results of the discussion on d.c. 1 and 2 clusters.

The bivariate radio luminosity function (RLF) of the cluster galaxies has been derived for the morphological types E, S0 and S+Irr, dividing these into 3 different classes of absolute optical magnitude. The RLFs of E and S0 galaxies in the present clusters do not differ from the corresponding ones for clusters of higher richness. There is a marginal evidence that S+Irr galaxies in cluster have a lower probability of being radio sources than field ones. The RLF of the E + S0 first cluster members with absolute optical luminosity less than -20 has been computed using all the d.c. 0, 1, 2 Abell clusters with $\delta < 15°$, except A407 and

A1185. There is an indication at 1.5 r.m.s. level that first cluster members have a higher probability of being radio sources than the other galaxies of the same optical luminosity range.

The proportion of resolved to unresolved sources in the present sample is similar to that of the WSRT survey of rich clusters. A plot of the maximum linear size as a function of the total radio power shows a clear trend indicating that more powerful radio sources have larger sizes. This size-power distribution does not differ from the corresponding distribution for low luminosity radio galaxies not belonging to rich clusters. Among the resolved sources, 3 show head-tail or wide-angle-tail type radio structure: one in A569, another in A576, the last in A2162. This is an indication of the presence of a dense intracluster medium (in the case of A576 at the periphery of the cluster).

Table 1

Name	D	R	N	Name	D	R	N
A71	3	0	0	A1569	5	0	-
A98	5	3	-	A1589	5	0	-
A154	3	1	1+(1)	A1654	5	0	-
A160	4	0	1	A1674	5	3	-
A179	3	0	1	A1760	5	3	-
A195	3	0	1	A1767	4	1	1
A262	1	0	8	A1781	3	0	1
A272	5	1	2	A1800	3	0	2
A278	3	0	2	A1831	3	1	2
A347	1	0	1	A1913	4	1	1+(1)
A397	3	0	1	A1927	4	1	1+(2)
A568	3	0	2+(1)	A1939	5	1	-
A569	1	0	2	A1940	5	3	-
A576	2	1	2	A1983	3	1	1
A608	5	1	-	A1990	5	3	-
A646	5	0	-	A1991	3	1	2
A655	5	3	-	A2065	3	2	1
A671	3	0	(1)	A2079	3	1	2
A779	1	0	0	A2089	4	1	1
A899	5	1	-	A2092	4	1	(2)
A963	5	3	-	A2100	5	3	-
A1177	4	0	1	A2107	4	1	(2)
A1213	2	1	3	A2148	3	0	1
A1228	1	1	1	A2162	1	0	5
A1254	3	1	1	A2244	5	2	..
A1267	3	0	1+(1)	A2301	4	0	2+(1)
A1268	5	2	-	A2572	3	0	3+(1)
A1377	3	1	1+(1)	A2625	3	0	1
A1413	5	3	-	A2630	3	0	0
A1425	5	1	-	A2666	1	0	0
A1500	3	0	1				

THE RADIO EMISSION OF INTERACTING GALAXIES

E. Hummel[1], J. M. van der Hulst[2], J. H. van Gorkom[3],
C. G. Kotanyi[4]

[1]University of New Mexico, Albuquerque, NM, USA
[2]University of Minnesota, Minneapolis, MN, USA
[3]NRAO, VLA Program, Socorro, NM, USA
[4]University of Groningen, The Netherlands

Gravitational interaction is a straightforward interpretation of some of the peculiar optical morphologies shown by galaxies. There have also been attempts to study the effects of a gravitational interaction on the radio continuum emission. Statistically, the central radio sources (inner 1 kpc) in interacting spiral galaxies are about three times stronger than in isolated spirals; on the other hand, the intensity of the extended emission does not seem to be affected (Stocke, 1978; Hummel, 1981). Peculiar radio morphologies are not a general property of interacting galaxies, since in the complete sample studied by Hummel (1981) of spirals with a probability ≥ 0.8 of being physically related to their companion, less than 5% have a peculiar radio morphology.

Even in the few cases where an unusual radio morphology has been seen, gravitational interaction is not always the most attractive explanation. We present here VLA observations of the radio structure in three galaxies where the peculiar radio morphology is well established. In at least one of these (NGC 4438) the radio structure is better interpreted in terms of interaction with a diffuse intergalactic gas rather than in terms of a galaxy-galaxy interaction.

NGC 4038/39 (Arp 244, the "Antennae"): The figure shows the distribution of the continuum emission at 20 cm. The bulk of the radio emission coincides with the dusty region between the galaxies. The two discrete radio sources near the optical positions are presumably nuclear sources or their remnants. The extended emission shows considerable structure. There are three separate components in the southern part and a kind of plateau that extends to the northern nuclear source. No emission was detected from the optical "tails" at a level of 1 mJy/beam.

NGC 4410/IC 790 (see photograph in Stocke et al. 1978): We show the continuum map at 20 cm convolved to a resolution of 20"x20". The radio emission is spread over an area of 40x100 kpc (D=74 Mpc, assuming H=100 km/sec/Mpc), while the associated galaxies are optically compact, their sizes are about 10 kpc. A nuclear radio source is associated with NGC 4410a, a ring galaxy with a small bulge within the ring. At high resolution this source is a double with a separation of 0.5 kpc at PA 90°. The eastern galaxy NGC 4410b shows no radio emission. The extended emission

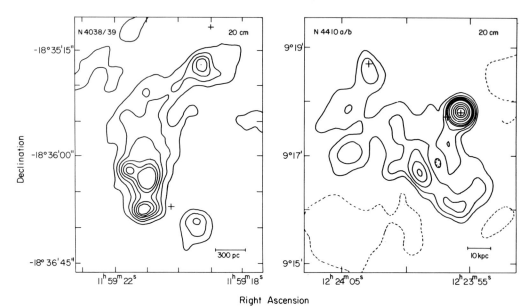

Figure: NGC 4038/39, cnt=2, 4, 6, 8, 10, 12 mJy/beam, beam=6"x6"; NGC 4410a/b, cnt=-1.5, 1.5, 2.5, 3.5, 4.5, 5.5, 7.5, 10, 12.5, 15, 17.5 mJy/beam, beam=20"x20". Optical positions of the galaxies are indicated by +.

shows a ridge of about 100 kpc in length displaced to the south of NGC 4410a/b by about 30 kpc. The emission associated with IC 790 suggests a head-tail morphology. A faint optical bridge, visible on the PSS, connects NGC 4410, IC 790 and an anonymous galaxy further to the northeast, suggesting a dynamical interaction involving these galaxies.

NGC 4438 (Arp 120): Observations with the Westerbork array showed that the centroid of the extended radio emission in this galaxy in the center of the Virgo cluster is displaced by about 4 kpc from the optical nucleus and central radio source, probably as a result of the interaction with the diffuse intergalactic medium in the cluster (Kotanyi et al., 1982). The high-resolution VLA map of the central source in the galaxy shows that this has a peculiar structure as well. The source is elongated in PA 124° and has a projected size of 1x0.3 kpc. The orientation shows no relation with either the optical major axis or the direction from the optical nucleus to the extended radio component (PAs of resp. 30° and 250°).

References

Hummel, E., 1981, Astron. Astrophys. 96, 111.
Kotanyi, C.G., van Gorkom, J.H., Ekers, R.D., 1982, Astron. Astrophys., in press.
Stocke, J.T., 1978, Astron. J. 83, 348.
Stocke, J.T., Tifft, W.G., Kaftan-Kassim, M.A., 1978, Astron. J. 83, 332.

STEPHAN'S QUINTET REVISITED

J. M. van der Hulst[1] and A. H. Rots[2]
[1]Department of Astronomy, University of Minnesota, Minneapolis, MN, USA
[2]National Radio Astronomy Observatory, Socorro, NM, USA

VLA observations at 1465 MHz of the Stephan's Quintet region reveal that the arc-shaped area of emission discussed by Allen and Hartsuiker (1972) breaks up into several components. The idea that NGC 7318b is a recent interloper in the group and that the interaction resulting from this event causes the enhanced activity at the east side of NGC 7318b is adopted as still the most reasonable explanation. The results are discussed in more detail in another paper (van der Hulst and Rots, 1981).

1. INTRODUCTION

When Allen and Hartsuiker (1972) originally discovered the radio emission between NGC 7319 and NGC 7318b, they proposed that the emission traces a galactic bow shock due to the gravitational accretion of gas by NGC 7319 as it moves through the group of galaxies. The more recent HI studies by Allen and Sullivan (1980) and Peterson and Shostak (1980) have somewhat modified this picture. These authors propose a collision between NGC 7319 and NGC 7318a, to explain the HI distribution. They also suggest that NGC 7318b had no part in the collision and is a recent interloper in the group.

2. OBSERVATIONS AND RESULTS

We observed Stephan's Quintet with the partially completed VLA in November 1979 at 1465 MHz. Maps were obtained with resolutions of 2.25", 6", and 24". Figure 1 shows the 6" resolution map superimposed on a IIIaJ photograph of Arp (1973). Various individual sources of radio emission were detected:

(i) The nucleus of NGC 7319. The flux density of this source is 20.5+1.0 mJy. It is slightly extended and shows a jet-like feature at a position angle of 207°, nearly perpendicular to the central bar in this galaxy.

(ii) Discrete source north of NGC 7318b. The northern part of

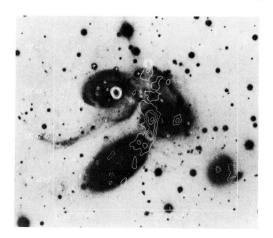

Figure 1. Contour diagram of the 6" resolution VLA map of Stephan's Quintet, superimposed on a IIIaJ photograph of Arp (1973). The contour values are -0.6 (dashed), 0.6, 1.2, 1.8, 2.4, 3.6, 4.8, and 9.6 mJy/beam.

Allen and Hartsuiker's (1973) arc of emission resolves into a discrete, slightly extended source. It probably is an unrelated background source.

(iii) NGC 7318b. The central part of the arc appears to curve to the east, rather than the west. On a global scale it coincides with spiral arm features and HII regions, although there are definite systematic displacements. There is a diffuse extension to the south, partly covering NGC 7320.

(iv) The center of NGC 7318a. A weak radio source was detected here with a flux density of 1.3+0.3 mJy.

3. DISCUSSION

Our observations lend support to the hypothesis that NGC 7318b is an interloper in Stephan's Quintet. The most likely scenario involves a burst of violent star formation at the eastern edge of NGC 7318b induced by its interaction with the group and the intergalactic medium, and resulting in the formation of giant HII regions and an enhanced rate of supernova events. The general coincidence of the radio emission with regions of optical emission, including bright HII regions as signposts of recent activity, supports this idea.

REFERENCES

Allen, R.J. and Hartsuiker, J.W. 1972, Nature 239, p. 324.
Allen, R.J. and Sullivan, W.T. 1980, Astron. Astrophys. 84, p. 181.
Arp, H.C. 1973, Astrophys. J. 183, p. 411.
Peterson, S.D. and Shostak, G.S. 1980, Astrophys. J. 241, p. L1.
van der Hulst, J.M. and Rots, A.H. 1981, Astron. J. (in press).

A MORPHOLOGICAL CLASSIFICATION OF CLUSTERS OF GALAXIES
FROM EINSTEIN IMAGES

C. Jones and W. Forman
Harvard-Smithsonian Center for Astrophysics

The earliest Uhuru observations showed that cluster X-ray sources were not associated with single individual galaxies but were extended sources (Gursky et al. 1971, Kellogg et al. 1972, and Forman et al. 1972). The detection of iron line emission from X-ray spectroscopic observations (Mitchell et al. 1976 and Serlemitsos et al. 1977) showed both that the dominant X-ray emission process was thermal bremsstrahlung and that the gas had been processed through stellar systems before being injected into the intracluster medium.

Various optical classification systems have been developed for clusters with the goal of providing dynamical information. Properties of cluster member galaxies have been suggested as indicators of dynamical evolution. For example, relaxed, evolved clusters tend to be characterized by lower fractions of spiral galaxies, higher velocity dispersions and larger central densities (see Bahcall 1977 for a review). Hausman and Ostriker (1978) suggested that the Bautz-Morgan system is directly related to cluster evolution with type I systems -- those with optically dominant cD galaxies -- the most highly evolved.

Although ten years have passed since clusters of galaxies were discovered to be X-ray sources, it has only been with the advent of the Einstein X-ray imaging observatory that a first look at cluster X-ray morphology and classification has been possible. The proposed classification system divides clusters into two families -- those with and those without X-ray dominant galaxies. Within each family, the dynamical indicators display a full range of values.

One subgroup of clusters is those whose X-ray emission is not regular and which do not contain an X-ray dominant galaxy. The luminosities of these irregular clusters are low by classic cluster standards (10^{42} to 10^{44} ergs/sec). One of the brighter, nearer and best studied members of this class is A1367. Bechtold et al. (1981) have discussed the Einstein observations of this cluster.

In the IPC image of A1367, Figure 1, the cluster fills most of the field and has an elongated central region. At the southeast is a bright elliptical galaxy which contains a 3C radio source with a radio tail (Gavazzi 1978) and a nuclear X-ray source (Elvis et al. 1981). Figure 2 shows contours of the X-ray emission on an optical photograph.

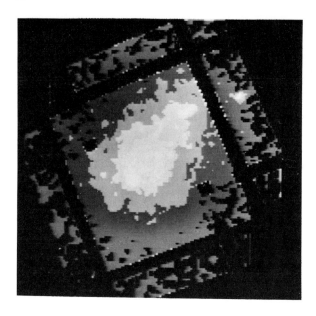

Figure 1: A1367 observed in two offset IPC fields in the energy range 0.5-3.0 keV.

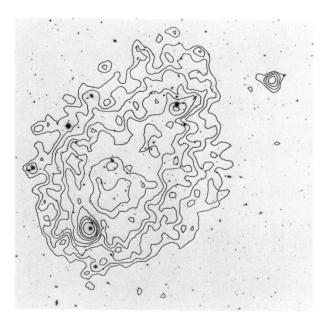

Figure 2: The X-ray iso-intensity contours of A1367 show the elongatation of the central region of the cluster and the galaxy associated with 3C284 in the southeast.

Using an isothermal sphere model Bechtold et al. obtained core radii of 0.80 Mpc for the long axis and 0.42 Mpc for the short axis. Carter and Metcalf (1980) have measured the cluster ellipticity from optical galaxy counts and find a ratio of .5 ± .1 with a similar position angle to the x-ray.

The IPC image was suggestive that the emission was not smooth

(Jones et al. 1979). Therefore, we performed a long (37K second) HRI observation of the central region. Since the background varies over the field of view due to the cluster emission, Bechtold et al. calculated the background locally in regions adjacent to the candidate source locations and searched the field with various detection cell sizes ranging from 12" to 2' on a side. For comparison, the same analysis was performed on the Cetus deep survey.

With the 1' cell Bechtold et al. detected ten extended regions in A1367 compared to one in Cetus. The one in Cetus is associated with a poor group of galaxies. To demonstrate that these clumps of emission were not an artifact of extended cluster emission, Bechtold et al. repeated the analysis on the Columbia Einstein observations of the Coma cluster. Both Coma and A1367 lie at the same distance and both are richness class 2, so their X-ray morphologies may be easily compared. In Coma two sources were detected, compared to the ten in A1367. One is a strong point source 5' from the cluster center with no obvious optical counterpart. The second corresponds to the peak of the cluster emission at the cluster center. Although clumped emission is observed in A1367 and not in Coma, clumps such as those found in A1367 could exist in Coma but would remain undetected due to the higher level of general cluster emission in Coma.

In A1367 about 5% of the total cluster emission is in blobs. Oemler (1980), in a magnitude limited sample counted 19 galaxies in the central region. The extended X-ray sources occupy 5% of the area and include six of Oemler's galaxies. The probability of finding this association by chance is 2×10^{-4}. A seventh galaxy, below Oemler's magnitude limit coincides with one additional source. Thus, there is a significant association of these extended sources and cluster galaxies. However, three sources are not near galaxies. One contains several faint objects and may be a distant cluster.

At the distance of A1367, the X-ray luminosities of these blobs are 10^{40} to 10^{41} ergs/sec. which is five to fifty times the X-ray luminosity of our galaxy. Bechtold et al. considered several scenarios for maintaining the hot corona. In the core of the cluster, mass loss rates from ram pressure stripping and evaporation are ten to one hundred times the maximum gas replenishment rate postulated for normal galaxies. But these excessive galactic winds are not required if the galaxies have massive halos.

While the observation of coronas around approximately one third of the core galaxies suggests that massive halos may be a common property of galaxies in A1367, for the remaining galaxies Bechtold et al. did not detect hot coronas. These galaxies appear to be of the same galactic types as those with coronas so that the mass injection rate which varies with galactic type is not apparently the deciding factor.

Observations suggest that A1367 is a young, dynamically unevolved cluster. Its irregular appearance and low central concentration both optically and in X-rays indicate that it has not fully relaxed. Its low X-ray temperature (Mushotsky and Smith 1980) implies a weak cluster potential, characteristic of a system in an early stage of collapse.

White (1976) has modelled the dynamical evolution of clusters. In his numerical simulation, the galaxies first separate out from the general Hubble expansion and form small groups. These groups may merge into two large subclusters which condense to form a relaxed cluster.

From the X-ray imaging observations, four clusters have been discovered to have double structure in their surface brightness distributions (Forman et al. 1980). Figure 3 shows a mosaic of fields around the southern double cluster SC0627-54. The extended source to the north is another cluster at the same redshift as the double cluster, separated by ~4 Mpc.

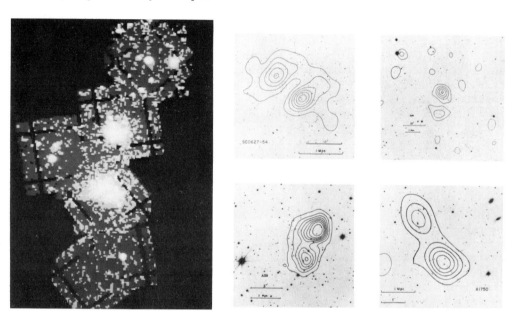

Figure 3: The 0.5-3.0 keV image around SC0627-54 showing the double cluster and a cluster to the north.

Figure 4: Iso-intensity contours for four double clusters superposed on optical photographs.

Figure 4 shows the X-ray iso-intensity contours for the four double clusters superposed on optical photographs. Each subcluster is centered on one of the bright cluster galaxies. Redshifts of each pair are consistent with the doubles being physically associated. Forman et al. have shown that all of the cluster components are extended with core radii similar to those of A1367. Their luminosities are also typical for rich clusters.

Unfortunately, there is no precise information for any of the dynamical indicators including spiral fractions, X-ray temperatures or accurate subcluster velocity dispersions. The observed subcluster separations and core radii agree with those expected from White's model for an intermediate dynamical phase. When the cluster survey is completed, the frequency of double clusters combined with numbers of clusters in each subgroup, should allow an estimate of how often the evolution proceeds through a double phase.

The Coma cluster is the archtype of an evolved, fully relaxed cluster. A Coma-type cluster A2256 is shown in Figure 5. The X-ray emission in these clusters is smooth and can be well-described by an isothermal sphere. The X-ray luminosities of these clusters are high and the X-ray temperatures are hot.

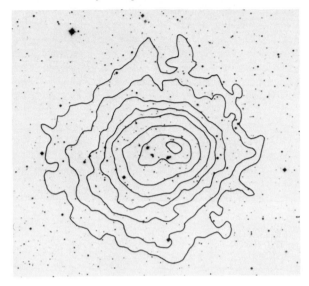

Figure 5: The X-ray iso-intensity contours of A2256 superposed on the PSS photograph.

In a parallel structure to the family of clusters from A1367 to Coma are those which contain a dominant galaxy. One such dynamically unevolved cluster is Virgo. Figure 6 shows the Virgo cluster in X-rays. Most of the bright sources in this figure are associated with

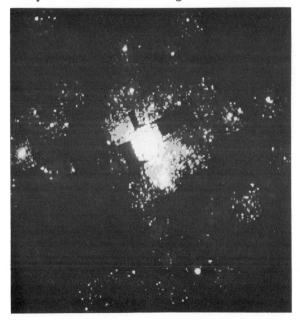

Figure 6: The 0.5-3.0 keV image of the Virgo cluster. X-rays from M87 dominate the central region. Each field is a single 1° x 1° IPC image.

galaxies. Dominating the central region is M87. Although much of the emission is associated with the gas halo around M87, there is extended cluster emission which is not symmetric about M87 but brighter to the north and west.

Fabricant, Lecar, and Gorenstein (1980) have used the IPC observations to trace the gravitational potential of M87 and thereby determine the mass of the galactic halo needed to bind the X-ray emitting gas. To a distance of ~230 kpc, the inferred galaxy mass is several x 10^{13} $M_\odot$ - ten times more massive than either the galaxy or the X-ray gas. Alternatively Binney and Cowie (1981) have suggested that the pressure of the cluster gas and a temperature gradient in the halo gas may allow the gas to be bound and slowly accreting with a less massive halo.

Two other ellipticals in the Virgo core are M86 and M84. Figure 7 shows an observation of these made with the IPC. In addition to extended emission centered on the galaxy, there is also an X-ray plume or tail. The spectra of both the galaxy and the plume can be characterized by thermal bremsstrahlung from 1-2 keV gas.

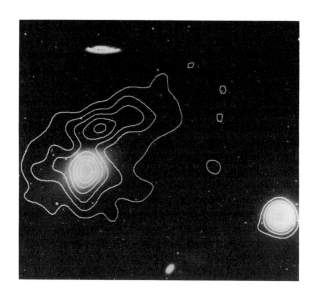

Figure 7: The X-ray image of M86 shows a central region with a detached plume to the north. M84 is to the west.

The key to understanding M86 is its high velocity of approach - nearly 1500 km/sec with respect to the Virgo cluster. Forman et al. (1979) suggested that the gas we presently see associated with M86 was accumulated during the intervals between the periodic passages of M86 through the Virgo core. M86 spends about five billion years between the time it exits the core, rises to its maximum height above the core and then falls back to the cluster center. M86 is now well into the Virgo core. The plume is gas that has been ram pressure stripped from the envelope around M86 by the intracluster medium but not yet heated and incorporated into that medium.

M84 is a less luminous X-ray source. Since it contains a 3C radio source, it was possible that the emission was associated with a compact nuclear source. However, HRI observations showed that the M84 X-ray source was extended on a scale of about 20 seconds, smaller than that around M86. M84 has a velocity which is equal to the mean of the Virgo cluster. Therefore, M84 appears to be a permanent resident of the cluster core where it is continually exposed to the intracluster gas which strips or evaporates its outer envelope. We observe only the core of the gas distribution which is tightly bound to the galaxy. As for A1367, these observations show that hot X-ray coronas are not an unusual feature of galaxies in dynamically unevolved clusters. The Virgo observations provide strong evidence that at least some of the intracluster gas is generated by ram pressure stripping of gas produced in the member galaxies.

In the family of clusters with dominant galaxies, we have in addition to young clusters like Virgo, dynamically evolved clusters, which are exemplified by A85 whose X-ray contours are shown in Figure 8.

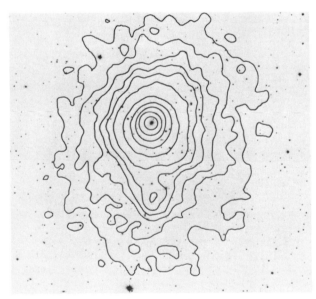

Figure 8: The X-ray iso-intensity contours of A85. The lowest contour is 3σ above background.

The emission is sharply peaked towards the center and coincides with the bright cD galaxy that sits at the bottom of the cluster's potential well. The X-ray gas in these clusters is hot and is associated with the cluster as a whole, not with individual galaxies. The x-ray emission is quite smooth in contrast to the irregular emission seen in A1367 type clusters. The X-ray surface brightness distribution of cD clusters is quite symmetrical suggesting relaxed systems.

Figure 9 shows the radial surface brightness profiles for four types of clusters. The first two are A1367 and A262 (like M87) which are the dynamically less evolved clusters and the last two - the Coma-type and the cD's are the more evolved clusters. Each evolved cluster has its progenitor - the M87-type for the cD clusters and the irregular A1367-type for the Coma-type clusters.

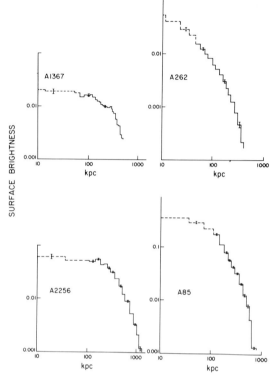

Figure 9: The 0.5–3.0 keV X-ray surface brightness profiles of A1367, A262, A85, and A2256. (The dashed lines designate where the IPC resolution effects the true profile.)

A85 and A2256 are at the same redshift so the detector resolution is the same for both. The surface brightness of the cD cluster falls much faster than that of the Coma type. Similarly with the young, unevolved clusters, the surface brightness profile of clusters with dominant galaxies falls faster than those without dominant galaxies. The clusters with steep slopes are both cD clusters and M87 type systems. The flatter distributions are both dynamically young A1367 systems and relaxed Coma-type clusters. The X-ray surface brightness distribution can be used to place the cluster in its family, then optical and X-ray indicators are used to determine the relative evolutionary stage.

Table 1 summarizes the dynamical indicators. The low X-ray temperature of A1367, its irregular X-ray and optical appearance and high spiral fraction indicate a relatively unevolved system. Similarly Virgo and A262 are in early stages. In contrast, Coma and A85 with their high X-ray temperatures, high central concentrations, and low spiral fractions are more evolved systems.

TABLE 1 - Cluster Classification

X-ray Dominant (XD) Systems	Non XD Systems
A262 Cool X-ray gas 2.4 ± 0.8 keV Emission around single galaxy Low velocity dispersion 478^{+183}_{-110} km/sec High spiral fraction 45%	A1367 Cool X-ray gas 2.8 ± 1.0 keV Clumpy emission around galaxies Low velocity dispersion 694 ± 75 High spiral fraction 40%
A85 Hot X-ray gas 6.8 ± 0.5 keV Smooth emission centered on single dominant galaxy Low spiral fraction $\leq22\%$	A2256/Coma Hot X-ray gas $7^{+3}_{-2}/7.9\pm.3$ keV Smooth,"isothermal sphere" emission High velocity dispersion $1274\pm250/900\pm63$ Low spiral fraction ?/13%

In addition to defining the X-ray classes of clusters, the percentage of clusters of each type is relevant. Although our samples are not complete, we can make some estimates. In the first family, the largest class is the irregular, low luminosity A1367-type clusters. Short observations of weak clusters make it difficult to determine if clumpy emission corresponds to an irregular cluster or a cluster with a double condensation. At present less than 10% of all clusters appear as clear doubles. Although Coma is thought of as the "classical" cluster, again less than 10% are like Coma.

In the second family, we have separated the two main classes based primarily on the presence of a cD galaxy although X-ray temperatures were used when available. About 20% of the clusters are unevolved with dominant M87 type galaxies. For the evolved dominant systems, Leir and van den Berg (1977) found that about 10% of the clusters were Bautz-Morgan types I and I-II. Since many of these appear as cD X-ray clusters, the fraction in this class is also about 10%.

The efficacy of the X-ray images in cluster classification studies derives from their sensitivity in mapping the mass distribution within clusters. The studies of gaseous corona around individual cluster galaxies suggest the existence of massive halos. This analysis applied to entire clusters gives masses in good agreement with the virial mass. Although only about 10% of the total cluster mass is in the form of hot X-ray gas, it is a remarkably sensitive tracer of the unseen material defining the cluster potential.

In summary the cluster images are useful in determining the cluster family and the cluster's dynamical state. However, further detailed analysis is necessary to understand the processes that occur as evolution proceeds.

We are indebted to K. Gilleece for her care in the preparation of this manuscript. This research was supported by NASA Contract NAS8-30751.

REFERENCES

Bahcall, N. 1977, Ann. Rev. of Astr. and Astrophys. 15, 505.
Bechtold, J., Forman, W., Giacconi, R., Jones, C., Schwarz, J., Tucker, W., and Van Speybroeck, L. 1981, Ap.J., in press.
Carter, D. and Metcalf, N. 1980, MNRAS, 191, 325.
Elvis, M., Schreier, E., Tonry, J., Davis, M., and Huchra, J. 1981, Ap.J., 246, 20.
Fabricant, D., Lecar, M., and Gorenstein, P. 1980, Ap.J., 241, 552.
Forman, W., Kellogg, E., Gursky, H., Tananbaum, H., and Giacconi, R. 1972, Ap.J., 178, L309.
Forman, W., Schwarz, J., Jones, C., Liller, W., and Fabian, A.C. 1979, Ap.J., 234, L27.
Forman, W., Bechtold, J., Blair, W., Giacconi, R., Van Speybroeck, L., and Jones, C. 1981, Ap.J., 243, L133.
Gavazzi, G. 1978, Astr.Ap., 69, 355.
Gursky, H., Kellogg, E., Murray, S., Leong, C., Tananbaum, H., and Giacconi, R. 1971, Ap.J., 167, L81.
Hausman, M. and Ostriker, J. 1978, Ap.J., 224, 320.
Jones, C., Mandel, E., Schwarz, J., Forman, W., Murray, S., and Harnden, F.R., Jr. 1979, Ap.J., 234, L21.
Kellogg, E., Tananbaum, H., Giacconi, R., and Pounds, K. 1972, Ap.J., 174, L65.
Leir, A. and van den Berg, S. 1977, Ap.J. Suppl., 34, 381.
Mitchell, R.J., Culhane, J.L., Davison, P.J.N., and Ives, J.C. 1976, MNRAS, 176, 29p.
Mushotsky, R. and Smith, B. 1980, Highlights of Astronomy, 5, 735.
Oemler, G. 1980, private communication.
Serlemitsos, P., Smith, B., Boldt, E., Holt, S., and Swank, J. 1977, Ap.J., 211, L63.
White, S.D.M. 1976, MNRAS, 177, 717.

RADIO AND X-RAY STRUCTURE OF CENTAURUS A

Eric D. Feigelson
Center for Space Research and Department of Physics
Massachusetts Institute of Technology
Cambridge, Massachusetts 02139

Abstract. Recent studies of the nearby radio galaxy Centaurus A with the Very Large Array and the Einstein X-Ray Observatory reveal complex radio and X-ray structures. A prominent one-sided jet comprised of resolved knots located 0.2-6 kpc from the nucleus is seen in both radio and X-rays. The X-ray emission is probably synchrotron, requiring in situ reacceleration up to $\Gamma \simeq 10^7$. Inverse Compton emission is not a likely explanation though a thermal model in which the nucleus ejects dense $10^5 M_\odot$ clouds cannot be excluded. An elongated X-ray region is also found near the "middle" radio lobe and optical HII regions $\sim$ 30 kpc NE of the nucleus. Conditions around the active nucleus, the absence of X-rays from the inner radio lobes, and X-ray evidence for a hot interstellar medium are briefly discussed.

1. INTRODUCTION

Centaurus A = NGC5128 first came to the attention of modern astronomers when Sir John Herschel (1853) described it as "an elliptically-formed nebula...cut asunder...by a broad obscure band" and associated it with three similar nebulae (Andromeda, the Sombrero, and NGC4565). The following century witnessed little progress in understanding this unusual object: it was catalogued as the 10^m star CPD-42°6250 (Kapetyn and Gill 1897) and was even considered to be a planetary nebula (Evans 1949). Cen A grew in importance as astronomical studies expanded into non-optical frequencies. It was one of the first radio, X-ray, and γ-ray sources to be identified with an external galaxy, and is now known to be the closest example of an "active galaxy" characterized by a luminous non-thermal nucleus and ejected radio lobes. Its extreme proximity (about 5 Mpc) provides a welcome opportunity to investigate intrinsically faint and small structures associated with such nuclear activity.

I will summarize here results from recent high-resolution observations of Cen A performed with the NRAO Very Large Array (VLA) and the Einstein X-Ray Observatory. The X-ray findings presented in preliminary form by Schreier et al. (1979), are discussed in detail by Feigelson

(1980) and Feigelson et al. (1981). The radio observations are described
by Schreier, Burns, and Feigelson (1981) and Burns, Feigelson, and
Schreier (in preparation). The following sections deal with several
identifiable radio and X-ray structures of the galaxy: the nucleus,
NE jet, inner radio lobes, "middle" NE lobe, and interstellar medium.
I omit discussion of the X-ray component associated with the dust lane
or disk, as we believe it to be of stellar origin. It should also be
noted that a recent X-ray map of 15° x 15° around Cen A by Marshall and
Clark (1981) demonstrates that, despite an earlier report to the contrary (Cooke et al. 1978), inverse Compton X-ray emission from the outer
radio lobes has not been detected to date. The average magnetic field
in the outer lobes must thus exceed 1.6 µG.

2. THE NUCLEUS

X-ray emission from the nucleus of Cen A has been extensively
studied over the past decade. The principal contribution of Einstein
observations is investigation of the gaseous environment immediately
around the nucleus. There is considerable evidence that it is enshrouded
in a cloud of gas and dust: the column density along the line of sight,
deduced from the X-ray spectrum, is very high ($N_H = 1 \times 10^{23}$ cm^{-2}); an
optical HII region coincident with the nucleus has been found (Gardner
and Whiteoak 1976); and there is excess IR and millimeter emission from
its vicinity that may be of thermal origin (see Mushotzky et al. 1978).
Seven percent of the X-ray emission is expected to be electron scattered
by this surrounding material (Fabian 1977). We have searched unsuccessfully for scattered or emitted soft X-radiation with the Einstein High
Resolution Imager (HRI). The nuclear source itself has a radius <0.3"
(1σ) or 8 pc, and less than a few percent of its flux (<1 x 10^{39} erg/s
in the 0.5-4.5 keV band) is scattered between 2" and 8" from the nucleus.

The VLA maps show that the inner knots of the jet lie only 10"-20"
from the nucleus. Their flux density exceeds that of the nucleus at
frequencies below about 1 GHz. The jet thus accounts for the confusing
radiation ~7" in size reported by Wade et al. (1971) and the core-halo
structure of the nucleus inferred by Slee and Sheridan (1975), and must
have caused overestimation of the nuclear flux density in a number of
earlier observations. A downward version of earlier low frequency data
casts doubt on the existence of the two synchrotron components invoked
by Grindlay (1975) to explain the nuclear X- and γ-ray flux by self-synchrotron Compton emission.

3. THE JET

The 1979 discovery of a jet of X-ray emission protruding from the
nucleus (Schreier et al. 1979) was particularly surprising as a radio
jet was not yet then known to exist. The first model for the X-ray
jet assumed a thermal origin. Now that radio emission with structure
very similar to that seen in X-rays has been found, nonthermal (in

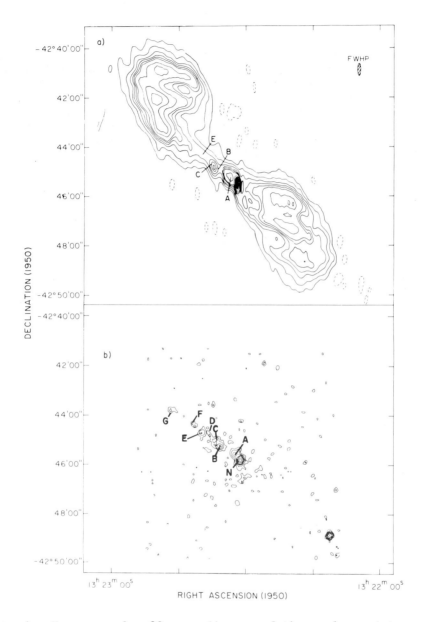

Figure 1. Upper panel: 20-cm radio map of the nucleus, jet, and inner lobes of Cen A. Observations made with the VLA C configuration, beam FWHM = 31" x 10" (Schreier, Burns, and Feigelson 1981). Positions of several X-ray knots are marked.
  Lower panel: X-ray map of the same region showing the nucleus, string of jet knots, and (at lower right) foreground dMe star. Observations made with the Einstein Observatory HRI, beam FWHM = 3" (Feigelson et al. 1981).

particular, synchrotron) models seem more likely though the thermal model has not yet been disproved.

Figure 1 (lower panel) shows a contour map of the HRI X-ray image of the nuclear region. The X-ray jet consists of several spatially resolved emission regions (called "knots" here) distributed in a straight line along P.A. 53° between 8" and 4' from the nucleus. There is additional evidence for excess emission between 2" and 8". The bright unresolved X-ray source 4.7' from the nucleus opposite the jet coincides with a 13^m dM2e star and is not a counter-jet. The locations, sizes, and flux densities of the X-ray knots are given in Table 1. Their 0.5-4.5 keV luminosities are $0.5 - 2 \times 10^{39}$ erg/s, comparable to the X-ray luminosity of our galaxy or M31. Each knot is resolved both along and perpendicular to the jet axis with diameters of 0.2'-0.3' (300-450 pc). Knot A is displaced 3" north of the principal jet axis, and has a bright elongated core about 6" long. Knot B, on the other hand, has a round center-filled appearance with FWHM = 12".

The C configuration VLA 20 cm map (Figure 1, upper panel) shows the brighter Knots A and B at levels of 2.3 Jy and 1.1 Jy respectively. The outer knots are not clearly evident, though they may be hidden in the ascending lobe emission. The radio spectral index of the knots appears to be similar to that of the lobe (~ 0.7), though this is at present poorly determined. Figure 2 shows a preliminary map of the inner knots obtained with the VLA in the A configuration. Knot A is now resolved into 4 small, misaligned subknots while Knot B is large and without structure. The radio and X-ray morphologies are thus quite similar though some differences in detail may be present. Interpretation of Knot A is confused by possible absorption of X-rays (but not radio) in the dust lane.

TABLE 1

X-Ray and Radio Properties of Nucleus, Jet, and Inner Lobe

Feature	Distance, PA from Nucleus	Size	S_{2keV} (μJy)	S_{20cm} (Jy)	α_{20cm}^{2keV}
Nucleus	--	< .3" (1σ)	83 (var)	3.37	0.54
Jet Knot A	17", 38°	18" x 15"	0.1	2.26	0.86
Jet Knot B	57 , 55	20 x 18	0.1	1.11	0.83
Jet Knot C	75 , 53	12 x 12	0.05	$\sim$0.75	
Jet Knot D	93 , 52	22 x 4	0.03		
Jet Knot E	113 , 54	12 x 12	0.05	$\sim$0.53	
Jet Knot F	140 , 54	12 x 8	0.03		
Jet Knot G	208 , 55	12 x 8	0.03		
Inner Lobe	3'-6'	--	<0.06 (3σ)	$\sim$120	>1.09

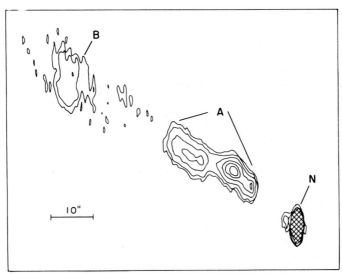

Figure 2. High resolution radio map made at 20 cm with the VLA A configuration showing the nucleus, Knot A, and Knot B. From Burns, Feigelson, and Schreier (in preparation).

The Cen A jet is of considerable interest purely by virtue of its radio properties. It appears to be narrowly collimated within 1 kpc of the nucleus, after which it expands laterally with an opening angle of about 22°. The bend within Knot A occurs on very small (0.1 kpc) spatial scales. But the paramount issue is to explain its X-ray emission. X-rays have now been detected in three radio jets (Feigelson 1980): Cen A, M87, and 3C273. There are three possible ways to produce the X-ray emission: thermal bremsstrahlung, inverse Compton, or synchrotron (Feigelson et al. 1981). In a thermal model, each knot of the Cen A jet is a 2×10^5 $M_\odot$ cloud with a temperature of several x 10^7K and density 0.2 cm^{-3}. The entire jet can be modeled by the ejection of such a cloud by the nucleus every 10^5 yr at a velocity between 2, and 10,000 km/s. The kinetic power of this "heavy" jet is 10^{43}-10^{44} erg/s, comparable to or exceeding the energy currently radiated by the nucleus. Within these constraints, the thermal model for the X-ray jet is possible. Faraday depolarization measurements, which will be forthcoming, may or may not indicate the presence of the required thermal densities. But such calculations depend on knowledge of the true magnetic field strength, and are thus always uncertain. The principal qualitative argument against the thermal model is that it does not explain the presence of the observed radio knots.

Inverse Compton explanations for the jet X-ray emission are not satisfactory without invoking exceptional conditions. Scattering of microwave background photons to the X-ray band implies a magnetic field of 0.2 µG for Knot B, far below the equipartition field of 30-40 µG. Inverse Compton scattering of starlight photons demands that the jet knots contain $\sim 10^{58}$ ergs in $\Gamma = 25$ electrons. This is 10^2 times the minimum energy of the inner lobe that is presumably the depository of all the energy in the jet.

Perhaps the most attractive model is that the X-ray jet, like the radio jet, is direct synchrotron emission. This accounts for the spatial coincidence and similar morphology of the X-ray and radio knots, and makes the Cen A jet closely analogous to the M87 jets where X-ray synchrotron emission is the only viable model (Schreier et al. 1981a). Both the Cen A and M87 jets have flat ($\alpha_r \sim 0.6$) radio spectra, and have a measured (M87) or inferred (Cen A) spectral break around the optical band to give $\alpha_r^x \sim 0.9$. Since X-ray emitting electrons would have radiative lifetimes of only 100 years, substantially shorter than the 1,000 - 20,000 LY travel time from the nucleus, reacceleration of particles within the knots would be necessary. Efficient acceleration up to the required $\Gamma = 10^7$ is not easily explained by Kelvin-Helmholtz induced turbulence or shocks, though resistance instabilities in a current-carrying jet may be feasible (Ferrari and Trussoni 1981). Direct confirmation for the synchrotron model would come from the detection of optical (or IR or UV), continuum knots coincident with the radio/X-ray knots.

4. THE INNER LOBES

The most striking finding regarding the inner lobes 3-5' NE and SW of the nucleus is that they are not seen in X-rays though they are 100 times more luminous than the jet in the radio. Whatever acceleration or thermal processes active in the jet knots are not present in the lobes. The first IPC images of Cen A showed excess X-ray emission 3'-5' NE and SW of the nucleus, which was interpreted by Schreier et al. (1979) to be inverse Compton emission from the inner radio lobes. Additional data revealed that the NE excess was due to jet and the SW excess to a foreground dMe star unfortunately located at the edge of the radio lobe. The current upper limit on inverse Compton emission requires $B > 3$ μG, consistent with an equipartition field strength of $B = 34$ μG derived from radio data. Furthermore, the synchrotron spectrum of the lobes must have $\alpha_r^x > 1.1$, and each lobe must contain $<3 \times 10^6$ $M_\odot$ in hot gas. It is interesting to note that amount of X-ray emission relative to radio emission decreases monotonically from $\alpha_r^x = 0.5$ in the nucleus to $\alpha_r^x = 0.9$ in the jet to $\alpha_r^x > 1.1$ in the lobe (see Table 1).

5. THE "MIDDLE" NE LOBE

Cen A is sometimes called a "double-double" radio source with an inner (4' NE/SW of the nucleus) and outer (3° N/S) pair of lobes. Additional radio structure is also present, however, including an enhancement at 25', P.A. 40° that we choose to call the "middle" NE lobe. Using beams of 7' FWHM, Cooper et al. (1965) report a flux density of 13 Jy at 1.4 GHz, $\alpha_r = 0.6$, and a half-power radius of 15' (23 kpc).

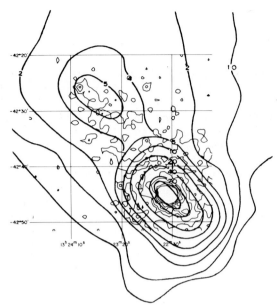

Figure 3. Superposition of Einstein IPC X-ray map (thin contours, from Feigelson et al. 1981) and 2.65 GHz radio map (thick contours, from Cooper et al. 1965) showing association between a faint X-ray feature and the middle radio lobe.

Its only notable property in the radio is the absence of a similar lobe to the SW. A line of X-ray emission about 5' long and 2' wide is detected 8σ above background in the Einstein Imaging Proportional Counter (IPC) coincident with the radio lobe (Figure 3). The feature has $L_x \simeq 1 \times 10^{39}$ erg/s and has an unusually soft X-ray spectrum, with almost all of the photons below 1.5 keV. Using standard assumptions, the observed emission is 30 times more luminous than expected from inverse Compton scattering of the microwave background. It could be direct X-ray synchrotron, as suggested for the X-ray jet above, with $\alpha_r^x = 1.1$. Both non-thermal explanations imply that the radio morphology is similar to that seen in X-rays. Radio maps with resolution $\leq 1'$ could determine if this is true.

The middle radio lobe is also accompanied by an unusual optical "jet" that makes a thermal model for X-ray emission more attractive than non-thermal models. A diffuse swath of HII regions, early-type stars, and shocked, turbulent gas (Graham and Price 1981) lies near the X-ray and radio structures. One can imagine that the ejected cloud of particles and magnetic field compressed and heated an ambient interstellar medium, giving rise to enhanced soft X-ray emission. The X-ray images provide independent evidence for a hot interstellar gas residing in the galaxy (see next section). The cooling time of the $\sim 10^6 M_\odot$ of $5 \times 10^6 K$ gas inferred from the X-ray data is short ($\leq 10^7$ yr), so it would soon condense into cold gas and young stars. Several radio galaxies have optical emission line regions near radio structures, though only Cen A is known to have associated X-ray emission as well.

6. A HOT INTERSTELLAR MEDIUM?

Both the HRI and IPC images of Cen A give indications of a faint diffuse X-ray component extending several arc minutes about the nucleus. In the HRI, the background level around the nucleus is higher than elsewhere in the field, though this may be partly or entirely due to instrumental background variations. In the IPC, a real diffuse region of X-rays is seen 5'-10' N-NW of the nucleus. We are unable, however, to determine whether emission is also present in other directions due to the presence of the jet and several unrelated X-ray sources. If this diffuse component is really present, it has L_x = 1-3 x 10^{40} erg/s and could be produced either by the integrated emission of dM stars in NGC5128 or by a hot interstellar medium. In the latter case about 2×10^8 $M_\odot$ of gas at 1×10^7 K would be present, exerting a hydrostatic pressure comparable to the equipartition pressure of the radio jet knots and the inner lobes. The Cen A jet thus might be confined by an external pressure.

Acknowledgements. I would like to thank Drs. Alan Bridle, Jack Burns, and Ethan Schreier for many stimulating discussions. This work is supported in part by NASA Grants NAS8-30751 and NAS8-30752.

REFERENCES

Cooke, B.A., Lawrence, A., Perola, G.C.: 1978, M.N.R.A.S. 182, 661.
Cooper, B.F.C., Price, R.M., Cole, D.J.: 1965, Aust. J. Phys. 18, 589.
Evans, D.S.: 1949, M.N.R.A.S. 109, 94.
Fabian, A.C.: 1977, Nature 269, 672.
Feigelson, E.D.: 1980, Ph.D. thesis, Harvard University.
Feigelson, E.D., Schreier, E.J., Delvaille, J.P., Giacconi, R., Grindlay, J.E., Lightman, A.P.: 1981, Ap.J. in press.
Ferrari, A., Trussoni, E.: 1981, ESLAB Symposium on X-Ray Astronomy, Amsterdam.
Gardner, F.F., Whiteoak, J.B.: 1976, Proc.As.Soc.Aust. 3, 63.
Graham, J.A., Price, R.M.: 1981, Ap.J. in press.
Grindlay, J.E.: 1975, Ap.J. 199, 49.
Herschel, J.F.W.: 1853, Outlines of Astronomy, 4 ed, Philadelphia.
Kapteyn, J.C., Gill, D.: 1897, Ann. Cape Obs., vol. 4.
Marshall, F.J., Clark, G.W.: 1981, Ap.J. 245, 840.
Mushotzky, R.F., Serlemitsos, P.J., Becker, R.H., Boldt, E.A., Holt, S.S.: 1978, Ap.J. 220, 790.
Schreier, E.J., Feigelson, E., Delvaille, J., Giacconi, R., Grindlay, J., Schwartz, D.A., Fabian, A.C.: 1979, Ap.J. 234, L39.
Schreier, E.J., Burns, J.O, Feigelson, E.D.: 1981, Ap.J. in press.
Schreier, E.J., Feigelson, E.D., Gorenstein, P.: 1981, in preparation.
Slee, O.B., Sheridan, K.V.: 1975, Proc.As.Soc. Aust. 6, 1.
Wade, C.M., Hjellming, R.M., Kellermann, K.I., Wardle, J.F.C.: 1971, Ap.J. 170, L11.

EMISSION REGIONS IN CENTAURUS A

R. M. Price
University of New Mexico

J. A. Graham
Cerro Tololo Inter-American Observatory

Centaurus A, at an estimated distance of five megaparsecs, is the closest radio galaxy. It presents the best opportunity to examine in detail the physical mechanisms and resulting structures that are to be found in radio galaxies. Centaurus was first studied in detail at radio wavelengths by Cooper, Price and Cole (1965), hence CPC. Many of the comments, interpretations, and conclusions recorded in that paper remain valid today and provide the broader framework in which the more detailed studies using today's more powerful instrumentation can be understood. Historically, it is also interesting to note that Centaurus A was the first extragalactic radio source in which linear polarization and Faraday rotation were discovered and extensively studied.

Recent work at radio, optical and X-ray wavelengths had dealt with the nuclear regions of NGC 5128 and the middle or NE radio lobe. The comparative or relative distribution of the different wavelength emissions has several noteworthy features. The optical "jet" first described by Blanco et al., 1975, is at a position angle of 58 degrees and a distance of 15 minutes from the center of NGC 5128. It is fairly closely aligned with the "nuclear" X-ray and radio jet which has a length of approximately four minutes (Schreier et al., 1979, 1981). The faint optical HII regions (25' from the nucleus) and the middle radio lobe are more closely aligned with the well-known inner double radio source at position angle 45 degrees. The X-radiation discussed by Feigelson in this volume is roughly coextensive with the radio (CPC) and the optical filaments studied in detail by Graham and Price (1981). The X-ray and optical emission lie closer to the S side of the lobe (unresolved), particularly with respect to the polarized emission from this region as reported by CPC. We believe this relative alignment could be a significant piece of information regarding the physical processes that have led to the formation of the HII regions in this area. This also could be indicative of the distribution of Faraday depolarization in the vicinity of the HII regions. The rotation measure reported by CPC appears uniform across this region but lacks sufficient detail to shed light on the current situation.

The degree of excitation in the HII regions was studied by Graham and Price. Using log [O III]/Hα as a useful indicator of HII region excitation, we noted that the degree of excitation clearly falls off with distance from the center of NGC 5128, although not monatonically (see fig. 7 in Graham and Price, 1981).

Another measure of the degree of excitation of HII regions was log Hα/[NII] compared to log [OIII]/Hβ. In the NE radio lobe of Centarus we note regions that show a much higher degree of excitation than is normally found in HII regions in spiral galaxies (see Searle, 1971; and figs. 7 and 8 in Graham and Price, 1981).

There are a number of conclusions and conjectures which can be made from an analysis of the information provided over the entire spectrum in the northeast lobe of Centaurus A:

• Ionized hydrogen regions are found at distances of at least 45 kiloparsecs from the center of the galaxy NGC 5128. Spectra of the HII regions are not consistent with a simple photoionization mechanism.

• The degree of excitation of the HII regions decreases as a function of distance from the center of NGC 5128 and indicates that the ultimate energy source lies in the nucleus of the galaxy.

• Some of the regions show higher excitation than for HII regions in normal spiral galaxies.

• The disparity in position angle between the nuclear radio and X-ray jet and the middle radio lobe requires precession of the channel through which energy flows from the nucleus of NGC 5128.

Finally, we stress the need for further radio polarization measurements in this region.

References

Blanco, V. M., Graham, J. A., Lasker, B. M., and Osmer, P. S. 1975, Ap. J. (Letters) 198, L63.
Cooper, B. F. C., Price, R. M., and Cole, D. 1965, Australian J. Phys. 18, 589.
Graham, J. A. and Price, R. M. 1981, Ap. J. 247, 823.
Schreier, E. J., Burns, J. O. and Feigelson, E. D. 1981, Ap. J. (in press).
Schreier, E. J., Feigelson, E., Delaville, J., Giacconi, R., Grindley, J., Schwartz, D. A., and Fabian, A. C. 1979, Ap. J. (Letters) 254, L39.
Searle, L. 1971, Ap. J., 168, 327.

X-RAY EMISSION FROM CENTAURUS A

James Terrell
University of California, Los Alamos National Laboratory
Los Alamos, New Mexico, U.S.A.

ABSTRACT: Observations of 3-12 keV X-ray emission from NGC 5128 (Cen A) were made by Vela spacecraft over the period 1969-1979. These data are in good agreement with previously reported data, but are much more complete. Numerous peaks of X-ray intensity occurred during the period 1973-1975, characterized by rapid increases and equally rapid decreases (in less than 10 days). Thus it seems probable that most of the X-ray flux from the nucleus of Cen A came from a single source of small size.

The two Vela 5 spacecraft launched in May 1969 were among the first to be capable of X-ray astronomy. Their collimated NaI detectors (6° FWHM) scanned the sky along a great circle as they rotated with a 64-second period, observing X-ray sources in the 3-12 keV range. The entire X-ray sky was surveyed every 56 hours, half of the orbital period. Reasonably complete data were thus obtained on many X-ray sources, until the last detector failed in June 1979. Among the sources observed were X-ray transients such as Cen XR-4 (Conner et al., 1969), X-ray bursts (Belian et al., 1976), and even gamma-ray bursts (Terrell et al., 1981).

Centaurus A (NGC 5128), the nearest active galaxy, was a relatively weak source in 1969, but from 1973 through 1975 its X-ray strength was much higher and showed considerable variability (Beall et al., 1978). The Vela 5 observations of Cen A during this period of peak activity are shown in Figure 1. The data are presented as 10-day averages, based on a weighted sum of all the observations within 8° of the source, corrected for background. Standard deviations shown represent counting statistics only, but systematic errors are believed to be small. The data in Figure 1 are new and have been obtained from a complete reprocessing of the Vela 5B data. Where other observations have been reported the agreement is very good. The sudden increase by a factor of 1.6 reported by Winkler and White (1975) may be seen around April 9, 1973. It was followed by a further large increase and then by a precipitous decline, by a factor of almost 3, to a very low level in June 1973. Other peak fluxes were attained in January 1974, January 1975, and June 1975. Often the increases and decreases in X-ray flux occurred in 10 days or less.

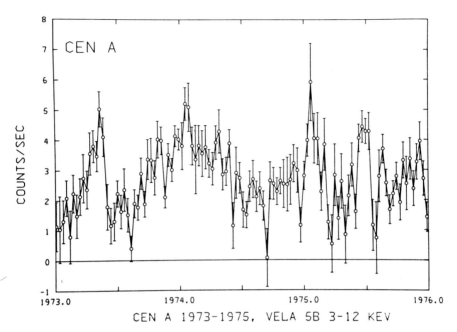

Fig. 1. Cen A 3-12 keV X-ray flux, 1973-1975 (Vela 5B, 10-day averages)

The counting rates in this graph may be converted to energy received in the 3-12 keV range by means of the factor 2.8×10^{-10} ergs/count. For NGC 5128, each count/sec represents $\sim 10^{42}$ ergs/sec emitted by the galaxy, so that the X-ray source was very intense, and doubtless very massive. The large amplitude and short time scale of the fluctuations suggest that most of the X-ray flux came from a single source of size at most a few light-days. It has been proposed that the nucleus of Cen A is a massive condensation ejecting matter at relativistic speeds, in order to account for the immense radio power emitted from regions much further out in the galaxy (Hoyle and Fowler, 1963; Terrell, 1966, 1967, 1975). The X-ray fluctuations reported here give some support to the idea of a massive, condensed, and active galactic nucleus in NGC 5128.

This work was supported by the U. S. Department of Energy.

REFERENCES:
Beall,J.H., et al.: 1978,Ap.J.219,pp.836-844.
Belian,R.D., Conner, J.P.,and Evans,W.D.: 1976,Ap.J.206,pp.L135-L138.
Conner,J.P., Evans,W.D., and Belian,R.D.: 1969,Ap.J.157,pp.L157-L159.
Hoyle,F., and Fowler,W.A.: 1963, Nature 197,pp.533-535.
Terrell,J.: 1966,Science 154,pp.1281-1288.
Terrell,J.: 1967,Science 156,p.265.
Terrell,J.: 1975,Nature 258,pp.132-133.
Terrell,J.,Fenimore,E.E.,Klebesadel,R.W.,and Desai,U.D.:1981,Ap.J.,subm.
Winkler,P.F., and White,A.E.: 1975,Ap.J.199,pp.L139-L142.

VLBI OBSERVATIONS OF THE NUCLEUS OF CENTAURUS A

R. A. Preston, A. E. Wehrle, D. D. Morabito
Jet Propulsion Laboratory, Pasadena, California

D. L. Jauncey, M. Batty, R. F. Haynes, A. E. Wright
CSIRO, Epping, Australia

G. D. Nicolson
CSIR, Johannesburg, South Africa

VLBI observations of the nucleus of Centaurus A have been made at three southern hemisphere observatories. Since Centaurus A is the nearest active galaxy, VLBI investigations are important because the physical processes in the nucleus can be studied in greater linear detail than in other similar galaxies. Previous VLBI observations of Centaurus A have been hampered by its southerly declination (-43°) and the sparsity of VLBI capability in the southern hemisphere, leading to only scattered single point u,v coverage. This paper presents results from the early stages of development of a southern hemisphere VLBI network.

The known radio structure of Centaurus A has three prominent features: i) a large double structure centered on the optical galaxy with a position angle of $\sim 0°$. These components extend over $\sim 10°$, or equivalently ~ 900 kpc; ii) A smaller double structure also centered on the optical galaxy with a position angle of $\sim 50°$, which is nearly perpendicular to the optical dust lane. The component separation is $\sim 7'$ (~ 10 kpc), which is comparable in size to the optical galaxy; iii) An extremely small nucleus positioned in the center of the optical galaxy. Wade et al. (1971) determined the size of the radio nucleus to be $\leq 0\rlap{.}''5$ (~ 10 pc), and VLBI nuclear structure has been detected at the $5 \times 10^{-4}{''}$ ($\sim 10^{-2}$ pc) level (Jauncey et al. 1981). Optical and x-ray jets have been detected within a few arcminutes of the nucleus, and are aligned with the nucleus with a principal axis that roughly matches (PA $\sim 60°$) the inner double radio structure (Dufour and Van den Bergh 1978, Schreier et al. 1979). A recent VLA map (Feigelson, this volume) also shows an inner radio jet extending inward to within a few arcseconds of the nucleus, with a strong correlation between radio and x-ray hotspots.

The Mark II VLBI observations were performed at 2.3 GHz on 1980 April 22-27 with right circular polarization. Participating observatories were located at Tidbinbilla and Parkes in Australia and Hartebeesthoek in South Africa, providing baselines with maximum resolutions of $\sim 0\rlap{.}''10$ and $\sim 0\rlap{.}''0027$. Complete u,v tracks were obtained.

Figure 1 shows the history of measured correlated flux densities as a function of interferometer hour angle on the Tidbinbilla-Parkes baseline. An elliptical Gaussian model of the nucleus was fitted to these results (see solid line in Fig. 1). The total flux density of the nucleus at 2.3 GHz was constrained to be 6.8 Jy by a separate VLBI measurement in April 1981 on a 25 km baseline at Goldstone, California. We note that the 2.3 GHz nuclear flux density is about double the 2.7 GHz value measured a decade earlier by Wade et al. (1971). The best fit elliptical Gaussian model had a major axis of $\sim 0''.05$ (~ 1 pc). The minor axis has a maximum value of $\sim 0''.03$ (~ 0.6 pc), but could be much smaller. The position angle along which the nucleus is elongated is $30° \pm 20°$, indicating a possible jet which is roughly aligned with the larger radio, optical, and x-ray jets and the inner double radio lobes, with the aligned structure extending over a dynamic range in size of $\sim 10^4$. The linear size of the elongated nucleus is typical of the sizes of nuclear jets found in other active galaxies by VLBI (Linfield 1981).

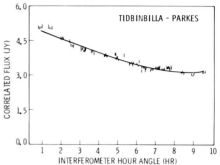

FIGURE 1

On the Australian-South African baseline, the nucleus was completely resolved, indicating there is no 2.3 GHz nuclear source $\lesssim 0''.001$ ($\lesssim 0.02$ pc) with a flux density $\gtrsim 20$ mJy. However, a previous VLBI measurement at 8.4 GHz by Jauncey et al. (1981) found a nuclear source of strength ~ 0.5 Jy with a size of $\sim 0''.0005$ (~ 0.01 pc). This 8.4 GHz component must be highly self-absorbed at 2.3 GHz to account for our non-detection. It is interesting to note that the detected radio structure in Centaurus A extends over a dynamic range in size of $\sim 10^8$.

REFERENCES:

Dufour, R. J., and Van den Bergh, S.: 1978, Astrophys. J. 226, 73.
Jauncey, D. L., Preston, R. A., Kellermann, K. I., and Shaffer, D. B.: 1981, submitted to Astrophys. J. Letters.
Linfield, R.: 1981, Astrophys. J. 244, 436.
Schreier, E. J., Feigelson, E., Delvaille, J., Giacconi, R., Grindlay, J., Schwartz, D. A., and Fabian, A. C.: 1979, Astrophys. J. Letters 234, L39.
Wade, C. M., Hjellming, R., Kellermann, K. I., and Wardle, J. F. C.: 1971, Astrophys. J. Letters 170, L11.

(This work was supported at JPL by NASA contract NAS 7-100.)

SYSTEMATICS OF LARGE-SCALE RADIO JETS

Alan H. Bridle
National Radio Astronomy Observatory
and
Department of Physics and Astronomy, University of New Mexico

ABSTRACT Radio jets occur in sources with a wide range of radio luminosities, and in 70% to 80% of nearby radio galaxies. There may be two basic types of large-scale (>1 kpc) jet -- $B_{\|}$-dominated one-sided jets in sources with luminous radio cores, and $B_{\perp}$-dominated two-sided jets in sources with weak radio cores. The large-scale jets that have been observed at high linear resolution are well collimated within a few kpc of their cores, then flare and recollimate further out. Their brightness-radius evolution is often "subadiabatic".

1. TERMINOLOGY

The extended radio lobes of many extragalactic sources are linked to small-diameter cores at the centers of the optical objects by narrow bridges of radio emission (e.g. Fig. 1). Such bridges are usually called *jets*, although there is no direct evidence for matter transport along them. The term *jet* is popular because the bridges occur where it has been *postulated* that collimated outflows of relativistic fluid transport energy from the cores to the lobes, as continuous *beams* (Blandford and Rees 1974) or as trains of discrete *plasmoids* (Christiansen et al. 1977).

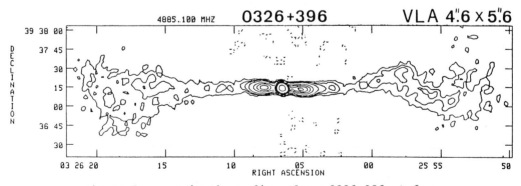

Figure 1. Jet in the radio galaxy 0326+396 at 6cm.

Use of the word *jet* prejudges the physics of the bridges. It presumes that they arise directly from radiative inefficiencies in the postulated fluid flows. In fact, the *direct* evidence for outflow from the cores of extragalactic sources (e.g., proper motions of VLBI components, asymmetries in optical emission line profiles) does not relate explicitly to material in the *large-scale* bridges. As the observers' terminology can bias thinking about the phenomena, it is dangerous to employ language which prejudges the physics without placing clear restrictions on its use. I call a feature a *jet* only if (a) its length is at least four times its width, (b) high-resolution maps separate it from other source components either spatially or by brightness contrast, and (c) it points from a radio core towards a lobe or a "hot spot".

2. OCCURRENCE OF RADIO JETS

Radio jets are now known in 69 extragalactic sources; this number is growing rapidly as high-resolution maps with good sensitivity and dynamic range are made of large samples of extended sources. Jets are found in sources with a wide range of radio luminosities, from weak ($P_{1400} \sim 2 \times 10^{23}$ W/Hz, $H_0 = 100$) radio galaxies such as M84 and NGC 3801 to powerful ($P_{1400} \gtrsim 10^{27}$ W/Hz) radio quasars such as 3C334 and 4C29.68 (Wardle et al., this volume) and 3C345. Jets occur in sources with both "edge-darkened" and "edge-brightened" overall morphologies.

Jets are detected most frequently in lower-luminosity sources. At least 15 of the 22 3CR galaxies at $z < 0.05$, $\delta > 10^0$ with $P_{178} > 10^{23}$ W/Hz (this luminosity cutoff excludes only M82) have detectable radio jets. Ekers et al. (1981) detect jets in 9 of 11 *well-resolved* sources in a complete sample of 40 B2 galaxies with $m_{pg} < 15.7$, $S_{408} > 0.2$ Jy and $24^0 < \delta < 40^0$. This 70% to 80% incidence of detectable jets in weaker radio galaxies may not apply to more powerful sources; only $\sim 25\%$ of the 3CR "complete sample" sources so far observed at the VLA have detectable jets. Jets may therefore emit a greater fraction of the total luminosity in intrinsically weaker extragalactic sources.

Radio jets exist on size scales ranging from parsecs to hundreds of kpc, e.g., the 250-kpc jet and 400-kpc counterjet in NGC315 (Willis et al. 1981) and the one-sided 0.7-pc jetlike feature extending from the core of the source towards the brighter large-scale jet (Linfield 1981). This review outlines some systematic properties of jets on scales >1 kpc, based on maps of the 69 presently known jets. These data are *not* from an unbiased sample, so trends reported here must be re-examined when homogeneously-observed "complete samples" of sources with radio jets become available.

3. MAGNETIC CONFIGURATIONS

Linear polarizations in radio jets are typically 10% to 40% at centimetre wavelengths (values as high as 60% were found at 20cm in the

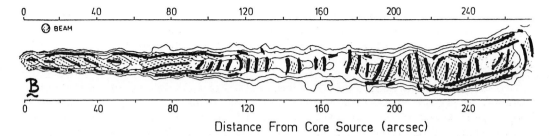

Figure 2. Projected magnetic field in the NGC6251 jet.

jets of NGC315 by Willis et al. 1981). Multi-frequency polarimetry can determine the rotation measures and hence the projected magnetic field configurations. Figure 2 illustrates the field configurations commonly found in jets. Close to the radio core the field is predominantly *parallel* to the jet. Further out the field on the jet axis is predominantly *perpendicular* to the jet while that on the edges may be *either* parallel or perpendicular. I will label field configurations $B_\parallel$ or $B_\perp$ according to which component prevails *on the jet axis*.

Figure 3 shows that the fraction of the jet length over which $B_\parallel$ dominates is strongly related to the radio core luminosity. This relation is strong enough to suggest that there are two basic types of jet distinguishable by their magnetic properties -- $B_\perp$-*dominated jets* associated with weak ($P_{5000} \lesssim 10^{23}$ W/Hz) cores and $B_\parallel$-*dominated jets* associated with more powerful cores. Note that although $L_\parallel/L_{jet}$ is independent of projection (in straight jets) and of redshift, there could be a selection effect present in Fig. 3. The higher-luminosity cores are in more distant sources, so the transverse resolution of the

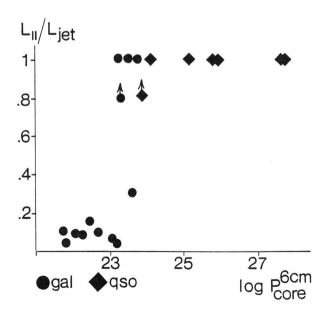

Figure 3. The fraction of the jet length over which $B_\parallel$ dominates, plotted against the logarithm of the 6cm core luminosity in W/Hz for 20 jets with known magnetic configurations.

field configuration decreases from left to right. This might bias the result if, for example, more luminous sources had stronger $B_\parallel$ *edges* to outer $B_\perp$ regimes. It will be important to test the relation in Fig. 3 in samples with similar linear resolution for sources of all powers.

The short $B_\parallel$ regions of $B_\perp$-dominated jets are those closest to the cores, as expected in flux-conserving expansions where $B_\parallel \propto 1/R^2$ and $B_\perp \propto 1/R$, R being the jet radius. The presence of $B_\parallel$ edges to some $B_\perp$ regions may indicate either that $B_\parallel$ is maintained by shearing at the edges of some jets, or that the field has an overall helical structure. Field strengths estimated by equipartition calculations range from $\sim 2 \times 10^{-4}$ gauss in some quasar jets to $\sim 10^{-6}$ gauss in galaxy jets.

4. JET-COUNTERJET ASYMMETRIES

Figure 4 is a logarithmic histogram of the intensity ratios between jets and their counterjets at corresponding distances on opposite sides of the parent objects. The symbols code the magnetic configurations in the jets where the intensities were measured. The ratios were taken from maps on which the jet widths were poorly resolved, to integrate over their transverse intensity profiles. The ratios plotted with triangles are *lower limits*, i.e., the jets were detected only on one side of the parent object. Most $B_\perp$ regions have side-to-side intensity ratios <4:1 while most $B_\parallel$ regions have ratios >4:1 (the exceptions are a symmetric $B_\parallel$ regime near the core of 3C449 (Cornwell et al., this volume) and the $B_\perp$ outer jet of NGC6251). The division of Fig. 3 into $B_\parallel$-dominated and $B_\perp$-dominated jets therefore parallels a division into *one-sided* (>4:1) and *two-sided* (<4:1) jets.

The short $B_\parallel$ regimes at the bases of $B_\perp$-dominated jets have side-to-side intensity ratios as high as those of $B_\parallel$-dominated jets. They usually occur at the base of the *brighter* jet in a two-sided system. Their high side-to-side ratios result in a longer "gap" between the core and the start of the jet on the *fainter* side of such systems.

In 7 of 10 sources with *both* a "VLBI" jet in the radio core *and* a large-scale jet system, the brighter large-scale jet is on the same side of the core as the VLBI jet. The asymmetries of the parsec and kpc scales in such sources may therefore be partly correlated.

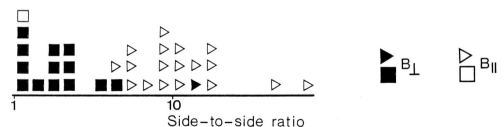

Figure 4. Jet-counterjet intensity ratios (bin width = × 1.259).

5. COLLIMATION

Eight of the nearer jets in radio galaxies (NGC315, 3C31, 0326+396, NGC6251, M84, 1321+319, Cen A and 3C449) have been mapped with adequate transverse resolution to show their lateral expansion directly. These jets are all $B_\perp$-dominated. The data for NGC315 exemplify the systematic properties that are becoming clear. Figure 5 plots the HWHM (R) of the brighter jet in NGC315 against the angular separation (z) from the core from 60" to 800" (15 to 190 kpc). The local *expansion rate* dR/dz varies from >0.23 near the core to -0.04 on the collimation "plateau" 45 to 100 kpc from the core. Beyond 100 kpc the jet re-expands with dR/dz $\sim$ 0.09. Similar variations of dR/dz occur in 3C31 (Bridle et al. 1980), 0326+396, NGC6251 (Willis et al. 1982), 1321+319 (Fanti and Parma 1981) and 3C449 (Perley et al. 1979). The *constant* expansion rate expected in simple freely-expanding supersonic jets is rare, and early attempts to estimate Mach numbers M for jets by putting $M = \overline{dz/dR}$ now appear to have been premature.

Recent VLA observations of the first few kpc of the jets in NGC315, NGC6251 (Willis et al. 1982), Cen A (Feigelson et al., this volume) and M84 (Laing and Bridle 1982) show a consistent pattern of collimation behavior closer to their cores. Figure 6 *(next page)* illustrates the pattern with data on NGC315. For the first 15" (3.5 kpc) the expansion is slow (dR/dz $\leq$ 0.09) but the jet "flares" to dR/dz $\sim$ 0.3 by 20" (5 kpc) from the core. The collimation properties of these large-scale jets are evidently *not* set once and for all at the sub-kpc scales characteristic of VLBI jets. Rather, the observed expansions imply that the jets "break out" from confinement on scales $\leq$1 kpc, "flare", and then *recollimate* on scales of tens of kpc.

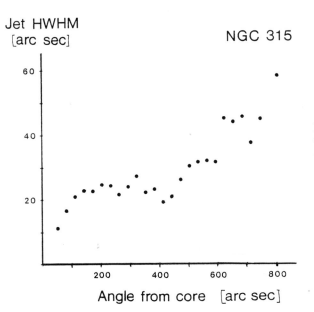

Figure 5.
The lateral expansion of the brighter jet in NGC315. 20cm WSRT data from Willis et al. (1981).

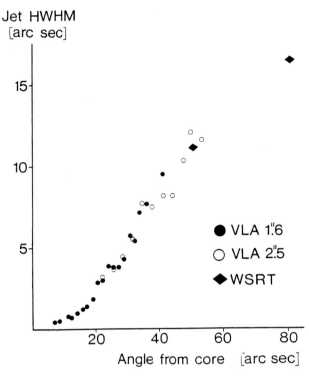

Figure 6.
The lateral expansion of the brighter jet in NGC315 (unpublished VLA 20cm data and inner WSRT data from Fig. 5). Note the excellent agreement between HWHMs measured at different resolutions.

In these nearby radio galaxies, the recollimation could be due to thermal pressures in extensive components of the galactic atmospheres (e.g., Bridle et al. 1981). The pressures needed to recollimate these jets could be provided by media whose X-ray emission is near the detection limits of the recent *Einstein* observations, if the pressures in the jets are near those of the radiating particles in equipartition. In this case the recollimation of *two-sided* jets should occur at similar distances from the core on both sides of the galaxy. This appears to be the case in NGC315, 3C31, 0326+396, 1321+319 and 3C449.

I have used the HWHM of the transverse intensity profiles of these jets to describe their expansion properties because most of the profiles are center-brightened and reasonably Gaussian (although flat-topped or asymmetric profiles *do* occur occasionally). No clear relation has yet emerged between the transverse profile shapes and other properties of the jets (such as the magnetic configurations or intensity gradients).

6. SPECTRA

The radio spectra of 15 jets have now been measured near 20cm. Most jets have very similar spectra; the mean spectral index is 0.6, with a standard deviation of 0.15. This is a spectral index frequently seen in SS433 (Hjellming and Johnston, this volume). Jets usually have slightly flatter spectra than the lobes they enter (Willis 1981).

There are usually only weak (<0.2) spectral index gradients along the jets and no *systematic* gradients across them. There is no discernible correlation between spectral index and expansion rate, contrary to the predictions of Benford et al. (1980).

7. BRIGHTNESS-RADIUS EVOLUTION

The surface brightnesses of jets rarely decrease as rapidly with increasing jet width as would be expected if they were steady, constant-velocity flux-conserving flows of unreaccelerated particles. Even if the *radiating* particles did no work in the lateral expansion of the jets, magnetic flux conservation would make the central surface brightness T_b decrease with HWHM R *at least* as fast as $1/R^{(\gamma+3)/2}$ in a jet satisfying these assumptions. (This is the dependence for a pure transverse magnetic field varying as $1/R$). For the mean jet spectral index of 0.6, the *slowest* "adiabatic" brightness decline would thus be $1/R^{2.6}$.

Figure 7 exemplifies the brightness-radius evolution seen in most nearby jets, using data for NGC6251. Close to the radio core T_b initially *increases* with R, then declines slowly ($1/R^{0.7}$ here). Such regions within a few kpc of the cores are where optical continuum emission was detected in the jets of 3C31 and 3C66B (Butcher et al. 1980). The optical-to-radio spectral indices are similar to the radio indices in these regions, requiring *in situ* particle replenishment unless the magnetic fields are far below their equipartition values.

In the "flaring" regions further from the cores the *mean* T_b-R relations in seven of nine well-studied nearby jets are significantly flatter than $1/R^{2.6}$, all seven being in the range $1/R$ to $1/R^{1.4}$. Only in 0326+396 and on the "collimation plateau" of NGC315 does the brightness fall off as steeply as, or steeper than, $1/R^{2.6}$ for a significant fraction of the length of the jet.

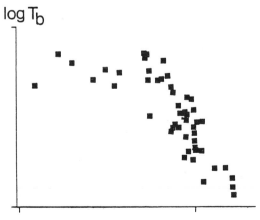

Figure 7.
Logarithmic plot of brightness temperature T_b on the axis of the jet in NGC6251 against the jet HWHM, over the first 3' (60 kpc) of the jet. The HWHM does not increase monotonically, due to knots in the jet, but the *mean* T_b-R relation is well-defined and evolves from $1/R^{0.7}$ to $1/R^{1.4}$ as the jet expands.
(VLA 1662-MHz data on arbitrary scales of T_b and R)

Such "subadiabatic" brightness-radius relations could arise in several ways, including (a) deceleration of the jet by entrainment of low-momentum material so that the jet is compressed longitudinally to compensate for its lateral expansion, (b) magnetic flux amplification, (c) relativistic particle acceleration or replenishment, or (d) decline of the power output of the core over the lifetime of the jet. In several jets the regimes of *faster* expansion are those of *slower* brightness-radius evolution. This may be consistent with interpreting them as turbulent mixing regions in which the emission of MHD waves and their damping by acceleration of relativistic particles is an effective source of viscosity on small scales (Henriksen et al. 1982).

Acknowledgements

I thank Jack Burns, Tim Cornwell, John Dreher, Ron Ekers, Eric Feigelson, Ed Fomalont, Robert Laing, Rick Perley, Larry Rudnick, Wil van Breugel, John Wardle and Tony Willis for showing me jet maps well in advance of publication. Ed Fomalont, Dick Henriksen, Robert Laing, Rick Perley, Tony Willis and Lorenzo Zaninetti have also sharpened my appreciation of jet systematics in numerous discussions. I am grateful to NRAO and to the University of New Mexico for their hospitality during the preparation of this review.

REFERENCES

Benford,G., Ferrari,A. and Trussoni,E. 1980. Ap.J., 241, pp. 98-110.
Blandford,R.D. and Rees,M.J. 1974. M.N.R.A.S., 169, pp. 395-415.
Bridle,A.H., Chan,K.L. and Henriksen,R.N. 1981. J.Roy.Astron.Soc. Canada, 75, pp. 69-93.
Bridle,A.H., Henriksen,R.N., Chan,K.L., Fomalont,E.B., Willis,A.G. and Perley,R.A. 1980. Ap.J.(Letters), 241, pp. L145-L149.
Butcher,H.R., van Breugel,W.J.M. and Miley,G.K. 1980. Ap.J., 235, pp. 749-754.
Christiansen,W.A., Pacholczyk,A.G. and Scott,J.S. 1977. Nature, 266, pp. 593-596.
Ekers,R.D., Fanti,R., Lari,C. and Parma,P. 1981. Preprint.
Fanti,R. and Parma,P. 1981. *Optical Jets in Galaxies* (Proc. 2nd ESO/ESA Workshop, 18-19 February 1981), pp. 91-95.
Henriksen,R.N., Bridle,A.H. and Chan,K.L. 1982. Ap.J., submitted.
Laing,R.A. and Bridle,A.H. 1982. In preparation.
Linfield,R.P. 1981. Ap.J., 244, pp. 436-446.
Perley,R.A., Willis,A.G. and Scott,J.S. 1979. Nature, 281, pp. 437-442.
Willis,A.G. 1981. *Optical Jets in Galaxies* (Proc. 2nd ESO/ESA Workshop, 18-19 February 1981), pp. 71-76.
Willis,A.G., Perley,R.A. and Bridle,A.H. 1982. In preparation.
Willis,A.G., Strom,R.G., Bridle,A.H. and Fomalont,E.B. 1981. Astron. Astrophys., 95, pp. 250-265.

RADIO AND X-RAY OBSERVATIONS OF LARGE SCALE JETS IN QUASARS

J. F. C. Wardle and R. I. Potash
Physics Department, Brandeis University
Waltham, Massachusetts

The first big radio jet to be found in a quasar was in 4C32.69 (z=.659) (Potash and Wardle, 1980). A new higher resolution map of this jet, made with the VLA at 5 GHz, is shown in Fig. 1. Observationally the important points are 1) it is very luminous, 2) it is very well collimated, 3) the magnetic field is parallel to the jet over its entire length, 4) it bends at least three times along its path to the outer radio lobe. Physically this is interesting because 1) it is a very lossy pipeline to the outer lobe, 2) the high degree of collimation suggests either the jet is confined by external pressure or it is highly supersonic. But the minimum internal energy density is very high ($>3\times10^{-10}$ erg cm^{-3}), so if it is confined the external pressure must also be very high. On the other hand, if the jet is expanding freely, it is easy to show that the momentum flux is enormous ($>4\times10^{38}$ dynes). Such a jet is very rigid. It is difficult to stop and difficult to bend, but evidently both of these things happen. A detailed discussion of these problems is given in Potash & Wardle (1980), who concluded that the jet cannot be expanding freely. They suggested that the jet might be confined either by the thermal pressure of external hot gas, or by a helical component of magnetic field due to currents in the jet.

We have recently discovered five more quasars with large scale (>100 kpc) radio jets, using the VLA at 5 GHz. In Fig. 2 we show a map of 3C334 (z=.555) as an example. The other quasars with jets are 4C39.27, 4C29.68, 4C22.26 and 4C24.02. These observations will be discussed in detail elsewhere.

The six jets were discovered among a total of only thirteen quasars that we have looked at with the VLA. It seems that large scale jets are a fairly common occurrence among quasars, as they are among galaxies. In all six cases the jet is visible on only one side of the nucleus, and in the three cases for which we have sufficient sensitivity to measure the linear polarization distribution, the magnetic field runs parallel to the jet along its entire length. These features are in agreement with the trends for high luminosity jets discussed by Bridle (this volume). Most important, five out of the eight largest angular diameter quasars from Schmidt's (1975) complete sample of 4C quasars contain jets. Sources selected by angular diameter lie preferentially close to the plane of the sky. The observed "onesidedness" of the jets is therefore unlikely to be due to the Doppler effect, and is probably an intrinsic feature of these structures.

Three of the six quasars (4C32.69, 4C39.27 and 3C334) have been detected in soft X-rays, using the IPC on board the Einstein Observatory. (The remaining three quasars have not been observed in X-rays.) The measured values of optical to X-ray spectral index, α_{ox}, are 1.12, 1.36 and 1.53 respectively.

The important question is whether the observed X-ray emission is consistent with the presence of enough hot gas surrounding the quasar to confine the jets. First we note that the measured values of α_{ox} are entirely typical of radio loud quasars (Zamorani et al, 1981), so there is no evidence for enhanced X-ray emission from these quasars. (Also, the images are unresolved in the IPC, <80 arc sec.) However, we shall consider a "best case" situation and assume a) <u>all</u> the X-ray emission comes from extended hot gas surrounding the quasar, b) the gas fills a uniform isothermal sphere whose extent is just enough to contain the observed radio structure, c) the gas is a pure H-He plasma, and line emission can be neglected. Since the IPC cannot yet determine good spectral parameters, we assume various temperatures and

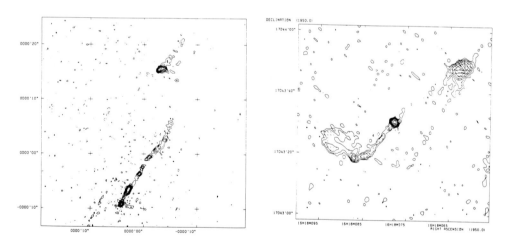

Fig. 1 (left)   The nucleus, jet and Np radio lobe of 4C32.69, observed at 5 GHz with a resolution of ~0.5 arcsec.

Fig. 2 (right) 3C334, observed at 5 GHz with a resolution of ~1.5 arcsec.

then calculate the ion density that would produce the observed X-ray emission, and the corresponding thermal pressure. The result is as follows. In all three cases, the ion density (which is not very sensitive to the assumed temperature) is close to 10^{-2} ions cm^{-3}, and the lowest temperature necessary to confine the <u>average</u> pressure inside the jets (which itself is a minimum number calculated from standard synchrotron theory) is in excess of 10 keV.

It is clear that the required temperatures are very high, even making the most favorable assumptions, and they are probably inconsistent with the raw counts in the IPC energy channels which indicate the X-ray spectra are in fact comparatively soft. As an absolute minimum these temperatures are somewhat higher than temperatures found in X-ray emitting clusters of galaxies (Gursky & Schwartz, 1977). We also point out that any gas that can confine the jets must itself be confined, requiring a gravitational potential well considerably deeper than provided by a typical rich cluster of galaxies. In fact there is no direct evidence that such a component of the X-ray emission exists at all, and we conclude that it is improbable that the observed quasar jets are confined by the pressure of a hot ionized gas.

However, we consider the arguments presented in Potash & Wardle (1980) in favor of confinement to be compelling. The simplest alternative appears to be models in which magnetic fields due to currents in the jet itself help prevent rapid expansion (Benford, 1978; Chan & Henriksen, 1980). In this picture, the breaking up of the jet in 4C32.69 (Fig. 1) into discrete blobs might be attributed to pinching instabilities, which could also drive the particle acceleration necessary to maintain the brightness of the jet as it expands.

REFERENCES

Benford, G. 1978, MNRAS, 183, p. 29.
Chan, K. L. and Henriksen, R. N., 1980, Ap. J., 241, p. 534.
Gursky, H., and Schwartz, D. A. 1977, Ann. Rev. A. & Ap., 15, p. 541.
Potash, R. I. and Wardle, J. F. C. 1980, Ap. J., 239, p. 42.
Schmidt, M. 1975, Ap. J., 195, p. 253.
Zamorani, G., et al, 1981, Ap. J., 245, p. 357.

DISCUSSION

BENFORD: If some of your jets do indeed display pinching or helical instability, you can use the theoretical wavelengths and growth lengths to eliminate some parameters--for example, the Mach number--and thus sharpen your arguments. Also, to display such instabilities, a beam must be confined.

DE YOUNG: I would just like to comment that pinch instabilities disrupt, but do not confine, these flows.

4C 18.68: A QSO WITH PRECESSING RADIO JETS?

Ann C. Gower,
Department of Physics, University of Victoria,
Victoria, B.C., Canada

J.B. Hutchings,
Dominion Astrophysical Observatory,
Victoria, B.C., Canada

The radio source 4C 18.68, identified with the 16.5 m QSO 2305+187 ($z = 0.313$), was observed at 20 cm and 6 cm with the VLA in its highest-resolution configuration (A) on April 25, 1981, as part of a programme to map the radio structure of quasars for which optical structure has been mapped using the Canada-France-Hawaii telescope. The radio structure is remarkable, showing complex, curved structure at 6 cm resolution embedded in a halo $\sim$ 60 kpc across ($H_0 = 100$ km/sec/Mpc, $q_0 = 0$) seen most clearly in the 20 cm map. The polarisation is low ($\sim$ 8% maximum at 6 cm) and follows the curved structure quite closely,

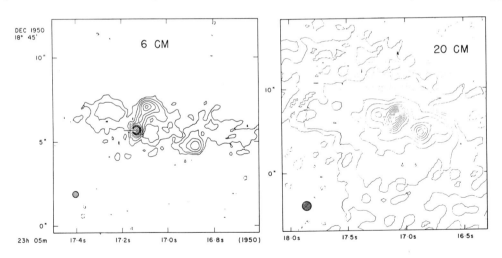

Figure 1: VLA maps of 4C 18.68 at 6 cm and 20 cm. Beam size as indicated. Scale: 1" = 2.8 kpc ($H_0 = 100$ km/sec/Mpc, $q_0 = 0$). Contour levels at: -1, 1, 2, 5, 10, 20, 30, 40, 50, 60, 70, 80, 90% of peak brightness. Peak brightness: 0.08 Jy/beam area at 6 cm and 0.13 Jy/beam area at 20 cm. Optical QSO position and error bars indicated on 6 cm map.

showing the magnetic field direction to be along the curved structure. The central source has a flat spectrum, while the spectrum of the structure generally steepens with distance from the centre.

We believe that this structure may represent the path traced out by the material flung out in a precessing or rotating beam inclined at a moderate angle to the line of sight. The large number of free parameters involved make it difficult to fit a unique model, but preliminary model-fitting has been done using a programme kindly provided by Dr. P.C. Gregory. We find, assuming that the emission has been roughly constant and equal in opposite directions (which is by no means necessarily the case), that the details of the geometry of the inner structure and the ratio of the intensities on the two sides of the source can be fit with a moderately relativistic jet velocity (0.5-0.7 c) at a cone angle of $\sim 20°$ about an axis inclined at about 50° to the line of sight. This jet velocity and the scale of the structure then give a period of $\sim 5 \times 10^4$ years, consistent with the possible presence of two interacting massive bodies in the central source.

The outer structure of the source is, however, not well explained by this model and if, in addition, the constraint of equal emission in opposing directions is removed, a wider range of jet velocities and inclinations is possible. A fuller account of this work is being published elsewhere.

This work was (partially) supported by a grant from NASA administered by the American Astronomical Society.

THE QUASAR JET 4C 32.69 at 1.4 GHz

J. W. Dreher
National Radio Astronomy Observatory
Socorro, New Mexico 87801

One of the most luminous radio jet sources known is 4C 32.69, identified with a z=0.67 QSO. It has previously been mapped at 5 GHz by Potash and Wardle (1980, PW hereafter). Here a 1.4 GHz VLA map of comparable resolution is presented and compared to the earlier map.

Figure 1 shows the source at 1.7" resolution with contours at 1, 2, 4,..., 64, 90% of the peak brightness of 57 mJy/beam. Apart from the prominent jet, the source has a typical high-luminosity-double morphology. Especially interesting is the bridge of emission spanning the outer hotspots and apparently surrounding the jet. Lower resolution maps show no sign of emission extending out beyond the hotspots. Using H=50 and q=0, the scale is 9 kpc/" and the source is 600 kpc across, much larger than average. Note that the central maximum on this map is not the core source but the bright hotspot 3" out along the jet.

For purposes of comparison, maps with identical beams (1.4"x2) were prepared from the 1.4 GHz data and the older 5 GHz data (PW). From these maps the following properties of the jet are evident:
1. The basal hotspot has α=-0.55 while the rest of the jet has α=-0.74 with no significant variations.
2. There appears to be a uniform foreground Faraday rotation of 47° consistent with the RM of -60 rad/m^2 suggested by PW. Any variations in rotation along the jet are $\leq$ 0.1 rad.
3. Most of the jet is not depolarized. At the core end, m(1.4 GHz)> m(5GHz), which is to be expected from the superposition of a flat-spectrum low-polarization core upon a steeper-spectrum, highly-polarized jet. Significant depolarization is seen at the far end of the jet where it begins to fade into the bridge, but the interpretation of this depolarization is unclear since some or all of it could be caused by the combination of polarized flux from the jet and the bridge. Generally, wherever the jet is brightest and the S/N best, the depolarization ratio is 1.0±1.

The presence of a radio bridge around the jet raises questions about the possibility of jet confinement by external pressure. Recent, high-resolution observations (Wardle, private communication) indicate a minimum energy density of 1.3×10^{-9} at the brightest part of the jet. For confine-

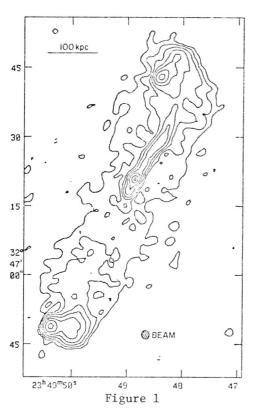

Figure 1

ment by a 10^8K plasma, $n \sim 3 \times 10^{-2}$ is required. Such a dense plasma in the bridge surrounding the jet should cause non-uniform Faraday rotation of the polarized emission from the jet. How large this rotation will be depends on the field geometry. The high degree of linear polarization normally seen in the bridges of other radio sources is usually taken to indicate that the fields in bridges are highly ordered, with at most a few independent, randomly oriented field regions along each line of sight. In this case, if N is the number of such regions, the bridge will cause fluctuations in RM along the jet of roughly $\sqrt{N}$(RM per cell)$\sim\sqrt{N}(.8\,n_e(B/\sqrt{3})l)$ where l is the length through each cell and $l \sim d/N$ if d is the width of the bridge, about 50 kpc. Using the equipartition field in the bridge of 7μG to estimate B, $N \sim 4$, and a limit on the variation in rotation of 0.1 radian, the upper limit on the thermal electron density is 2×10^{-5}, very much less than the density estimated above for pressure confinement.

The density needed for pressure confinement of the jet may be made consistent with the lack of rotation in three ways. Firstly, the confining plasma may be supposed to be extremely hot. For $n_e < 2 \times 10^{-5}$, a temperature $> 10^{11}$K would be needed to exert sufficient pressure. Although difficult to rule out observationally, postulating such a relativistically-hot plasma seems implausible without a more compelling argument. Secondly, the bridge could be far out of equipartion. If $n_e \sim 10^{-2}$ then B must be $<10^{-2}$μG. Not only would this lead to an absurdly high total energy for the bridge ($\sim 10^{10} M_\odot c^2$), but the consequent inverse-Compton x-ray flux would be >10 times the observed x-ray flux (see Wardle and Potash, in this Symposium). Thirdly, the magnetic field geometry may be such that there are many reversals of $B_\parallel$ along each line of sight with a resulting decrease in the net Faraday rotation. In the extreme, Laing (1981) has recently shown that for some field configurations a high degree of linear polarization may be coupled with no Faraday rotation at all. It is not clear, however, that the special field configurations required for this type of effect are likely to occur in actual radio sources.

Laing, R.A.: 1981, Ap. J., 248, pp 87 - 104.
Potash, R.I., and Wardle, J.F.C.: 1980, Ap. J., 239, pp 41 - 49.

BENT JETS IN RADIO QUASARS

Susan G. Neff
Astronomy Department, University of Virginia and
National Radio Astronomy Observatory

How is energy transported out from the central engine in quasars and radio galaxies to the distant radio lobes? This problem has been around since the early discovery of classical double radio sources, and is still not answered in detail. The idea of relativistic beams was first suggested by Martin Rees as a means of transporting plasma out of the nucleus (Rees, 1971, Blandford and Rees, 1974). This idea gained support first from the discovery of hot spots in the radio lobes of these large classical double sources, and later by observations of the beams themselves in radio galaxies. As more jets were observed, it became obvious that they were often curved, serpentine, or even sharply bent. This behavior has been modeled as precession of the central nozzle (Bridle et al., 1976, Ekers et al., 1978), as nuclear refraction (Henriksen et al., 1981), as a growing plasma instability (Hardee, 1981) and as various combinations of the above. At the present time, it seems safest to conclude that there are some examples of each of these processes known.

Here I present four examples of quasars with bent jets that were found during a more extensive study of quasars with known compact nuclear radio cores. The observations shown were made at one of two frequencies, using the "snapshot" mode of the partially completed NRAO Very Large Array. The 4.885 GHz observations were done in October, 1979 using up to 16 antennas, and the 1.635 GHz observations were done in June, 1980 using up to 21 antennas. The data were carefully edited, maps were made and cleaned, and finally the self-calibration technique was judiciously applied. In the figures shown below, a strong point source has been removed from the core of each quasar. The maps of 0742+318 and 1217+023 have been tapered slightly to enhance the low surface brightness emission.

Work is currently in progress to test various theories of bent jets using these observations, but already it is probably possible to rule out those theories that require the jets to be nearly parallel to the line of sight. As in the case of the giant radio galaxy 3C 236 (Willis et al., 1974), these quasars are quite large even if they lie precisely

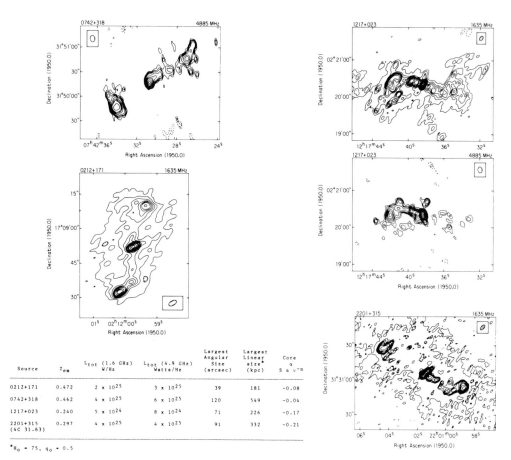

Source	z_{em}	L_{tot} (1.6 GHz) W/Hz	L_{tot} (4.9 GHz) Watts/Hz	Largest Angular Size (arcsec)	Largest Linear size* (kpc)	Core $S_\alpha \nu^{-\alpha}$
0212+171	0.472	2×10^{25}	3×10^{25}	39	181	-0.08
0742+318	0.462	4×10^{25}	6×10^{25}	120	549	-0.04
1217+023	0.240	5×10^{24}	8×10^{24}	71	226	-0.17
2201+315 (4C 31.63)	0.297	4×10^{25}	4×10^{25}	91	332	-0.21

*$H_0 = 75$, $q_0 = 0.5$

in the plane of the sky; tilting them very far out of the plane of the sky merely exacerbates travel time and velocity problems. All of the sources shown are known (from Preston's VLBI Survey, Preston, 1979) to have structure on the scale of a few tens of parsecs. Observations have been proposed to study the detailed small scale structure in those sources with the brightest radio cores.

REFERENCES

Blandford, R. D., and Rees, M. J., 1974, Mon. Not. Roy. Astr. Soc. 169, 395.
Bridle, A. H., Davis, M. M., Meloy, D. A., Fomalont, E. B., Strom, R. G., Willis, A. G., 1976, Nature 262, 179.
Ekers, R. D., Fanti, R., Lari, C., Parma, P., 1978, Nature 276, 588.
Hardee, P. H., 1981, preprint.
Henriksen, R. N., Bridle, A. H., and Chan, K. L., 1981, Ap. J. (submitted).
Preston, R., 1979, preprint.
Rees, M. J., 1971, Nature 229, 312 (Erratum, Nature 229, 510).
Willis, A. G., Strom, R. G., Wilson, A. S., 1974, Nature 150, 625.

THE JETS IN 3C 449 REVISITED

T. J. Cornwell, NRAO and R. A. Perley, NRAO

Introduction: Bridle (this volume) has summarized the overall characteristics of the jets found in numerous low-luminosity and some high-luminosity radio sources. Previous observations made with the partially completed VLA at wavelengths of 6 and 20 cm indicated that 3C449 was an archetypal radio source obeying all the "rules" summarized by Bridle. New observations with the VLA of the polarization structure at 6 and 20 cm have destroyed this simple picture and identify 3C449 as a "rogue" jet source.

Intensity Structure: The new observations indicate several interesting facets of the intensity structure of the radio jets:

1. The jets are very symmetrical at both wavelengths, the ratio of the total flux being less than 1.8 at 6 cm.

2. The spectrum of the jets is fairly constant over their whole length, the northern jet being slightly steeper ($\alpha = 0.67$) than the southern jet ($\alpha = 0.62$).

3. The collimation data is consistent with two regimes of simple linear expansion, first an opening angle of about 20 degrees then at about 6 degrees.

4. Profiles of total intensity are almost Gaussian with no signs of limb-brightening.

5. Both peak flux and total flux across the jet show no signs of decreasing along the jet as rapidly as adiabatic expansion predicts.

Polarization Structure: Both jets show unusual features in the polarization at 6 and 20 cm.

1. 6 cm: Both jets possess a region, relatively weak in total intensity and close to the core, where the inferred magnetic field direction is parallel to the jet axis before twisting off to become

perpendicular farther away from the core. The total emission rises as the field twists to become perpendicular. Almost no shearing of the B-field at the jet boundaries is observed, the only exception being at the first turn in the northern jet.

2. 20 cm: The jets are heavily depolarized showing gradients in percentage polarization both along and across the jets. The depolarization ratio (6cm/20cm) varies from as low as 1 to greater than 10, in some places over a distance of less than a jet diameter.

What Does It All Mean? A number of interesting conclusions arise:

A. The luminosity per unit length is approximately constant along the jet as predicted for constant opening angle by the model of Henriksen, Chan and Bridle (1981) in which turbulence provides the necessary particle accelaration. However, other mechanisms to preserve the luminosity per unit length exist (see Bridle, this volume).

B. Differences in smoothness of emission between the northern and southern jets suggest that any variations are intrinsic to the jet and are not due to fluctuation in the core luminosity.

C. Two regimes of expansion are seen, consistant with collimation by a two component atmosphere (e.g. Bridle, Chan and Henriksen 1981). The region of recollimation seen in sources such as NGC 315 on the scale of >10 Kpc may have been masked by the disruption of the jets on a scale of about 20 Kpc.

D. The depolarization is quite unlike that seen in any other jet source. Any explanation, either placing it internal or external to the jet, appears extremely contrived. The unexpected complexity of the polarization structure illustrates the dangers of using a simple depolarization model to interpret low resolution data.

E. Hardee (1981) has suggested that the kinks in the large scale structure are due to helical instabilities disrupting the jets. Our observations show that the amount of thermal matter required in the lobes is too high by an order of magnitude.

We acknowledge useful conversations with Alan Bridle, Robert Laing, Dick Henriksen and Lorenzo Zaninetti.

References:
Birkinshaw, Peacock and Laing, M.N.R.A.S. (in press) 1981.
Bridle, Chan and Henriksen, preprint 1981.
Hardee, preprint 1981.
Henriksen, Chan and Bridle, preprint 1981.
Perley, Willis and Scott, Nature, 281, pp 437, 1979.

RECENT WSRT AND VLA OBSERVATIONS OF THE JET RADIO GALAXY NGC 6251

A.G. Willis[1], R.G. Strom[1], R.A. Perley[2] and A.H. Bridle[2]
[1] Netherlands Foundation for Radio Astronomy
[2] NRAO

NGC 6251, a 14th mag elliptical galaxy, was shown by Waggett et al. (1977) to have large-scale radio emission features with a total angular extent of ~ 1.1°, which corresponds to a projected linear size of about 1.7 Mpc (H_0 = 75 km s^{-1} Mpc^{-1}). A bright radio jet links a central core source embedded in NGC 6251 to the extended emission on the northwest side of the galaxy.

We have observed the radio emission associated with NGC 6251 at 49cm wavelength with the Westerbork Synthesis Radio Telescope. The resulting radio map, made with a 55 arcsec beam (FWHM), is displayed in Figure 1. This map clearly indicates that in addition to the main jet oriented in position angle 296°, there is a faint counter jet extending to the southeast from the central core source. The cross-sectional integrated emission from the counter jet is only about 1/60 that of the main jet, which explains why the counter jet has not been detected previously. The 49cm map also indicates that the main and counter jets have rotational symmetry, changes in position angle of the main jet on its way to the northwest hotspot marked "A" being opposite to those of the counter jet on its way to feature "B". However, while hotspot A marks the end of collimated emission associated with the main jet, the counter jet appears to have additional loosely collimated emission extending some 13.9 arcmin (~360 kpc) from feature B to the outer hotspot marked "C". Note that hotspots A and C at the ends of the jets have rather similar surface brightnesses (to within a factor 2).

The bright inner 4 arcmin regime of the main jet seen in Figure 1 has been the subject of extensive VLA observations. These observations show that the Faraday rotation measure (RM) distribution and the projected magnetic field structure of the inner main jet are rather unusual. While the RM over the jet (and the northwest outer lobe) at distances greater than 90 arcsec (38 kpc) from the central core source is roughly constant with values lying between -40 and -60 rad m^{-2}, at distances of less than 40 arcsec (17 kpc), the RM has values ranging from -10 rad m^{-2} (at ~40 arcsec distance) to -130 rad m^{-2} (at ~18 arcsec distance). These variations of the RM over the inner part of the NGC

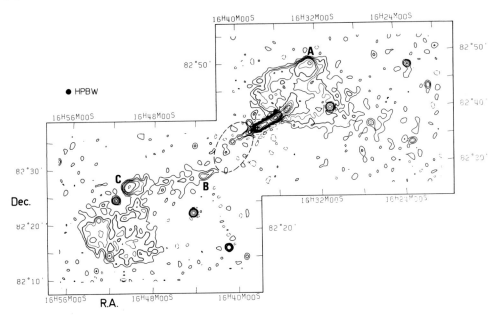

Fig. 1. NGC 6251 at 49 cm wavelength. The region inside the ellipse has contours at -2 (dashed),2,3.75,6,9 mJy/beam. In the rest of the map the contours are -3 (dashed),3,6,9,13.75,20,40,60,80,100,150,250,400,500 mJy/beam. A cross marks the position of the central core source.

6251 jet are displayed in Figure 2. We believe that a rather constant fraction of this RM, ~ -50 rad m^{-2}, is due to the foreground Faraday screen of our own galaxy. The remaining variations might be due to one or a combination of the following possibilities. Firstly, about ± 30 rad m^{-2} of the RM might occur within the jet but in a significantly ordered magnetic field so that RM variations across the jet would be expected.(An estimate of |30| rad m^{-2} RM internal to the jet is derived from the fact that the degree of polarization seen at 21cm wavelength is typically 0.6 of that seen at 6cm wavelength.) Secondly there may be ionized gas outside the jet but near or within NGC 6251 itself which produces the observed RM variations. The mass of ionized gas involved could then be as large as 2 x 10^9 solar masses.

That the RM variations must at least partially occur within the jet is suggested by the rough alignment between the RM contours over the distance ~20 to 30 arcsec from the core source in Figure 2 and the projected magnetic field structure also displayed there. Between 20 and 40 arcsec distance from the core the projected magnetic field has a diagonal orientation with respect to the jet extension as do the RM contours over 20 to 30 arcsec distance from the core.

This diagonal orientation of the projected magnetic field with respect to the main axis of the jet is also seen over the distance range 180 arcsec (77 kpc) to 260 arcsec (111 kpc). Elsewhere the

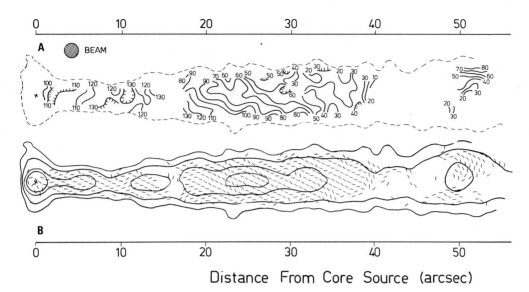

Fig. 2. (a) Contours of Faraday rotation measure (RM) over the inner part of the NGC 6251 jet. Negative signs have been omitted from all contour labels, i.e. a contour label of 100 implies a RM of -100 rad m^{-2}. The outer dashed line corresponds to a total intensity contour of 0.25 mJy/beam at 1662 MHz where the beam is 1.65 arcsec (FWHM). (b) Vectors indicating the orientation of the projected magnetic field in the jet superimposed on total intensity contours of 0.25, 1, 3 and 5 mJy/beam at 1662 MHz. Note that especially near the outer edges of the jet scatter in the vector orientations occurs because of low signal-to-noise ratio in the polarized flux density measurements used to generate this map.

magnetic field is aligned either parallel or perpendicular to the jet extension, a configuration already seen in many other radio jets such as that of 3C 31 (Fomalont et al., 1980).

The Westerbork Radio Observatory is operated by the Netherlands Foundation for Radio Astronomy with the financial support of the Netherlands Organization for the Advancement of Pure Research (Z.W.O.). The NRAO VLA is operated by Associated Universities Inc., under contract with the National Science Foundation.

REFERENCES

Fomalont, E.B., Bridle, A.H., Willis, A.G., Perley, R.A., 1980, Astrophys. J. 237, 418.
Waggett, P.C., Warner, P.J., Baldwin, J.E., 1977, Monthly Notices Roy. Astron. Soc. 181, 465.

DISCUSSION

LAING: I belive that your results show that the E-vector PA is proportional to λ^2 over 300° of rotation close to the nucleus. One cannot get more than 90° out of a simple slab model and this implies that most of the rotation is caused by gas associated with NGC 6251, but located in front of the jet. The depolarization may be due either to : (a) thermal matter within the jet, or (b) foreground gas clumped on a scale much smaller than the beam.

WILLIS: I agree with your remark that the E-vector roation proportional to λ^2 for over 300° of rotation indicates that a large part of the rotation must be due to gas outside the jet. However, I think we must be careful in the application of the simple slab model to jets; the magnetic field may be quite complicated.

NGC 4258: A BENT JET IN A SPIRAL GALAXY

R. H. Sanders
Kapteyn Astronomical Institute

The remarkable continuum arms in the spiral galaxy NGC 4258 are suggestive of some form of ejection from the nucleus of this galaxy (Van der Kruit, Oort and Mathewson, 1972). To summarize the observations (see Oort, Figure 2, this volume), the "anomalous spiral arms" are clearly distinct from the normal spiral arms, although wound in the same sense; there is a sharp gradient of the continuum emission on the leading edge of the arms, and an indication that the arms split on the western side; the arms go directly into the nucleus and coincide with $H\alpha$ emitting filaments (Courtes, Viton and Veron, 1965).

The only model which has been suggested for the anomalous arms involves highly directional ejection of 10^7 to 10^8 $M_\odot$ of gas at velocities from 800 km/s to 1600 km/s. Interaction of the clouds with the differentially rotating gaseous disk of the galaxy accounts for the observed shape of the arms. We may obtain an alternative model by assuming that the anomalous arms are more or less steady-state jets which are bent both by a pressure gradient in the interstellar medium and by the ram pressure of rotating extended atmosphere. From the balance of centrifugal force wthin the bent jet against ram pressure and pressure gradient forces, one may derive a differential equation for the shape of the jet (Begelman, Rees and Blandford, 1979). I have assumed a rotating isothermal atmosphere (or gaseous halo) in a Hubble low potential with constant angular velocity on cylindrical shells. In this model there are five free physical parameters:

(1) $f = V_{mx}^2/(\sigma^2 + V_{mx}^2)$.
 Here V_{mx} and σ are respectively the maximum rotational velocity and the velocity dispersion of the gas in the halo.
(2) θ_e = the initial angle of the jet with respect to the rotation axis of the galaxy.
(3) $\lambda = \rho_a/V_a^2/\rho_j V_j^2$.
 This is the initial ram pressure in terms of the momentum flux in the jet.
(4) S_o = the distance of the sonic point in the jet from the galactic nucleus.
(5) R_o = the nozzle radius.

In addition there are two projection angles θ_1 and θ_2, but one of these is the known inclination of the galaxy.

There is one very necessary constraint on such a model. Ram pressure bends a jet into a spiral, but it is a leading spiral. However, the anomalous spiral ams have the same sense of winding as the normal spiral arms, and, presumably, the normal spiral arms are trailing. This means that, in the context of this model, the anomalous arms cannot lie in the plane of the galaxy. If our line of sight is between the plane of the galaxy and the plane of the anomalous arms then both the trailing normal arms and the leading anomalous arms will project into spirals with the same sense of winding.

Figure 1 shows the optimum bent jet model compared to recent VLA 20-cm observations of NGC 4258.

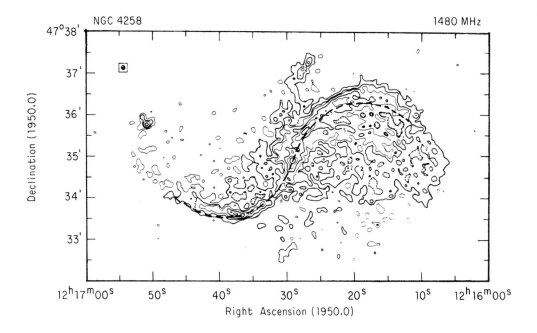

Figure 1. The bent jet projected on to the plane of the sky (dashed line) compared with the VLA 20-cm continuum map of Van Albada and Van der Hulst.

The parameters of this model are $f = 0.8$, $\theta^e = 80°$, $\lambda = 1.0$, $S_o = 250$ pc, $R_o = 100$ pc with projection angles of 198° and 72° inclination). We see that the fit to the shape of the anomalous arms is reasonable. Moreover, assuming a jet velocity of 1000 km s^{-1}, the mass loss rate is fairly mild (10^{-3} $M_\odot$/year).

This model has a very explicit observational consequence. To get the proper projection of the anomalous arms, the northwestern arm must lie behind the galactic plane. Therefore, we should see 21-cm absorption against the continuum arm.

NGC 4258 is an example of a normal spiral galaxy which may contain a jet. If jets do exist in normal spirals, perhaps the nuclei of most normal galaxies contain the same engines that are found in the powerful radio galaxies.

I am very grateful to G. D. van Albada and J. M. van der Hulst for kindly allowing me to use their 20-cm VLA map of NGC 4258 in advance of publication.

REFERENCES

Begelman, M. C., Rees, M. J., Blandford, R. D.: 1979, Nature, 279, 770.
Courtes, G., Viton, M., Veron, P.: 1965, Quasi Stellar Sources and Gravitational Collapse, eds. Robinson, Shild, Shucking, Univ. of Chicago Press, p. 307.
Van Albada, G. D.: 1978, Ph.D. Dissertation, University of Leiden.
Van Albada, G. D., van der Hulst, J. M.: 1981, in preparation.
Van der Kruit, P. C., Oort, J. H., Mathewson, D. S.: 1972, Astron. Astrophys., 21, 169.

DISCUSSION

OORT: I see several objections to your model. The first is that the normal arms of NGC 4258 show a pronounced weakening where according to the model of Van der Kruit, Oort and Mathewson the ejected gas would have swept them away. The second is that the inner parts of the anomalous arms, where they are seen in $H\alpha$ emission, show that the general, presumably disk-emission in the region preceding the arms is much stronger than in the region behind, indicating that these parts of the arms, which reach out to at least 6 kpc from the centre, are likewise situated in the disk. This is further corroborated by the fact that they participate in the disk's rotation.

SANDERS: The outer spiral arms, at least the southern arms, do become less intense closer to the center of this galaxy. From visual inspection of photographs of this galaxy, the weakening seems to me to be entirely consistent with what is seen in other normal Sb systems (M81, for example). As you point out, the general $H\alpha$ emission is certainly weaker in front at the southern anomalous arm; however, the region of this low intensity does coincide with a large dust patch. My impression from Van der Kruit's work on the $H\alpha$ filaments associated with the anomalous arms is that the kinematic features differ rather conspicuously from that of the general disk

KILOPARSEC SCALE STRUCTURE IN HIGH LUMINOSITY RADIO
SOURCES OBSERVED WITH MTRLI

Peter N. Wilkinson
Nuffield Radio Astronomy Laboratories, Jodrell Bank
University of Manchester

1. INTRODUCTION

The new Jodrell Bank Multi-Telescope-
Radio-Link Interferometer system (MTRLI) began
operating in January 1980 with four telescopes
and with its full complement of six telescopes
in December 1980. The location of the
telescopes and the baseline lengths (in km)
thus obtained are shown in Fig. 1. However,
before describing some of the exciting new
results it has yielded on high luminosity
radio sources it is important to outline the
capabilities of the instrument since the maps
are produced in a rather different way to
those from conventional synthesis instruments.

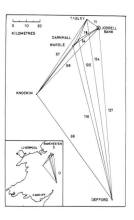

2. OUTLINE OF THE MTRLI MAPPING CAPABILITY

Fig.1 The MTRLI

The technical details of the MTRLI have
been outlined elsewhere[1] and need not concern us here since, regardless of the equipment, the phase stability on such long baselines as
these is fundamentally limited by the variations in the atmospheric/
ionospheric delay over each telescope. And while a few observations
have been made using compact sources within the primary telescope
beams as phase references to calibrate out these phase variations so
far nearly all the MTRLI maps have been produced using 'closure' phase[2];
thus they should more properly be termed "hybrid" maps[3].

Notwithstanding this reliance on closure phase which limits us to
mapping only those sources which give fringes above the noise on all
baselines, the mapping capability of the system has far exceeded our
initial expectations. The methodology we have adopted to analyse the
data strongly resembles that of the 'self-calibration' technique now
in use for improving VLA maps[4]. We assume that, apart from fixed

offsets, all the errors in the amplitude and phase of the fringes are associated with the individual telescopes. An algorithm (CORTEL)[5] has been written to correct these telescope errors via an iterative scheme involving CLEAN and we are now routinely obtaining maps whose noise levels approach the limit set by thermal noise (typically ∼1 mJy/beam). This has resulted in maps whose dynamic range is >1000:1 on strong sources dominated by a single compact component.

In the number of telescopes employed and in the use of 'closure' methods the MTRLI should be very like present-day VLBI arrays in its mapping capabilities. Why then are our maps markedly superior in noise level/dynamic range to any VLBI maps published so far? To first order it is because the fringe amplitude calibration on MTRLI (1→2% r.m.s.) is typically five times better than in VLBI where there are rarely a priori unresolved sources to normalise each baseline to the same flux scale. Secondly we believe that the CORTEL algorithm is more sophisticated than the early hybrid mapping methods used in VLBI. The (u,v) coverage obtained with MTRLI is only somewhat better than in many current VLBI experiments and so this cannot be a major contributory factor to the superior dynamic range of MTRLI compared with current VLBI.

Extensive simulations on artificial data have shown that the (u,v) coverage allows us to map linear sources (i.e. Virgo or Cygnus A-like) whose dimensions are ≲50 CLEAN beams. Larger sources must be convolved with a bigger beam until the ∼50-beam criterion is satisfied. At the major observing frequencies of 408, 1666 and 5000 MHz the conventional CLEAN beam is nearly circular at all declinations above 10° and has a FWHM of 1.0, 0.25 and 0.08 arcsec respectively. Thus a rule of thumb is that the MTRLI beam is a factor four smaller than the VLA 'A' configuration beam at the same frequency. Fig. 2 shows a grey

Fig. 2   Four-telescope map of Virgo A at 408 MHz

scale map of Virgo A at 408 MHz made with only four telescopes and a minimum baseline of 23 km. It shows the compact features in the source, i.e. the nucleus and the knotty radio jet, very well but of course is not sensitive to the smooth extended structure mapped by the VLA[6].

3. RECENT RESULTS ON HIGHLY LUMINOUS SOURCES

Naturally enough in its first year of operation the MTRLI has concentrated on mapping the strongest compact sources available to it. On average these are intrinsically (but see later) the more luminous sources in the sky. No complete unbiassed sample has yet been fully observed so here I will merely try to show some of the more interesting maps we have made which we hope contain new pointers towards the source physics.

Two of the strongest sources, both apparently and intrinsically, are 3C196 (QSO, $Z = 0.87$) and 3C295 (galaxy, $Z = 0.46$) both of which are doubles. However since the component separation in each is only ~ 4 arcsec no detailed maps of either object have been available until now. It is important to map many such powerful "normal" doubles, with similar linear resolutions, in order to determine whether at high redshifts there are any differences in structure ascribable either to intrinsic differences in the sources themselves or more likely to the modifying effect of the intergalactic medium whose properties are probably redshift-dependant. The 1666 MHz map of 3C295 shown in Fig. 3 reveals a structure in which the disposition of ridges and hotspots strongly resembles that in the low redshift ($Z = 0.056$) source Cygnus A as mapped by Cambridge 5 km telescope at 5 GHz[7]. The projected linear size of 3C295 is about half that of Cygnus A, a difference which could merely be due to the effect of projection.

In contrast the 408 MHz map of 3C196 shown in Fig. 4 reveals a very different structure. This can most easily be interpreted as the result of the rotation of its major axis leading to extended (and in fact steep spectrum) trails and hot spots marking the present direction of the putative beams[8]. Until now such rotation symmetries have only been seen in low luminosity objects (see the contribution by R. Ekers) where they are usually ascribed to precession of the rotation axis of the central engine.

The unique characteristics of the MTRLI, i.e. subarcsecond resolution at low frequencies with excellent dynamic range, are particularly apposite for the study of the extended steep spectrum emission near compact flat spectrum components. Many such sources have now been mapped and in the vast majority of them the extended emission is dominated by a one-sided jet often containing barely resolved knots. An illustrative example is shown in Fig. 5 which is the 1666 MHz map of 3C454.3 (QSO, $Z = 0.86$). Here the core contains $\sim 95\%$ of the total flux and a dynamic range of $\sim 2000:1$ is needed to

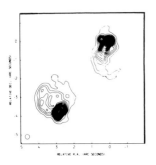

Fig. 3. 3C295 @ 1666 MHz

Contours @ 1,3,5 →25,35→95% of Peak. Bottom Contour ≡ 25 mJy/beam.

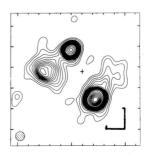

Fig. 4. 3C196 @ 408 MHz

Contours at 1,2,4→24,30→90% of peak. Bottom contour ≡ 82 mJy/beam.
The ⌐ shows 2 arcsec.

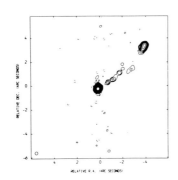

Fig. 5 3C454.3 @ 1666 MHz

Contours at 0.05, 0.1, 0.2→51.2% of peak.
Bottom contour ≡ 4.2 mJy/beam

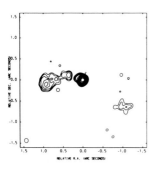

Fig.6a. 3C309.1 @ 1666 MHz

Beam = 0.1 arcsec
Contours @ 0.5,1.0,2.0→64% of peak
Bottom contour ≡ 18 mJy/beam

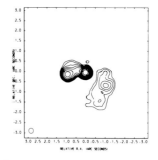

Fig.6b. 3C309.1 @ 1666 MHz

Beam = 0.25 arcsec
Contours @ 0.1,0.2,0.4→51.2% of peak.
Bottom contour ≡ 3.8 mJy/beam

delineate the jet properly; the knot or hotspot at the end of the jet has a peak brightness ≲3% of the core.

However, while one-sided jets may dominate the extended emission, high dynamic range maps often reveal components on the opposite side of the flat spectrum core. This is illustrated in Fig. 6a, the 1666 MHz map of 3C309.1 (QSO, Z = 0.90) which incidentally has a resolution of 0.1 arcsec since it was made with European VLBI as well as MTRLI data. This map is typical of many we have obtained of overall flat spectrum sources. However 3C309.1 is a steep spectrum source and comvolving the map to the standard MTRLI resolution (0.25 arcsec) improves the dynamic range (to ~1000:1) and reveals the full extent of the emission to the west of the core (Fig. 6b). It is only because of the relative strength of the steep spectrum emission in 3C309.1 that we are able to see this low brightness structure - a point to bear in mind when trying to interpret similar dynamic range maps of core-dominated sources.

Nearly all the overall flat spectrum objects are quasars and some turn out to have particularly peculiar, twisted, structures. Maps of three such objects 1636+473, 1823+568, and 3C418 are shown in Figs. 7a, 7b, 8 & 9. While it is clear that in 1823+568 and 3C418 the extended structure is physically associated with the core in 1636+473 the association can only be said to be statistically very probable, there being no evidence for a tell-tale bridge of emission between the core and the emission to the North.

I must not leave the impression that all the flux density is accounted for in these maps of core-dominated sources[9]. Typically 5-10% of the total flux is missing from the maps and must arise from diffuse low brightness emission with a characteristic scale ≳5 arcsec. The spectral index of this diffuse emission is rather steep ($\alpha \gtrsim -1$ with $S \propto \nu^{\alpha}$).

The crucial question about these core-dominated objects (and the halfway-house cases like 3C309.1) is whether they represent a separate type of radio source to the normal doubles, which also seem to exist at high redshifts vide 3C295, or whether the clear morphological differences between the two types are more apparent than real. For example if there are relativistic bulk motions in the radio-emitting material (and we know that this is almost certainly the case in the nuclei of the superluminals) then the combined effects of relativistic beaming and projection may greatly distort the apparent structure of a source as seen by instruments with limited sensitivity or dynamic range.

Regardless of the detailed physics involved there are a few simple observational facts which must have an important bearing on this question.

1) There must be some doubles seen "end-on" and these must be found among the apparently compact sources.

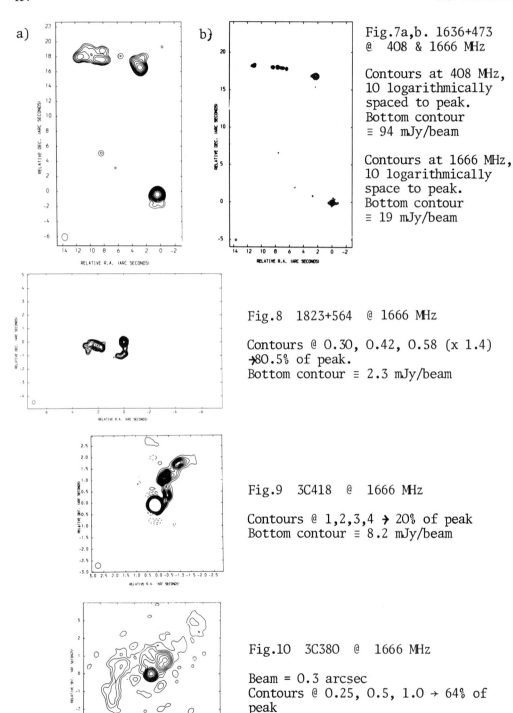

Fig.7a,b. 1636+473 @ 408 & 1666 MHz

Contours at 408 MHz, 10 logarithmically spaced to peak. Bottom contour ≡ 94 mJy/beam

Contours at 1666 MHz, 10 logarithmically space to peak. Bottom contour ≡ 19 mJy/beam

Fig.8 1823+564 @ 1666 MHz

Contours @ 0.30, 0.42, 0.58 (x 1.4) →80.5% of peak.
Bottom contour ≡ 2.3 mJy/beam

Fig.9 3C418 @ 1666 MHz

Contours @ 1,2,3,4 → 20% of peak
Bottom contour ≡ 8.2 mJy/beam

Fig.10 3C380 @ 1666 MHz

Beam = 0.3 arcsec
Contours @ 0.25, 0.5, 1.0 → 64% of peak
Bottom contour ≡ 9.0 mJy/beam

2) We do not observe any compact sources as amorphous "blobs" corresponding to a simple superposition of the outer lobes.

3) All compact sources in fact contain bright cores and have coherent extended structure, often jet-like and one-sided, as well as faint, diffuse "haloes".

4) We do not see "disembodied" jets i.e. without an accompanying core.

From these very general statements it seems almost inescapable first that some aspect-related amplification of the core flux is occurring, this is presumably due to relativistic beaming. Secondly point 4) implies that if the core fluxes are boosted then so, at least to some degree, must be those of the jets otherwise the jets would be swamped and invisible in limited dynamic range maps. Relativistic beaming of the jet emission is the obvious explanation since it also naturally accounts for the one-sidedness which is so prevalent.

A simple working hypothesis to explain the apparently diverse forms seen in highly luminous radio sources is therefore that there is really only one type of object and its appearance is critically dependent on the bulk velocity of the emitting material and its angle to the line of sight. However this pleasing synthesis is almost certainly a gross over-simplification of the real situation a view which is strengthened by the 1666 MHz map of 3C380 shown in Fig. 10. This steep-spectrum source seems to manifest virtually all types of structure and it is difficult to see merely a classical double source lurking amidst this tangle of emission. Further doubts will remain about this unified scheme until 1) maps with sufficient dynamic range ($>10^4:1$) are available to trace out the diffuse emission regions in core-dominated sources to see whether they really are the faint "ghosts" of the outshone outer lobes 2) VLBI techniques have advanced sufficiently in sensitivity to measure the supposed relativistic motions in the extended jets. At least we may have confidence that these observational goals will be met in the forseeable future.

I thank all my colleagues at Jodrell Bank who contributed towards making the MTRLI, and therefore this paper, possible. Among those people who helped me produce these maps or who allowed me to use their maps prior to publication were Mike Charlesworth, Ron Clarke, Marshall Cohen, Tony Foley, Colin Lonsdale, Tom Muxlow, Mark Orr and Althea Wilkinson. I am grateful to them all and to Ian Browne with whom I have had many illuminating conversations.

REFERENCES

1. Davies, J.G., Anderson, B. & Morison, I. 1981. Nature, 288, 64.
2. Rogers, A.E.E. et al. 1974. Astrophys.J., 193, 293.
3. Baldwin, J.E. & Warner, P.J., 1976. M.N.R.A.S., 175, 345.
4. Schwab, F.R., 1980. Proc. 1980 Int. Optical Computing Conference.
5. Cornwell, T.J. and Wilkinson P.N., 1981. M.N.R.A.S., 196, 106
6. Owen, F.N., Hardee, P.E. & Bignell, R.C., 1980. Ap.J., 239, L11.
7. Hargrave, P.J. & Ryle, M., 1974. M.N.R.A.S., 166, 33.
8. Lonsdale, C.J. and Morison, I., 1980. Nature, 288, 66
9. Browne, et al., 1981. M.N.R.A.S. (in press).

A SUGGESTED CLASSIFICATION AND EXPLANATION FOR HOTSPOTS IN SOME POWERFUL RADIO SOURCES

P.P. Kronberg
University of Toronto, Scarborough College and David Dunlap Observatory
and
T.W. Jones
Department of Astronomy, University of Minnesota

1. A CLASSIFICATION SCHEME FOR HOTSPOT COMPLEXES IN POWERFUL DOUBLE RADIO SOURCES

High resolution (1"→0".1) maps of the outer complexes of some "well formed" powerful radio sources suggest that we can now distinguish two physically distinct types of outer hotspots. We denote them as type "A" and "B" and describe them as follows: Type A hotspots, illustrated in Figure 1, occur at the outer leading edge, and have a cusp-like, or otherwise elongated shape. This strongly suggests that their shape and energy density are determined by the ram-pressure interaction between the end of a beam or momentum flux "pipeline", and the ambient i.g.m. Magnetic fields appear well-ordered <u>along</u> the cusp (Laing, 1981). The surface brightness of well-resolved Type A hotspots leads to a velocity of advance (V_a) which is typically $10^3 \rightarrow 10^4 (\rho_{ig_{-27}}^{-1})$km/s (ignoring the ion energy).

As the schematic illustration in Fig. 1 shows, Type B hotspots generally lie off the A hotspot-galaxy/QSO axis, and are also <u>behind</u> the Type A hotspots. In at least some cases, such as illustrated in Fig. 1, they are <u>more compact</u> and have a higher minimum energy density than the outer, Type A hotspots. Synchrotron lifetimes in the B hotspot of 3C351N are $\lesssim 10^4$ yrs, which is less than the light travel time from A- to the B-hotspot. The B-hotspots in 3C351N and CygAW appear to

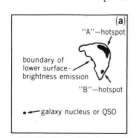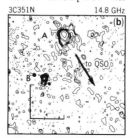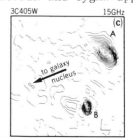

Fig. 1. "A" and "B" hotspots illustrated schematically (a), and in λ2cm maps of 3C351N (Kronberg <u>et al</u>. 1980) and CygAW (Hargrave & Ryle 1976) (b and c).

occur near the lobe-i.g.m. interface. In both of these cases the resolution is not yet sufficient to define their morphology or magnetic field structure in detail. Higher resolution ($\lesssim$0.1 arcsec) observations will be useful in further elucidating the structure of B-hotspots, in particular the question as to how they are confined.

While a number of strong extragalactic source lobes exhibit the A- and B-hotspot characteristics just described, we must emphasize that many others do not. Examples of the latter are sources having structure like 3C31 and 3C449, head-tail sources, and sources having complex or z-shaped structure. In other cases, eg. 3C9 at 1" resolution, one lobe appears to have at least a cusp-shaped A-hotspot, whereas the opposite radio lobe has an entirely different structure.

The nature of the B-hotspots is intriguing. The short radiative lifetimes of their constituent electrons imply that particle acceleration *is occuring in the B hotspots*. Prima facie ram pressure arguments suggest that, if the B-hotspots are at the end of the beam, they ought to have advanced fastest, hence furthest from the galaxy/QSO - which is contrary to observation in the examples we have cited. In the following section we propose an explanation for the A-B hotspot phenomenon.

2. A SUGGESTED MODEL FOR A AND B HOTSPOTS

We begin with the assumption that the A hotspots occur at the end of the beam which deposits the energy from the galaxy/QSO. The B hotspots arise from instabilities which form along or near the lobe-i.g.m. interface as the A hotspot advances into the ambient medium. The observed circumferential geometry of magnetic field lines in A hotspots provides a natural channel for rapid transport, backwards and off axis, of relativistic electrons from the A hotspot. These are then trapped and accelerated in instabilities at the lobe-i.g.m. interface, and in this way form B hotspots. The relatively infrequent appearance of strong B hotspots within a given radio lobe can be explained if they form and dissipate on timescales short compared with the dynamic lifetime of the radio lobe. Their relative proximity to the A hotspots ($\lesssim$20% of A-to-QSO) suggests that they are causally connected to the A hotspot phenomenon, and further that conditions for B hotspot formation are favourable off-axis and not too far from the leading edge of the outer lobe.

Since we can now resolve A hotspots and, hence estimate the minimum internal energy density (ε) of the relativistic gas, pressure balance with the i.g.m. of density ρ gives an estimate of the velocity of advance, $V_a \simeq 5770 \, (\varepsilon_{-9}/\rho_{-27})^{\frac{1}{2}}$ km/s. For typical sources this is comparable to the Alfvén speed, derived from the assumption of equipartition, at the ejecta-i.g.m. interface ($V_A = 8920 \, B_{-4} (\rho_{-27})^{-\frac{1}{2}}$ km/s), and the ion sound speed, $V_s \simeq 1200 \, T_8^{\frac{1}{2}}$ km/s. It is also not much greater than typical i.g.m. orbital or turbulent velocities $100 \lesssim V_t \lesssim 1000$ km/s (ε, ρ, B, and T in c.g.s.). The similarity of all these numbers (which we expect a priori if the Mach number of the beam flow is not very large) suggests that the shear velocities near the interface behind the A hotspot are of the requisite magnitude to stimulate the growth of the Kelvin-Helmholtz type instabilities. In their classical form these will propagate with wave number $k_z(=2\frac{\pi}{\lambda}\cos\theta)$ and amplitude a, and will grow to a limit given by $ka\sim 1$ (Gerwin 1968). The characteristic growth time is $\sim (\lambda/\Delta V)\sqrt{\rho_{ig}/\rho_{lobe}}$, where ΔV is the velocity differential which will stimulate suitable wave growth. As the field is amplified due to being "stretched" (the surface area of the interface is increased by the instability), electrons will gain energy by betatron acceleration on approximately the same timescale, and thus produce a B-hotspot. Taking $\Delta V<0.1c$, $\lambda<500$ pc we find that the kinetic energy available ($\rho_{ig}(\Delta V)^3 \lambda^2$) is comparable to the inferred particle energy in the B-hotspot. Once formed, however, the B hotspots are probably short-lived. Since instability growth slows as $a\sim\lambda$, acceleration of electrons should become less efficient (eventually due primarily to stochastic processes). Then short radiative lifetimes will cause the spot to fade. We might therefore expect to see the faded, fainter, remains of B hotspots at lower surface brightness levels more commonly than the "hot" B hotspots in Fig. 1. At the University of Minnesota this work was supported by NSF grant AST79-00304, and at the University of Toronto by NSERC grant No. A5713.

REFERENCES

Gerwin, R.A.: 1968, Rev. Mod. Phys. 40, pp. 652-658.
Hargrave, P.J., Ryle, M.: 1976, Monthly Notices Roy. Astron. Soc. pp. 481-488.
Kronberg, P.P., Clarke, J.N., van den Bergh, S.: 1980, Astron. J. 85, pp. 973-980.
Laing, R.A.: 1981, Monthly Notices Roy. Astron. Soc. 195, pp. 261-324.

DISCUSSION

LAING: I have observed several examples of hot-spots (e.g., in 3C 20, 133, 196) whose morphologies are inconsistent with the predictions of your model. The main problem is that the limb-brightened cusp in the diffuse subcomponent always points away from the compact subcomponent, whereas in your model one would expect it to point away from the associated galaxy or quasar. In addition, in 3C 20 and 3C 196, the compact subcomponents are on the source axis. (R. A. Laing, this volume.)

KRONBERG: We agree that 3C 196 (like Hydra A and other probably similar sources) do not appear to have our canonical form of A-hotspot, and are presumably subject to a somewhat different set of physical conditions. In our suggested model, however, the A-cusp axis could deviate from the axis to the QSO (as is apparent in Cyg A W) due to the effect of local i.g.m. velocities at the outer lobes (V_t) which are non-negligible relative to V_a.

DE YOUNG: I have difficulty seeing how the nonlinear Kelvin-Helmholtz instability will always produce just one compression region which gives the enhanced synchrotron emission.

JONES: First, we would like to emphasize that, in our suggested model, approximately one young B hotspot is seen at a given time; over a larger sample, we don't expect always to see exactly one B hotspot. Since synchrotron emissivity is such a strong function of the field amplification ($\epsilon \sim B^{7/2}$ in the simplest case) the enhanced spot should be strongly concentrated in a region (our B hotspot) where the largest instabilities go non-linear ($a \sim \lambda \sim r$). Larry Smarr's numerical calculations, shown at this meeting, support the idea that this region is fairly clearly defined.

HOT-SPOTS IN LUMINOUS EXTRAGALACTIC RADIO SOURCES

R. A. Laing
National Radio Astronomy Observatory

Compact hot-spots in luminous extragalactic radio sources are often double on the scale of a few kpc (Laing 1981a). Examples are shown in Figures 1 – 3; the maps were made with the A and B configurations of the VLA. The general features are as follows:
(a) It is usually possible to recognize a compact, "active" subcomponent within a radio lobe. This has a size of <1 kpc and may only contain a small fraction of the total flux of the hot-spot. It need not be at the leading edge.
(b) More diffuse regions are often grossly offset from the source axis (e.g. 3C 196). In some cases, there is apparent inversion symmetry about the optical identification.
(c) The characteristic morphology of a diffuse subcomponent is best illustrated by the eastern hot-spot of 3C 20, which is limb-brightened, with a circumferential magnetic field. The bright edge is on the side furthest away from the compact subcomponent.
(d) The polarization structure in the diffuse subcomponents, like that in most extended radio lobes, can be explained if the magnetic field has been sheared so as to be tangential to the surface, but is otherwise random (Laing 1980; 1981b).
(e) The most obvious explanation for the multiple structure is that the compact subcomponents represent the points of impact on the surrounding gas of twin beams from the associated galactic nucleus; if these alter their direction, then diffuse remnants may be formed. This idea would be consistent with the inversion symmetry seen, for example, in 3C 196.
(f) A problem with this model is posed by the detailed morphology of the diffuse subcomponents: why should the bright limb be opposite the compact subcomponent, rather than at the leading edge of the source? An alternative picture, in which the diffuse subcomponents are formed by material escaping from the active regions, should also be considered.

REFERENCES
Laing, R. A., 1980. M.N.R.A.S., 193, 439.
　　　　　　 1981a. M.N.R.A.S., 195, 261.
　　　　　　 1981b. Ap. J., 248, 87.

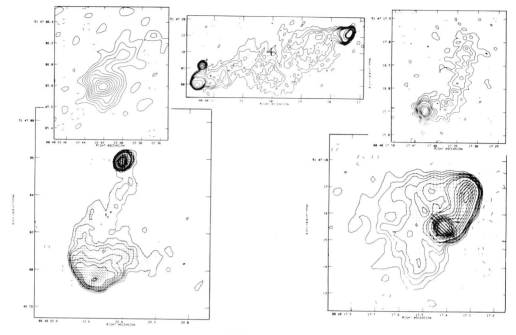

FIGURE 1   3C20

Centre: 4885 MHz, 1".1 beam

Bottom right and left: 4885 MHz, 0".35 beam

Top right and left: 15035 MHz, 0".14 beam

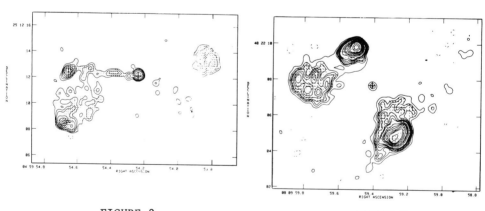

FIGURE 2

3C133   4885 MHz

Beam 0".4

FIGURE 3

3C196   4885 MHz

Beam 0".35

MORPHOLOGY AND POWER OF RADIO SOURCES

P.A.G. Scheuer
Mullard Radio Astronomy Observatory,
Cavendish Laboratory, Cambridge, U.K.

I want to make two points:

1. Observations suggest that the hot-spots move about either because the beam precesses (e.g. Ekers et al. 1978; Lonsdale & Morrison 1980) or more discontinuously, as in sources like 3C351 that have multiple hot-spots (Laing 1981). The natural interpretation is that the hot-spot at the end of the beam slides over the inner surface of a 'cavity' filled with very hot dilute ex-hot-spot material (Scheuer 1974), extending the cavity at various places at different times. Such a 'dentist's drill' model has various consequences:

(i) The mean speed (length/age) at which the cavity elongates may be considerably smaller than the instantaneous speed V of the hot-spot, estimated in the customary way from pressure balance:

$$\frac{1}{3}\binom{\text{minimum energy density for}}{\text{observed radio emission}} = (\text{external density}) \times V^2$$

(ii) Remark (i) raises the possibility that giant radio sources have become so long because their jets are unusually constant in direction. NGC6251, which is over 2 Mpc long, has a jet whose direction has not wavered by more than a few degrees in $\gtrsim 10^7$ years (Saunders et al. 1981).

(iii) We should not be surprised by sources whose hot-spots lie some way behind the extreme ends of the source; this may merely mean that the beam at present impinges on one side of the 'cavity'. 3C285 is a typical example; a more persuasive case is the Np component of 3C219 (Perley et al. 1980).

2. Most of the radio emission of most really powerful radio sources comes from their hot spots. Contrariwise, straightforward equipartition calculations on models lead to at least as much emission from the 'cavity' as from the hot-spots (Scheuer 1974). While there are several possible explanations, one in particular is suggested by Laing's (1981)

extensive data on 40 sources. He lists the synchrotron lifetimes in the hot-spots for electrons radiating at 15 GHz, and while these exceed the light crossing times, they do so by only one or two orders of magnitude. One expects flow through the hot spots at some fairly small fraction of the speed of light; the synchrotron loss cutoff in the hot-spot is then expected to appear in the mm or infra-red region. Expansion of the hot-spot material into the cavity shifts the loss cutoff to a much lower frequency. If the expansion is adiabatic (certainly an oversimplification) the ratio of loss cutoff frequencies is equal to the ratio of energy densities. We cannot use this relationship directly in the more interesting cases, as we cannot estimate the energy density (or even the size) of a 'cavity' that we cannot observe. However, we can estimate the cutoff frequency within the hot-spots (for some assumed streaming speed), and Figure 1 shows that the most powerful hot-spots have loss cutoff frequencies in the 10-100 GHz range, implying cutoff frequencies in the 'cavity' that could well be below 1 GHz.

The predictions of this model differ from those for synchrotron loss within the 'cavity' in that the latter gives a progressively shorter tail to each hot spot at higher frequencies, while the model sketched here can abolish the whole of the radio bridge. This model also accounts for the strong correlation found by Jenkins & McEllin (1977) between the power of a radio source and its 'compactness', i.e. the fraction of its flux density originating in hot-spots, for the synchrotron loss cutoff depends more strongly on the power of the hot-spot than on its size.

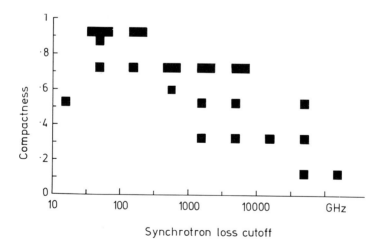

Figure 1. Compactness versus synchrotron loss cutoff on the hot-spots. Compactness is defined, following Jenkins & McEllin, as the fraction of flux within two 15 kpc regions, and measured by taking the point source flux equivalent to the maxima on a map convolved to a circular beam 15 kpc FWHP (H = 50, Ω = 0). The synchrotron loss cutoff ν_{loss} here

means the frequency at which the emission is expected to fall to 0.1 of the power law extrapolation of the radio observations, assuming equipartition, and a flow speed v = 0.25 c through a hotspot of observed radius r and radio power P; $\nu_{loss} \propto P^{-6/7} r^{4/7} v^{-2}$.

The sources form a very incomplete sample at present, and are mostly taken from Laing (1981).

The source at low ν_{loss} which apparently violates the otherwise strong correlation is 3C280.1, which has a hot-spot on one side of the nucleus, but a long jet on the other.

REFERENCES

Ekers, R.D., Fanti, R., Lari, C. & Parma, P.: 1978. Nature 276, 588.
Jenkins, C.J. & McEllin, M.: 1977. Mon. Not. R. astr. Soc., 180, 219.
Laing, R.A.: 1981. Mon. Not. R. astr. Soc., 195, 261.
Lonsdale, C.J. & Morison, I.: 1980. Nature, 288, 66.
Perley, R.A., Bridle, A.H., Willis, A.G. & Fomalont, E.B.: 1980. Astr. J., 85, 499.
Saunders, R., Baldwin, J.E., Pooley, G.G. & Warner, P.J.: (in press).
Scheuer, P.A.G.: 1974. Mon. Not. R. astr. Soc., 166, 513.

THE RADIO JET OF 3C273

R.G. Conway
University of Manchester
Nuffield Radio Astronomy Laboratories
Jodrell Bank

Most radio sources are two-sided, like Cygnus A. A minority, however, are one-sided, and the first-known and brightest example is 3C273 (see Fig. 1), a high-luminosity QSO, showing 'super-luminal' proper motions in the core. The explanation of such one-sided sources may follow one of two lines (and it seems that both schools of thought are represented at the present meeting): on the one hand, the ejection of material from the central object may truly be one-sided, while on the other hand the ejection may be two-sided but at a relativistic speed, so that the receding half is hidden by Doppler beaming.

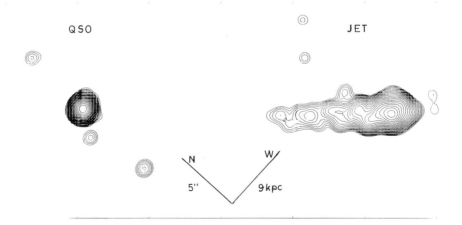

Fig. 1   Map of Radio brightness at 408 MHz of 3C273, with resolution 0.9 arc sec, tilted so that the jet in p.a. $223°$ is shown horizontal. The contours are logarithmic, with three contours to a factor 2 in brightness. The unresolved core (centre-left) coincides with the QSO. The radio jet coincides accurately in position with the optical jet. No radiation is detected from the opposite side of the QSO.

The radio map shown in Fig. 1, which was made with the Jodrell Bank MTRLI at 408 MHz ($\lambda 73.5$ cm), and has a resolution of 0.9 arcs, enables quantitative parameters to be estimated for testing these models. Fig. 1 shows the bright core coincident with the QSO, within which are found the superluminal proper motions. Three artificial sidelobes, to North and South of the core, may be ignored. Along the radio jet, the brightness increases by more than a factor 100, reaching 240×10^8 K at the outer head. No radiation is detected from the opposite side, indicating that the brightness of the postulated counter-jet must be <1/100 of the brightness of the visible jet.

If this ratio is due to Doppler beaming, the whole jet must be moving quasi-relativistically, at >0.7c. A simple calculation of the ram-pressure in front of the head shows that this motion would be halted by the I.G. ambient medium unless the number density <0.7 m^{-3}. Since more than one argument suggests that such a low density (< $\frac{1}{20}$ closure density) is implausible, it appears that in 3C273 the ejection from the central object is genuinely on one side only.

Question (by anonymous participant) Have you considered whether the further radio lobe might be hidden by absorption?

Answer By straightforward free-free absorption is not possible, since the EM of the absorbing matter would then make it visible in the optical region. An idea to play with is that the ejection might be sometimes in one direction, sometimes in the other, and of course conventional double sources could be the time-average of such a flip-flop scheme. Seen in this regard, our result shows that the ejection from the centre QSO has remained in p.a. 223° for at least the last 10^6 years.

RELATIVISTIC BEAMING AND QUASAR STATISTICS

I.W.A. Browne and M.J.L. Orr
Nuffield Radio Astronomy Laboratories
Jodrell Bank, Macclesfield, Cheshire

ABSTRACT

The predictions of a scheme which attributes the observed differences between flat and steep spectrum quasars to projection and the effects of relativistic beaming are explored. We conclude that the statistical properties of quasars are entirely consistent with such a scheme provided the mean Lorentz factor in the central components of quasars is ~ 5.

INTRODUCTION

Many maps of core-dominated sources have been shown by Peter Wilkinson and Rick Perley at this conference. One possible interpretation of these maps is that what we are seeing in core-dominated sources are just normal doubles viewed along their axes. If this is so the statistics of flat spectrum (core-dominated) and steep spectrum (normal double) sources are not independent. In this contribution we use a simple model of a quasar consisting of a compact relativistically beamed core, spectral index zero, and unbeamed lobes, spectral index -1, to predict the proportion of flat spectrum sources in flux limited samples selected at different frequencies. Also using the same model quasar we construct the flat spectrum number/flux density counts from the observed steep spectrum counts. Our aim is 1) to see if quasar statistics are consistent with such a unified scheme of flat and steep spectrum objects, and 2) to see if we can put useful constraints on the Lorentz factors in quasar cores. This work is described in more detail in Orr & Browne (1981).

THE QUASAR MODEL

We will neglect the contribution of jets because at high frequencies they are nearly always very much weaker than the cores. It will be assumed that the core emission is Doppler boosted with a Lorentz factor γ and that the rest of the emission (i.e. that from the lobes) is unbeamed.

We take the ratio

$$R_T = \frac{\text{Core strength perpendicular to the line of sight}}{\text{Lobe strength}}$$

to be a constant. In any particular source the observed ratio

$$R = \frac{\text{Core strength at angle } \theta \text{ to the line of sight}}{\text{Lobe strength}}$$

will depend on θ and γ (Scheuer & Readhead, 1979).

RESULTS

The distribution of R in a sample of sources selected without reference to beamed properties can be predicted, since in such a sample θ will be randomly distributed. 3CR quasars approximate closely to such a sample, and from the distribution of R amongst these we deduce that $R_T \simeq 0.025$. The distribution of R for a small sample of quasars is, however, not a good way to determine γ (c.f. Scheuer & Readhead, 1979). This is because predicted distributions for various γs, only differ significantly for sources with high values of R and these are expected to be very rare, requiring a high degree of alignment.

A better way to estimate γ is by predicting the expected numbers of flat and steep spectrum quasars in flux limited samples selected at different frequencies and comparing the predictions and reality. Such predictions are possible because the condition for a quasar to have a flat spectrum can be re-expressed as a condition on R. In other words if $R > R_C$, some critical value, the overall radio spectrum will be flat. Table I shows the observed fraction of flat spectrum quasars in various surveys and the Lorentz factors required to produce that fraction. It is clear from the table that $\gamma \sim 4.5$ is consistent with the statistics of nearly all the samples.

Another way in which our simple quasar model can be used to check the consistency of the unified scheme is to predict the number/flux density counts of flat spectrum quasars from those of steep spectrum quasars. (We make the assumption that quasar counts at 408 MHz are essentially free from the effects of beaming). Fig. 1 shows the 5 GHz differential source counts for flat spectrum quasars and the predicted counts for γ = 4, 5 and 6. No one value of γ gives a perfect fit, but considering the simplicity of the model and the uncertainties in the starting 408 MHz counts, we think the agreement is encouraging. In particular the overall shape is roughly right and the very small number of flat spectrum quasars at low flux densities (Condon & Ledden, 1981) is successfully predicted.

Table I The fraction (F) of flat spectrum quasars in various surveys and the core Lorentz factor (γ) required to predict F. ν is the frequency and S_0 the flux density limit of the survey.

Sample	ν(MHz)	S_0(Jy)	F	γ
3C	178	10	8 ± 4	3.2 ± 0.9
4C	178	2	17 ± 6	5.8 ± 0.8
Bologna(B2)	408	0.9	18 ± 5	4.9 ± 0.7
Jodrell Bank	966	1.0	21 ± 4	3.7 ± 0.3
Parkes	2695	0.35	60 ± 8	5.2 ± 0.5
Kuhr et al.	5000	1.0	82 ± 6	5.3 ± 0.3
NRAO Deep Survey	5000	0.1	56 ± 18	4.5 ± 1.1

A number of trials were conducted using a distribution of γ rather than a constant value. In practice it was found that approximately the same predicted curves were produced as for a constant γ provided the centre of gravity of the distribution $\bar{\gamma} \simeq \gamma$.

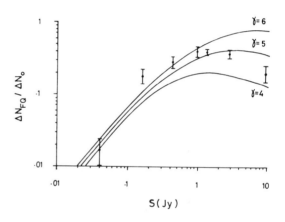

Fig. 1 The observed and predicted differential number counts of flat spectrum quasars

DISCUSSION AND CONCLUSION

Independent estimates of γ can be obtained from observations of superluminal sources (see contribution by M. Cohen). These all come out to be in the range 3 to 10 (if $H_0 \sim 100$ km s^{-1} Mpc^{-1}) and are within the range of the present results.

We conclude that a simple relativistic beam model of quasars (with $\gamma \simeq 5$) is entirely consistent with the statistical properties of quasars.

ACKNOWLEDGEMENTS

We thank Jim Condon for providing the results of his 5 GHz deep survey identifications prior to publication.

REFERENCES

Condon, J.J. and Ledden, J.E., (1981). preprint.
Orr, M.J.L. and Browne, I.W.A., (1981). In preparation.
Scheuer, P.A.G. and Readhead, A.C.S. (1979). Nature 277, 182.

Discussion

J.F.C. Wardle. Does your model predict $\left\langle \frac{V}{V_m} \right\rangle \sim 0.5$ for flat spectrum quasars?

I.W.A. Browne. Yes it does predict $\left\langle \frac{V}{V_m} \right\rangle \sim 0.5$ for intermediate flux density quasars. For the strongest quasars the predicted Log N/Log S slope is steeper than the Euclidian value indicating $\frac{V}{V_m} > 0.5$. This has what has been found to happen in reality by Peacock and his collaborators.

THE RADIO CORE IN 3C 236

E. B. Fomalont and A. H. Bridle
National Radio Astronomy Observatory
and
G. K. Miley
Sterrewacht Leiden

ABSTRACT: The two-kpc steep-spectrum radio core in the giant radio galaxy 3C 236 has been mapped with 0.″1 resolution using the VLA. The core morphology is substantially different from other radio cores and suggests that the flow of energy from the galactic nucleus may be continuous on one side and "blobby" on the other side.

1. THE RADIO STRUCTURE OF THE CORE

The radio map of the 2" radio core at 22.5 GHz with 0.″1 resolution is shown in Figure 1. The lowest contour level is 1.0% of the peak of 0.4 Jy and the diagonal line shows the direction to the outer radio structure of size 40 arcmin. From comparison with maps at 15 GHz with 0.″15 resolution and at 5 GHz with 0.″25 resolution, the features of the radio core are:

A. A nuclear core $<0.″05$ (<80 kpc) with $\alpha = 0.15$ ($S = \nu^\alpha$);
B. A jet extending to the SE with $\alpha = -0.70$;
C. A "blob" extended perpendicular to the major axis, about one kpc to the NW of the nuclear core, with $\alpha = -0.75$;
D. Some additional low level emission.

The map is in good agreement with an 0.″3, 1.6 GHz map (Schilizzi et al. 1981). They show that the SE jet breaks into several components and does curve slightly away from the source axis defined by the large double. They completely resolve out the NW blob but do separate an additional component about 150 pc (0.″1) NW of the nuclear core.

Sufficient signal-to-noise for the linearly-polarized maps were obtained with 0.″25 resolution at 4.9 and 15 GHz. They show that the electric-vector rotation between 4.9 and 15 GHz is about 30° counter-clock-wise except for the region in the southern part of the NW blob, where the rotation is about 90° CCW. The degree of polarization is about 5 to 15 percent except near the nuclear core and some depolarization is present at 5 GHz. Tentative conclusions about the magnetic field orientation suggest a longitudinal B field along the SE jet and a B field along the periphery of the NW blob. The differential

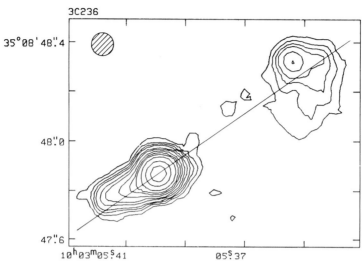

Figure 1: Radio map of the core of 3C 236 at 22.5 GHz with 0.1 arcsec resolution. The peak is 0.4 Jy with contours at 1, 2, 3, 4, 6, 8, 12, 16, 20, 30, 50, 75 percent of the peak. The clean beam is shown in the upper left corner of the map.

Faraday rotation and depolarization suggest (using the usual equipartition arguments) a thermal density of about 3×10^{-3} cm^{-3} in the NW blob.

2. CONCLUSIONS

   A. An opaque radio core of 0.4 Jy lies at the nucleus of 3C 236.
   B. The radio emission has a different morphology on each side of the radio core; jet-like to the SE and blob-like to the NW. This suggests that the more collimated, continuous flow of energy is occurring towards the SE. This is consistent with the morphology in the extended 40 arcmin radio emission, which shows a hot spot in the SE suggestive of the termination of a flow. The extended emission to the NW has little fine-scale structure suggestive of little collimated flow (Strom and Willis, 1980).
   C. The higher thermal electron density in the NW blob may account for the disruption of energy flow in this direction.
   D. Although the extended 40 arcmin source is straight (<3° bend) the two-kpc radio core shows deviations from linearity of over 10°. This suggests that the ultimate collimation of the extended source is not controlled by the inner few kpc but by the outer regions of the galactic environment.

REFERENCES

Schilizzi, R. T., Miley, G. K., Janssen, F.L.J., Wilkinson, P. N., Cornwell, T. J. and Fomalont, E. B., 1981, Optical Jets in Galaxies, ed., B. Battrick and J. Mort, 107.

Strom, R. G. and Willis, A. G., Astron. Astroph., 1980, 85, 36.

THE ARCSECOND MORPHOLOGY OF COMPACT RADIO SOURCES

R. A. Perley
NRAO, Socorro, New Mexico

As part of the VLA calibration program, 404 small angular size sources have been observed in the "A" array at both 6 and 20 cm with resulting resolutions of 0.4" and 1.2" respectively. Use of self-calibration techniques has allowed a search for associated extended structure to a level of ~0.3% of the peak. Here we report preliminary analysis of the results.

Fifty-two sources had spectral indices $\alpha > 0.5$ (defined as $S = C\nu^{-\alpha}$), and seven others were missed at one frequency from the survey. The analysis is based on the remaining 345 flat-spectrum sources. The search for secondary structure was conducted within a window of radii of 0.3" to 3.2" at 6 cm, and 1.2" to 16" at 20 cm. Although not complete, the sample includes almost all flat spectrum sources with $\delta > -40°$ and $S_6 > 1$ Jy, as well as ~150 with $S_6 < 1$ Jy. The only bias used in selection was spectral shape, so the sample should be a fair representation of bright flat spectrum objects as a whole.

The basic results are:
1. All 345 sources are dominated by an unresolved core.

2. 166 (48%) contain unambiguous evidence of secondary structure. The following results pertain to these 166 sources.

3. The angular and physical size distributions are both roughly exponential. The median source size is ~20 Kpc with a maximum size of ~120 Kpc (using $q_o = 0$, $H_o = 75$).

4. The brightness ratio distribution at 20 cm between the secondary structure and the core is also exponential, with a median of ~4%. There is no correlation between the brightness ratio distribution and flux of the core.

5. The brightness ratio distribution between detected lobes (or between the lobe and noise for asymmetric sources) is very different from steep spectrum sources. Two-thirds of the sources have only one

detected lobe, with a median brightness ratio lower limit of ~5:1. For those sources with two lobes, the brightness distribution has a median of 2:1.

6. Seventy-nine sources had secondary structure detected at both bands. For every source, the secondary has a normal, steep spectrum. In only one are the spectral indices of the core and secondary similar.

The morphology of the secondary structures is remarkably simple. Although this is surely partially due to limited dynamic range, more detailed studies (see below) indicate that the simplicity remains after higher dynamic range is achieved. The sources can be separated into two simple classes:

1. "Blobs". There is a single (rarely, two) often unresolved secondary. A bridge (often curved) of emission is frequently found linking the secondary to the core, but the brightest secondary structure is always at the end. About 70 to 80% of the 166 sources are in this group. The "typical" sources are 3C273, 3C345 and 3C454.3.

2. "Jets". Much more varied in structure, with bending commonly found. Diffuse emission enveloping core and brighter secondary structure is sometimes present. These sources are generally of lower luminosity. "Typical" souces are 3C371 and 0716+714.

It must be emphasized that <5% of the 166 sources show "classical double" structure.

Twenty-one sources with exceptional secondary structure have been observed in greater detail with particular emphasis on mapping all angular scales. The detailed results will be published elsewhere, but the basic results are:

1. The lobe brightness ratios are extremely high for 3C454.3, 3C345 and 3C273, exceeding 70:1 for the latter.

2. No depolarization is found in any "jet" or "blob". The B-field is always parallel to the main axis of elongated emission.

3. The spectral indices of all secondary emission is typical of extended steep sources. The secondary component of 3C273 has $\alpha = 0.85$. No spectral index gradient is found in any "jet" or "bridge".

Perley, Fomalont and Johnston (submitted to Ap.J.) have argued that the extreme asymmetry of the secondary structure implies that the flat-spectrum soucres are fundamentally different than normal, well resolved, steep spectrum doubles. Either the flat-spectrum sources are intrinsically one-sided, or relativistic beaming has caused or enhanced the observed imbalance. The latter explanation is favored as the arcsecond structure is often linked to superluminal, milliarcsecond scale structure (Perley, Johnston and Fomalont, in preparation).

HIGHLY POLARIZED EMISSION FROM THE E-HOTSPOT IN DA240

S.C. Tsien and Richard Saunders
Mullard Radio Astronomy Observatory, Cambridge.

The hotspot in the eastern lobe of the nearby giant radio galaxy DA240 (z=0.0356) provides a rare opportunity to examine the detailed polarization structure of a hotspot. Maps have been made with the Cambridge 5-km telescope at 2.7 and 5.0 GHz. The 5.0-GHz maps are shown in Figs 1 and 2. The main characteristics are: (a) The hotspot has an overall size in the 2.7-GHz map of 10x18 kpc^2. It blends smoothly at its outer edge into the background of the extended lobe. The 5.0 GHz total intensity map shows a yet more compact region (subcomponent A). It has not been fully resolved in the direction of the minor axis and has a size <1x2.5 kpc^2. There is a second much weaker and diffuse region (subcomponent B). Both subcomponents are superimposed on a broad plateau of emission. The hotspot has spectral index $\alpha(0.61-5.0)=0.52$, minimum total energy $E_{min}=3\times10^{56}$ ergs and equipartition magnetic field $B_{eq}=2\times10^{-5}$ G, values typical of other hotspots, although its projected distance from the nucleus is very large (0.65 Mpc);

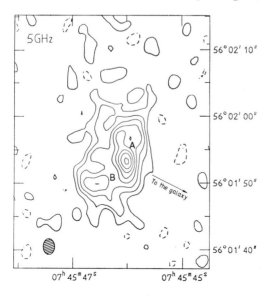

Fig.1. 5.0-GHz map of total intensity. First contour and contour interval are 4 mJy/beam.

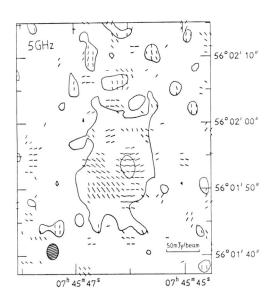

Fig.2. 5.0-GHz map of polarized intensity. The lines indicate the electric vectors.

(b) The hotspot is strongly polarized at both frequencies. At the peak of the compact subcomponent the polarization is about 20% and it increases towards the southeast, reaching 50-60% at the outer edge. The electric vectors in the polarized emission are well aligned over regions large compared with the beamwidth. The integrated polarizations in the hotspot are 24% and 28% at 2.7 and 5.0 GHz. Compared with 22% at 0.61 GHz, these values imply little depolarization at low frequencies;
(c) The rotation measure derived from the available data on the integrated polarization for the whole source is very small, about 2.4 rad m^{-2}, and most of it may be galactic in origin. The projected magnetic field may thus be taken as perpendicular to the E-vectors in Fig.2. The magnetic fields in subcomponents A and B are therefore nearly parallel to the major axes of these components, although neither parallel nor perpendicular to the axis of the source as a whole (at PA 63°). The mean direction in the hotspot, however, is almost perpendicular to the source axis.

The very high percentage polarization seen in some parts of the hotspot does not necessarily imply that the magnetic fields in these parts are nearly perfectly aligned. Hotspots are usually regarded as regions of interaction between the advancing energetic beam and the external medium, and the magnetic fields in hotspots are probably initially irregular. The production of high polarization may be due to shearing and compression of these irregular fields into a thin slab or shell-like structure of field.

We are indebted to Drs J.E. Baldwin, G.G. Pooley and J.R. Shakeshaft for helpful discussions.

SEYFERT GALAXIES

Andrew S. Wilson
Astronomy Program, University of Maryland

ABSTRACT. Observations of sample of Markarian Seyferts with the VLA indicate that a large fraction possess linear radio structure on a scale of a few hundred parsecs to a few kiloparsecs. The radio components generally straddle the optical nucleus and several sources are simple doubles. Similar structures are seen in the classical Seyferts NGC 1068, 4151, and 5548. NGC 4151 is probably best interpreted as a jet. A few sources (e.g. Mark 315, NGC 7469) exhibit diffuse, non-aligned radio structure on a scale similar to that of the linear sources. The radio axis in linear sources is misaligned with respect to the rotation axis of the galaxy disc by a large angle. The linear sources are discussed in terms of a model of a supersonic beam or jet which is "disrupted" by interaction with interstellar gas in the inner part of the galaxy (often a spiral). Two aspects of this interaction are emphasised. Firstly, the curved shape of the radio sources in NGC 1068 and NGC 4151 is ascribed to beam bending by the ram pressure of the rotating interstellar medium. Simple models of this process are shown to be consistent with the observations. Secondly, it is suggested that the broadened forbidden lines in Seyferts originate in part from interstellar gas accelerated outwards by the beam. This picture accounts for some of the empirical correlations found between radio and optical forbidden line properties.

1. INTRODUCTION

When investigating Seyfert galaxies, astronomers find a quite different form of active galactic nucleus to radio galaxies, the major topic of this meeting. The radio luminosity of a Seyfert is typically 10^{39-41} erg s^{-1} which, although greater than a normal spiral, is orders of magnitude weaker than is characteristic of most radio galaxies. Furthermore, the luminosity of a Seyfert nucleus in the radio band represents only a small fraction of its total electromagnetic luminosity, which is dominated by infra-red, X or gamma radiation (10^{43-46} erg s^{-1}). For these reasons, it has until recently been unclear whether the non-thermal emission from these objects is a consequence of an enhancement of the processes found in normal spirals (such as pulsars, supernova remnants and disk interstellar cosmic rays and magnetic fields) or whether it represents activity related to, but weaker than, the kind responsible for extended double radio galaxies (ejection of radio emitting gas from a compact nucleus). I hope to demonstrate in this paper that for a large fraction of Seyfert galaxies the second point of view is appropriate i.e. the radio sources may be crudely termed "mini radio galaxies".

2. RADIO STRUCTURE OF A "COMPLETE" SAMPLE

In studying statistical properties of radio sources, it is desirable to deal with well defined samples in order to avoid bias introduced by observational selection effects. Unfortunately a complete sample of Seyferts is not readily available. In the meantime, the best that can be done is to use the Seyfert galaxies in the lists of Markarian, but we must bear in mind that these lists are incomplete in apparent magnitude and discriminate against edge-on or otherwise obscured systems, since UV excess is the basis selection criterion.

Of the 41 Seyferts (Huchra and Sargent 1973) in the first 4 lists of Markarian, 16 have radio flux densities S(1415 MHz) > 10 mJy (de Bruyn and Wilson 1976). These 16 have all been observed at the VLA at 4885 MHz and some at 1465 or 15035 MHz. The present section is a preliminary report on the structure of this sample (Ulvestad and Wilson, in preparation). I divide the radio maps into 4 structural classes:
U: 7 galaxies (Mark 1, 110, 176, 231, 268, 279, 374). The radio emission is too compact for the structure to be defined. The source is either slightly resolved or unresolved.
L: 5 galaxies (Mark 3, 6, 34, 78, 270). The galaxies show discrete radio components, or a distribution of emission, lying along a line and straddling the optical continuum nucleus. These are mostly double or triple sources.
D: 1 galaxy (Mark 315). The radio emission is diffuse and not linear in structure.
A: 1 galaxy (Mark 348). The source structure cannot be classified into one of the above categories unambiguously.
A/L: 2 galaxies (Mark 79, 273). The radio emission may fall in the linear (L) category, but further observations are needed. Thus, of the 9 well resolved galaxies, at least 5 (56%) and possibly as many as 7 (78%) exhibit aligned components straddling the optical continuum nucleus. Maps of some of the sources in this sample are given in Fig. 1.

Statistical studies of the original sample of 41 and the 16 with structural information indicate (assuming $H_o = 50$ km s^{-1} Mpc^{-1}):
a. P(1415 MHz) ranges between <3×10^{20} and 1.5×10^{23} W Hz^{-1} Ster^{-1}.
b. Type 2 Seyferts tend to be more radio luminous than type 1´s (Sramek and Tovmassian 1975; de Bruyn and Wilson 1978).
c. The linear sizes of the radio sources range from below a few hundred parsecs to a few kiloparsecs, comparable to the classical Seyferts.
Very tentative results from this study, which require a larger sample for a definite conclusion include:
d. A higher fraction of the most radio luminous sources may be of linear (L) type; however aligned structure may also be found in intrinsically weak sources (like Mark 270 and NGC 5548).

SEYFERT GALAXIES

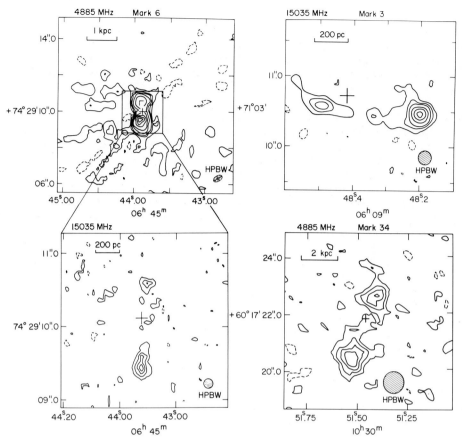

Figure 1. VLA maps of double radio sources in Markarian Seyfert galaxies. The cross marks the optical position (Clements 1981); the r.m.s. errors are typically $\simeq 0\overset{"}{.}1$ and are represented by the arms of the cross (except in the top left diagram). Left hand panel: Markarian 6 at 4885 (top) and 15035 MHz (bottom). In the top left diagram, contours are at -2 (dotted),2,4,6,10,20,40,60,80% of the peak brightness of 24.1 mJy (beam area)$^{-1}$. In the bottom left diagram, contours are at -20,20,40,60,80% of the peak brightness of 5.2 mJy (beam area)$^{-1}$. Top right is Markarian 3 at 15035 MHz; contours are at -10,10,20,30,50,70, 90% of the peak brightness of 24.8 mJy (beam area)$^{-1}$. Bottom right is Markarian 34 at 4885 MHz; contours are at -20,20,40,60,80% of the peak brightness of 1.2 mJy (beam area^{-1}).

e. The radio sources in type 1 Seyferts may be physically smaller, on average,than those in type 2´s.

3. THE CLASSICAL SEYFERTS

The "classical" Seyferts are much closer than those found in Markarian´s lists and may be mapped in correspondingly more detail.

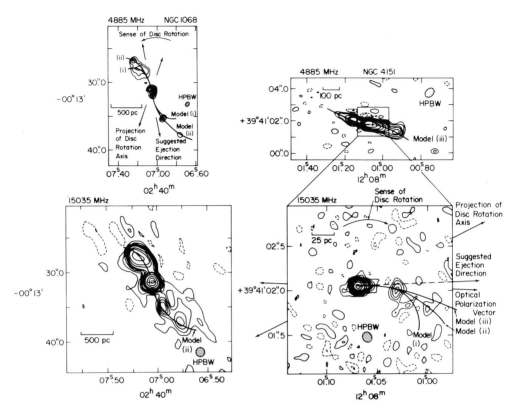

Figure 2. VLA maps of NGC 1068 and NGC 4151. The cross represents the optical nucleus in each case. The lines represent the "bent beam" models of Section 5.1. Left-hand panel is NGC 1068. Top left at 4885 MHz; contours at -2,2,4,6,10,15,20,25,30,35,40,45 times 6.83 mJy (beam area)$^{-1}$ (= 1000K). Bottom left at 15035 MHz; contours at -2,2,4,6,8,10,15,20,30,50,70,90% of the peak brightness of 161 mJy (beam area)$^{-1}$. Right hand panel is NGC 4151. Top right at 4885 MHz; contours at -5,5,10,15,20,30,40,50,60% of the peak brightness of 35 mJy (beam area)$^{-1}$. Bottom right at 15035 MHz; contours at -12,-6,6,12,18,24,30,40,50,70,90% of the peak brightness of 12.2 mJy (beam area)$^{-1}$.

Generally speaking, however, the structures of the two groups are quite similar. Both linear (L) and diffuse (D) class sources may be found and brief notes on individual cases follow.

NGC 1068. Early aperture synthesis maps by Crane (1977) and Wilson and Willis (1980) showed a triple source with angular size ≃ 13" (1.37 kpc). Maps with the completed VLA are shown in Fig. 2. Top left is a "snapshot" observations in the ´A´ configuration at 4885 MHz by Condon et al. (1981) while bottom left is the result of a longer integration in the ´C´ configuration at 15035 MHz (Wilson and Ulvestad, in preparation). Both maps show aligned structure in p.a. ≃ 33°; at least four components may be recognised. The brightest component coincides with the optical continuum nucleus and is resolved at 4885 MHz, with size 0″.7x0″.3 in p.a. 28° (Condon et al. 1981). This direction is similar to that of the outer structure.

NGC 3227. A VLA observation at 4885 MHz by Ulvestad et al. (1981) shows an unresolved point source plus emission extending over 3-4" (290-390 pc).

NGC 4151. Fig. 2 gives two maps of this famous galaxy. Top right is from the incomplete VLA at 4885 MHz (Johnston et al. 1981) while bottom right is a VLA map of the central region at 15035 MHz (Wilson and Ulvestad, in preparation). A map at 1666 MHz has also been made by Booler et al. (1981) with the MTRLI (see talk by R.D. Davies in this volume). These maps show a highly elongated structure of length $\simeq$ 3″.5 (340 pc) in p.a. 77°-92°, with the optical continuum nucleus lying close to the brightest radio peak. Comparison of the 1666 and 15035 MHz maps shows that the brightest component has a flatter spectral index (α = 0.43, $S \propto \nu^{-\alpha}$) than the one 0″.46 (44 pc) to the west (α = 0.90), supporting its association with the true nucleus. The brightest component is resolved at 15035 MHz with (deconvolved) size 0″.11 (11 pc) in p.a. $\simeq$ 272°.

NGC 5548. VLA maps reveal an unresolved component coincident with the optical continuum nucleus and two extended lobes straddling it (Wilson and Willis 1980; see also paper by Ulvestad et al. in this volume).

NGC 6764. This galaxy exhibits faint, diffuse asymmetric radio structure on a scale of $\simeq$ 10" (2.6 kpc); see Wilson and Willis (1980) and Ulvestad et al. (1981).

NGC 7469. VLA maps of this object show an unresolved (<0″.3 or <0.15 kpc in p.a. 62°) core coincident with the optical continuum nucleus, plus a halo of total extent $\simeq$ 10" (4.9 kpc); see Ulvestad et al. (1981) and Condon et al. (1981).

4. ORIGIN OF THE RADIO EMISSION

The spectral indices of the total radio emission from Seyfert galaxies lie in the range 0.4 < α < 1.1, although a few have flat spectrum cores with $\alpha \simeq 0$ (e.g. de Bruyn and Wilson 1978). The radiation mechanism is, therefore, non-thermal and presumably synchrotron. The general lack of polarization in these sources at 1415 MHz (de Bruyn and Wilson 1976) may be ascribed to Faraday depolarization by the large quantities of thermal material responsible for the optical emission lines. Calculations of the thermal radio radiation expected on the basis of the optical emission line properties confirms it is negligible at 4885 MHz (Ulvestad et al. 1981).

"Starburst" models are clearly relevant to some non Seyfert spirals, especially those whose light is dominated by hot young stars and HII regions. However, I feel there are serious difficulties in application of this picture to Seyferts, such as the very high supernova rates needed ($\sim 10^2$ yr^{-1} for NGC 1068, see Condon et al. 1981, Ulvestad 1981), the large fraction of linear sources, the general lack of evidence for large quantities of young stars and the need for photoionization by power law, rather than stellar, continua to account for the optical line spectra.

On the other hand, a number of arguments favor nuclear ejection as the mode of generation of the linear sources.

1. The classical Seyferts with linear radio structure are viewed close to "face-on" (the angle between the plane of the disc and that of the sky is i = $39°$ for NGC 1068 (Burbidge et al. 1959), i = $21°$ for NGC 4151 (Simkin 1975) and i = $28°-45°$ for NGC 5548 (Simkin et al. 1980)). Furthermore the radio axis does not seem to correlate with the projected major axis of the optical disc (Section 4). These considerations rule out the suggestion by Condon et al. (1981) that the elongated radio structures in these galaxies are a consequence of a disc of radio emitting material coplanar with the stellar disc and viewed obliquely. Their suggestion probably is valid, however, for non-Seyfert spirals with strong radio sources. Although the precise orientations of the more distant Markarian Seyferts are often difficult to determine, the high fraction of these galaxies containing linear, often double or triple sources (Section 2), again indicates that the structures are truly aligned and are not caused by projection effects.

2. The radio components in double sources are commonly elongated along the radio axis (e.g. Mark 3, 6, 270).

3. The radio structure of NGC 4151 (Fig. 2) is strongly suggestive of quasi-continuous nuclear ejection (Ulvestad et al. 1981), probably in the form of a jet (Johnston et al. 1981; Booler et al. 1981).

4. The direction of elongation of the radio source in NGC 4151 (p.a $\simeq 272°$ in the center) is close to the polarization vector of the optical continuum (p.a. $\simeq 268°$, Schmidt and Miller 1980). If this agreement is not coincidental, a connection between an inner (light months, the timescale of the optical variability) nuclear axis and the much larger scale (380 pc) shape of the radio source is implied. Furthermore, this alignment is similar to that found for radio-loud, double lobed quasars (Stockman et al. 1979).

5. There may be a continuity, both in radio luminosity and in the correlation between radio luminosity and forbidden line width (Wilson and Willis 1980; Heckman et al. 1981) between Seyfert galaxies and the kpc size cores of radio galaxies (e.g. 3C 236, 3C 293, 3C 305).

6. The "ridge-lines" of the radio sources in NGC 1068 and NGC 4151 describe a curve which "leads" the rotation of the galaxy. As will be argued in Section 5, this shape is readily accounted for by the bending of a supersonic jet under the ram pressure of the rotating interstellar gas disc. On the other hand, models of propagating star formation would be expected to yield "trailing" ´S´ shapes.

Some simple consequences of the nuclear ejection model bear mentioning.

a. Because double radio sources are not, in general, found outside the optical isophotes of Seyfert galaxies (de Bruyn and Wilson 1976), the beam or plasmoids must not escape the nuclear environment (~ kpc scale). The most natural means of achieving this is to "disrupt" the beam or "halt" the plasmoids by friction with the dense interstellar medium in the inner part of the spiral (Wilson and Willis 1980). This interaction is further discussed in Section 5.

b. A large fraction of Seyferts appear to form aligned radio sources and

Seyferts comprise ≈ 1-2% of all galaxies (Woltjer 1959). Thus the ability to eject radio components seems to be present in at least ≈ 1% of all galaxies.

c. Spiral galaxies as well as ellipticals and lenticulars can form linear radio sources. It should be borne in mind that, of the sources with unambiguous linear structure, only NGC 1068 (Rev. Morph. type (R)SA(rs)a) and NGC 4151 ((R´)SAB(rs)b) show, without doubt, spiral arms, although NGC 5548 ((R´)SA(r)0/a) and Mark 34 may do. Mark 79, which is probably a linear source, shows clear spiral arms.

d. The evidence for activity on an ultracompact scale in type 2 Seyferts is less convincing than for type 1 Seyferts and radio galaxies. The existence of double radio sources in such galaxies suggests (but does not imply) that they too are fuelled from a compact object.

e. A comparison of the radio axis in linear sources with the minor axis of the outer optical isophotes (which will represent the projected rotation axis in simple disc systems) shows no apparent relation. This result suggests that <u>if</u> the radio sources are ejected perpendicular to the plane of an accretion disc, this disc is misaligned with the outer stellar disc by a large angle (Ulvestad et al. 1981). Some caution must be exercised in applying this result since the beam, which is presumed to power the radio source, may be bent by pressure gradients or ram pressure in the interstellar gas, so that the observed radio axis may differ from the true direction of ejection.

5. THE INTERACTION OF THE BEAM WITH THE INTERSTELLAR MEDIUM OF THE GALAXY

5.1 The Shapes of the Radio Sources

Examination of direct photographs of NGC 1068 (NGC 4151) indicates that the galaxy rotates in a counter-clockwise (clockwise) sense, if the spiral arms are trailing. Figure 2 shows that the "ridge line" of the SW side of the radio source in NGC 1068 curves towards the west with increasing distance from the optical nucleus. For NGC 4151, the initial direction of ejection is P.A. ≃ 272° (as judged from the elongation of the central source) which changes to P.A. ≃ 257° at ≃ 2" from the nucleus. These bendings are in the sense expected if they are consequence of the ram pressure of the rotating gas in the discs of the galaxies, as noted independently for NGC 4151 by Booler et al. (1981). The sharp boundary of the radio emission on the SE side of the SW ridge in NGC 1068 and the diffuse boundary on the opposite (NW) side support this interpretation. In order to explore this possibility quantitatively, I have developed a simple model to describe the jet shape, making assumptions similar to those of Begelman et al. (1979) in their model of head-tail radio galaxies. The flow along the jet is assumed to be mass conserving, adiabatic and of constant speed (v_j). Presuming the rotation velocity of the interstellar gas to be supersonic, there will be a cylindrical stand-off bow shock associated with the jet. By equating the acceleration of jet material, $g = v^2/R_j$ (where R_j is the instantaneous radius of curvature of the jet) to the ram pressure per unit length per unit mass of material, $-p/\rho_j h$ (where p is the pressure, ρ_j the jet density and h its scale height), a second order differential equation is obtained for the jet shape. This equation is then integrated numerically to derive the trajectory in x,y coordinates. The situation differs from the head-tail case in that the interstellar gas is taken to perform purely rotational motion, with velocity as a function of distance from the center (r) appropriate to a typical spiral galaxy rotation curve. An assumption about the variation of interstellar gas density with r must also be incorporated. At present, only ejection in the plane of the galaxy has been considered and gravitational forces on the jet are neglected.

Some illustrative results are given as the solid lines in Figure 2. Neither NGC 1068 nor NGC 4151 can be fitted by a rotation curve which is solid-body over the whole jet region with a constant interstellar density (model i). The reason is that if the parameters are chosen to fit the inner parts of the source, the rapid rise in interstellar ram pressure with r causes more rapid bending than is observed in the outer parts. The problem may be alleviated by flattening the rotation curve and/or allowing the interstellar density to decrease with r. Model (ii) has an initial solid body portion plus a flat part for the rotation curve, along with constant interstellar density; it provides a good fit to NGC 1068. Model (iii) incorporates the same rotation curve as (ii) but the density drops as r^{-2} beyond the turnover. Even these assumptions do not provide a particularly good fit for NGC 4151, in which the jet seems to bend rapidly in the inner 1" and thereafter moves almost straight. Possibly the ejection in this galaxy is at an angle to the plane; such would render a rapid density fall off more plausible.

Models of this type can be checked by finding more objects with curved radio structure in spirals. As long as the ejection is close to the plane of the galaxy, the ´S´ shape defined by the radio emission should curve in the <u>opposite</u> sense to that defined by the normal spiral arms. In this picture, the shape of the radio source represents a steady-state situation in which a constant velocity, continuous beam is bent by ram pressure. It may be contrasted with the related picture advanced by van der Kruit et al. (1972, see also talk by J. H. Oort in these proceedings) for NGC 4258 in which the ejection is essentially instantaneous but covers a wide range of velocity. In this case, the "radio arms" show the current position of the ejected material after deflection by gravitational and ram pressure forces; they curve in the <u>same</u> sense as the normal spiral arms.

5.2 Acceleration of Interstellar Clouds

There are a number of correlations between the radio and forbidden optical emission line properties of Seyfert galaxies. These relations include a similarity in overall spatial scale and possibly shape (de Bruyn and Wilson 1978; Ulvestad et al. 1981) and correlations between radio continuum and forbidden line powers and between radio continuum power and forbidden line width (de Bruyn and Wilson 1978; Wilson and Willis 1980; Heckman et al. 1981). Although the precise relation between the beam which fuels the radio emission and the thermal gas in its vicinity is probably very complicated, I shall briefly discuss one aspect, namely the possibility that the high velocity thermal gas (broadened forbidden lines) represent interstellar clouds "blown outwards" by the beam.

The equation of motion of an interstellar cloud (mass M_c, velocity V_c, scale height h) immersed in a particle beam (density n_b, velocity V_b) is:

$$M_c dV_c/dt = C_F n_b m (V_b - V_c)^2 h^2 \qquad (1)$$

where C_F is a constant and m is the mean mass per particle in the cloud (Blandford and Königl 1979). For h = constant, the time for the cloud to be accelerated to $V_c (\ll V_b)$ is

$$t_{acc} = M_c V_c V_b / 2 C_F L_b$$
or
$$t_{acc} = 2 \times 10^5 \ (M_c/10^6 M_\odot)(V_c/500 \text{ km s}^{-1}) \qquad (2)$$
$$\times (V_b/10^4 \text{ km s}^{-1})(C_F/8.11)^{-1}(L_b/10^{42} \text{ erg s}^{-1})^{-1} \text{ years}$$

where $L_b = n_b m h^2 V_b^3 / 2$ and represents approximately the total power of the beam incident on the cloud. Interstellar clouds are presumably carried into the beam by the general galactic rotation, are accelerated radially outwards while in it and leave it with both a rotational and radial component of velocity.

A stand-off bow shock is formed in the beam on the upstream side of the cloud, where roughly half of the bulk kinetic energy incident on the cloud is converted into internal energy. In models of this type, the radio radiation represents a fraction C_L of this dissipated power:

$$L_{rad} \simeq C_L C_F \{(V_b - V_c)/V_b\}^3 L_b \qquad \text{or} \qquad (3)$$
$$L_{rad} \simeq 10^{40}(C_L/10^{-2})(C_F/8.11)\{2(V_b-V_c)/V_b\}^3 (L_b/10^{42} \text{ erg s}^{-1}) \text{ erg s}^{-1}$$

Thus, for a beam luminosity of $L_b \simeq 10^{42}$ erg s^{-1}, an efficiency factor $C_L \simeq 10^{-2}$ would yield $\simeq 10^{40}$ erg s^{-1} of radio emission, a typical value for a Seyfert galaxy.

There are a number of advantages in this type of model:
a. It accounts for the similarities in the spatial distribution of radio emission and thermal clouds since radio emission can only be generated when the beam impinges on a cloud.
b. With the parameters indicated by equations (2) and (3), it is possible for the observationally derived cloud masses ($\sim 10^6 \ M_\odot$) to be accelerated to the observed velocities in reasonable time scales (10^5–10^6 years).
c. Models of this kind lead naturally to a relation between L_{rad} and V_c since higher cloud velocities will be associated with faster or more luminous beams, which also generate more radio emission. The exact exponent in the L_{rad} vs V_c relation is model dependent but behavior close to the observed relation $L_{rad} \propto V_c^{3.5-4.5}$ is not unreasonable (Wilson 1981).

ACKNOWLEDGEMENTS

I thank A. G. de Bruyn, A. G. Willis, R. A. Sramek and especially J. S. Ulvestad for collaboration. K. Turpie assisted with computations. This research was supported by the National Science Foundation under grant AST 79-24381.

REFERENCES
Adams, T.F. 1977, Astrophys. J. Suppl., 33, 19.
Begelman, M.C., Rees, M.J. and Blandford, R.D. 1979, Nature, 279, 770.
Blandford, R.D. and Königl, A. 1979, Astrophys. Letts., 20, 15.
Booler, R.V., Pedlar, A. and Davies, R.D. 1981, Mon. Not. R. astr. Soc., (submitted).
de Bruyn, A.G. and Wilson, A.S. 1976, Astron. & Astrophys., 53, 93.
de Bruyn, A.G. and Wilson, A.S. 1978, Astron. & Astrophys., 64, 433.
Burbidge, E.M., Burbidge, G.R. and Prendergast, K.H. 1959, Astrophys. J., 130, 26.
Clements, E. 1981, Mon. Nots. R. astr. Soc. (in press).
Condon, J.J., Condon, M.A., Gisler, G. and Puschell, J.J. 1981, Preprint.
Crane, P.C. 1977, Ph.D. Thesis, Massachusetts Institute of Technology.
Heckman, T.A., Miley, G.K., van Breugel, W.J.M. and Butcher, H.R. 1981, Preprint.
Huchra, J. and Sargent, W.L.W. 1973, Astrophys. J., 186, 433.
Johnston, K.J., Elvis, M., Kjer, D. and Shen, B.S.P. 1981, Astrophys. J. (Letters), (submitted).
Kruit, P.C. van der, Oort, J.H. and Mathewson, D.S. 1972, Astron. & Astrophys., 21, 169.
Schmidt, G.D. and Miller, J.S. 1980, Astrophys. J., 240, 759.
Simkin, S.M. 1975, Astrophys. J., 200, 567.
Simkin, S.M., Su, H.J. and Schwarz, M.P. 1980, Astrophys J., 237, 404.
Sramek, R.A. and Tovmassian, H.M. 1975, Astrophys. J., 196, 339.
Stockman, H.S., Angel, J.R.P. and Miley, G.K. 1979, Astrophys. J. (Letters), 227, L55.
Ulvestad, J.S. 1981, Ph.D. Thesis, University of Maryland.
Ulvestad, J.S., Wilson, A.S. and Sramek, R.A. 1981, Astrophys. J., (in press, July 15 issue).
Wilson, A.S. and Willis, A.G. 1980, Astrophys. J., 240, 429.
Wilson, A.S. 1981, In "Optical Jets in Galaxies", Proceedings of the 2nd ESO/ESA Workshop on the use of the Space Telescope and coordinated ground based research, 18-19 Feb. 1981, Munich, p. 125, (ESA SP-162).
Woltjer, L. 1959, Astrophys. J., 130, 38.

DISCUSSION

HENRIKSEN: Should the Seyfert jets be directed out of the plane of the disc, pressure gradient effects should be at least as important as the rotation ram.

WILSON: The present model for the shapes of the radio sources assumes that the ejection is in the plane of the galaxy. Under these circumstances only ram pressure need be included.

RADIO CONTINUUM OBSERVATIONS OF THE NUCLEI OF NEARBY GALAXIES

R.D. Davies, A. Pedlar and R.V. Booler
University of Manchester, Nuffield Radio Astronomy Laboratories
Jodrell Bank, Macclesfield, Cheshire U.K.

A survey has been carried out at Jodrell Bank of the continuum radio emission from nearby galaxies. The objects include normal galaxies, Seyfert galaxies and others with nuclei active at optical and radio wavelengths.

In particular, a systematic study has been made of 100 Sbc (T=4) galaxies taken from the Second Reference Catalogue of Bright Galaxies. Each has also been measured in the HI λ21 cm line, so that good estimates of systemic velocity, hydrogen mass and total mass are available for correlation with continuum and optical data. 30 of the Sbc galaxies show emission with flux densities >50 mJy at λ18 cm on the MK IA-MK II interferometer (baseline 2500λ). These were subsequently observed at λ18 cm on short tracks with the Multi-Telescope Radio-Linked Interferometer (MTRLI) and a number were found to have flux densities >20 mJy; the resolution covered by the various baselines of the MTRLI was 0.25 to 2 arcsec. Several of these objects are intense enough to be mapped with the MTRLI at λ75 cm with a resolution of 1 arcsec.

Another group of galaxies mapped with the MTRLI at λ18 cm and λ75 cm are those with active nuclei. These have been selected from the surveys of Hummel (1980) and others. Objects already observed with the MTRLI at λ18 cm and/or λ75 cm include NGC 1068, 1275, 3690 and Mkn 3.

In this short report we will discuss our observations of NGC4151 in some detail. NGC4151 is a type 1.5 Seyfert galaxy at heliocentric velocity of 990 km s^{-1} (i.e. a distance of 10 Mpc) where 1 arcsec is 50 pc. The integrated flux density of NGC4151 seen with the MTRLI at λ18 cm is 280 mJy and accounts for most of the known flux density in the nucleus. It consists of central double source and extended features either side extending 2 arcsec from the nucleus. The central double source has a separation of 0.45 + 0.02 arcsec at a position angle (pa) of $83°.5 \pm 1°.0$; it is coincident, within the combined measurement errors, with the optical nucleus. It is likely that the brighter and

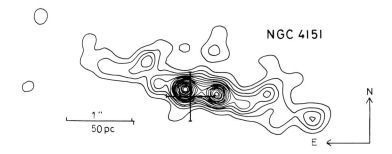

Fig.1 1665 MHz MTRLI map of the nucleus of NGC4151; the angular resolution is 0.25 arcsec

more compact (<0.15 arcsec) eastern component is associated with the optical nucleus which has a diameter <0.08 arcsec (Schwartzschild 1973). Low brightness elongated structure extends either side of the nucleus along pa = 77° ± 2°, a value significantly different from that of the central double source.

We consider that the extended structure in the nuclear regions of NGC4151 is indicative of a jet-like phenomenon. Its linear scale is 200 pc, an order of magnitude smaller than the jet in Virgo A or in radio galaxies and quasars. Further evidence for the similarity with the processes in quasars is the observation (Schmidt & Miller 1980) of optical polarization at pa = 88° which is close to that of the radio continuum. Furthermore, there is ample evidence from optical spectral line studies that expansion motions of up to 550 km s^{-1} exist within the forbidden-line region (5 x 2 arcsec) and possibly extends out to ∿20 arcsec from the nucleus.

It is of interest to describe the radio and optical activity in the nucleus of NGC4151 in terms of a single model. Despite the observed rotation in the inner part of the galaxy, the inner and outer forbidden-line regions are at the same pa, thus implying a stationary source. Furthermore, the relativistic electrons producing the radio continuum and the expansion motions in the forbidden-line region both have lifetimes of 2 x 10^5 to 10^6 yrs. The radio jets are conceived of as arising from relativistic particle beams fixed in space. Forbidden-line clouds which rotate with the general interstellar medium in the centre of NGC4151 are envisaged as entering the jet region and then being accelerated outwards as described by Blandford & Königl (1979). An outwards motion for the clouds of 550 km s^{-1} can be achieved in 10^5 years with realistic cloud parameters. More details of the observations and the model are given in Booler, Pedlar & Davies (1981).

Blandford, R.D., and Königl, A.: 1979, Astrophys.Lett. 20, 15.
Hummel, E.: 1980, Astron.Astrophys.Suppl.Ser. 41, 151.
Schmidt, G.D., and Miller, J.S.: 1980, Astrophys.J. 240, 357.

RADIO EMISSION FROM THE SEYFERT GALAXY NGC 5548

J. S. Ulvestad[*], A. S. Wilson[**], and D. G. Wentzel[**]
[*]NRAO - Charlottesville
[**]Astronomy Program, University of Maryland

ABSTRACT. Weak radio emission from the type 1.5 Seyfert galaxy NGC 5548 has been mapped with high resolution at the VLA at both 1465 and 4885 MHz. The galaxy contains the largest (5.9 kpc) triple radio source known in a Seyfert galaxy. The central component of that triple is unresolved (<0.39x0.15 kpc) and has a flatter spectrum than the well-resolved outer lobes. In addition, the field surrounding NGC 5548 and two of the sources in that field have been mapped at 1465 MHz; the field sources are unlikely to be physically associated with NGC 5548.

NGC 5548 is a Seyfert galaxy whose optical spectrum shows relatively narrow forbidden line emission and both narrow and broad emission components in the permitted lines. The galaxy, which lies at a distance of 100.7 Mpc (assuming H_0 = 50 km s^{-1} Mpc^{-1}), is a strong source of both x-rays and infrared radiation. It is a somewhat peculiar spiral having an inner disk and outer rings (Su and Simkin 1980) and has an apparent axial ratio of ~0.8.

An early VLA map of NGC 5548 at 4885 MHz was made by Wilson and Willis (1980). It showed the radio source to be triple, with two extended lobes straddling an unresolved component coincident with the optical continuum nucleus. This "nuclear radio source" has a total extent of 12" (5.9 kpc) in p.a. 165°. Van der Kruit (1971) and de Bruyn and Wilson (1978) also noted the presence of two additional sources lying outside the optical galaxy and aligned roughly E-W across its nucleus. One source lies 1!7 to the west (p.a. 270°) and the other is 4!1 to the southeast (p.a. 113°) of the nucleus.

NGC 5548 and the surrounding field were observed in a total of 4 runs on the partially completed VLA during 1979 and 1980. High-resolution maps of the nuclear radio source were made at 1465 and 4885 MHz. These maps show an unresolved central component together with two well-resolved outer lobes. Comparison of 1465 and 4885 MHz maps convolved with the same beam shows a spectral index of ~0.4 ($S_\nu \propto \nu^{-\alpha}$) for the central core and ~0.7 to ~1.0 in the outer lobes. The total

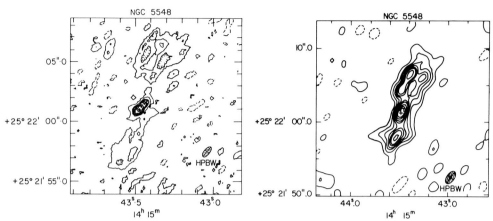

Figure 1. NGC 5548 at 4885 MHz (left) and 1465 MHz (right).

flux densities for the nuclear source, which is shown in Figure 1, are 23 and 8 mJy at 1465 and 4885 MHz, respectively. The corresponding powers are 2.2×10^{21} and 7.6×10^{20} W Hz^{-1} sr^{-1} at the two frequencies. Thus NGC 5548 contains not only the largest, but one of the weakest double or triple radio sources yet detected in a Seyfert galaxy.

Radio sources such as the nuclear source in NGC 5548 have been attributed to ejection of material from the inner nucleus or to bursts of star formation followed by multiple supernovae. In the first case, matter flowing from the inner nucleus may generate the radio emission in shocks where optical line-emitting clouds are encountered. Alternatively, a supernova rate of ~3-4 per year would be necessary to produce the amount of radio emission observed in NGC 5548.

A 20'-square field centered on NGC 5548 was mapped at low resolution at 1465 MHz, and the two field sources were mapped at high resolution. The field contains no low surface brightness features that might indicate a relation between the external sources and NGC 5548. The western and eastern field sources are extended and have total 1465-MHz flux densities of 34 and 21 mJy, respectively. Since neither source can be identified optically, the possibility of association with NGC 5548 cannot be ruled out, but seems unlikely.

We thank R. A. Sramek for participating in the observations and A. G. de Bruyn and S. M. Simkin for supplying optical photos of NGC 5548.

REFERENCES

de Bruyn, A. G., and Wilson, A. S.: 1978, Astron. Astrophys. 64, pp. 433-444.
Su, H. J., and Simkin, S. M.: 1980, Astrophys. J. 238, pp. L1-L5.
van der Kruit, P. C.: 1971, Astron. Astrophys. 15, pp. 110-122.
Wilson, A. S., and Willis, A. G.: 1980, Astrophys. J. 240, pp. 429-441.

WESTERBORK OBSERVATIONS OF LOW LUMINOSITY RADIO SOURCES

P. Parma
Istituto di Radioastronomia, Bologna, Italy

The two radio galaxies 0326+39 and 1321+31 are both part of a complete sample of low luminosity radio galaxies which was obtained by identifying B2 radio sources with galaxies from the Zwicky Catalogue. Both sources have a very symmetrical double jet, like many other sources of the sample (Ekers et al. 1981a).
What makes 0326+39 and 1321+31 unusual is their morphology: in other sources (for example 3C 31 and 3C 449) the jet widens and changes into the extended components, but in 0326+39 and 1321+31 the jet is surrounded by an extended structure right from the core (Figs. 1-2).
The source 1321+31 has been studied in detail (Ekers et al. in prep.) and a tentative explanation for its unusual morphology is presented. It is assumed that the jet is supersonic and freely expanding. The depolarization data and the dependence of the brightness profile on the transverse dimension of the jet are both suggestive of a decreasing flow velocity of the jet. Considering that the opening angle is constant and taking account of the Bernoulli equation we are led to the conclusion that most of the energy is flowing out of the jet to form the extended structure.

References

1) Ekers, R.D., Fanti, R., Lari, C., Parma, P.: 1981a, Astron. Astrophys., in prep.
2) Ekers, R.D., Fanti, R., Lari, C., Parma, P.: in prep.

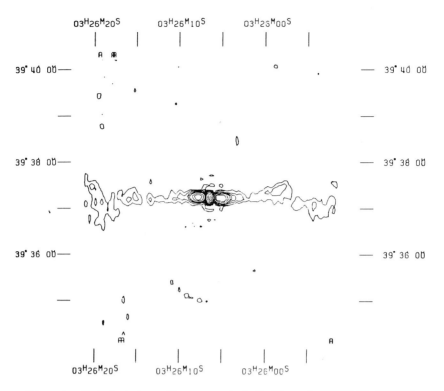

Fig. 1. Total intensity map at 5.0 GHz of 0326+39 (WSRT).

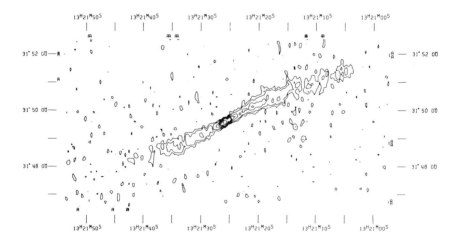

Fig. 2. Total intensity map at 5.0 GHz of 1321+321 (WSRT).

RADIO OBSERVATIONS OF MARKARIAN 8

D. S. Heeschen
NRAO, Charlottesville, VA
J. Heidmann
Observatoire de Meudon, Meudon, France
Q. F. Yin
NRAO and Beijing University, Beijing, China

Markarian 8, a clumpy irregular galaxy (Casini et al. 1979), was observed with the VLA at 20 cm (Mar. 19, 1981; 26 ant.) and 6 cm wavelengths (June 12, 1980; 18 ant.) The structure is alike at the two wavelengths, consisting of 3 distinct clumps imbedded in a diffuse envelope of about 40 arcsec extent. Figure 1 shows the 6 cm structure. At higher resolution the clumps break up into several components. The 20 cm structure is shown in Figure 2, which also compares the optical and radio morphologies. There is excellent general agreement, suggesting a common origin of emission in the clumps.

Radio continuum fluxes of the clumps were obtained by fitting gaussian models to the maps. Total fluxes were derived from low resolution (20"x20") maps. The envelope flux density is the difference between total flux and sum of the clump fluxes. Results are given in Table 1. The derived fluxes of individual clumps may be uncertain by as much as 25%. However, the sum of the clump fluxes is less uncertain, and it is clear that the spectra of the clumps are flat, suggesting thermal emission from an optically thin gas. The spectrum of the envelope is significantly steeper than that of the clumps and indicates non-thermal emission in this component.

Table 2 gives values of electron density, ionized mass and Lyman continuum flux derived for the three radio emitting clumps on the assumption of thermal emission. The last column gives the number of O stars that could produce the derived Lyman continuum flux. A distance to the galaxy of 48.5 Mpc was adopted for these calculations.

Benvenuti et al. (1980) have concluded, on the basis of optical and uv observations, that the clumps are massive HII regions containing more than 10^4 O and early B stars. The radio morphologies, spectra and fluxes, and the quantities in Table 2 derived from them, strongly support that conclusion. Markarian 8 may indeed be a galaxy in which there has been a recent extensive burst of star formation.

Table 1
Flux Densities and Spectral Indices

Feature	Flux Density (mJy) 20 cm	6 cm	Spectral Index, α ($S \sim \nu^{-\alpha}$)
total	12.6	8.2	0.36
A	2.8	2.2	0.20
B	0.9	0.6	0.34
C	0.7	0.7	0.00
A+B+C	4.4	3.5	0.19
Envelope	8.2	4.7	0.47

Table 2
Derived Parameters

Clump	N_e (cm^{-3})	M (10^8 M$_\odot$)	L_c (10^{53} photons s^{-1})	No. (10^4 O8V stars)
A	6.6	2.0	4.6	5.4
B	5.7	0.8	1.5	1.7
C	4.7	1.3	2.7	2.2

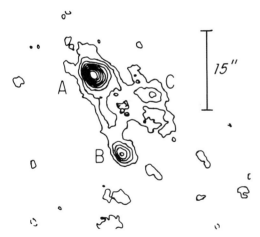

Fig. 1. 6-cm map, resolution 3".3 x 2".4 in p.a. 36°. Lowest and highest contours are 0.14 and 1.3 mJy beam area.

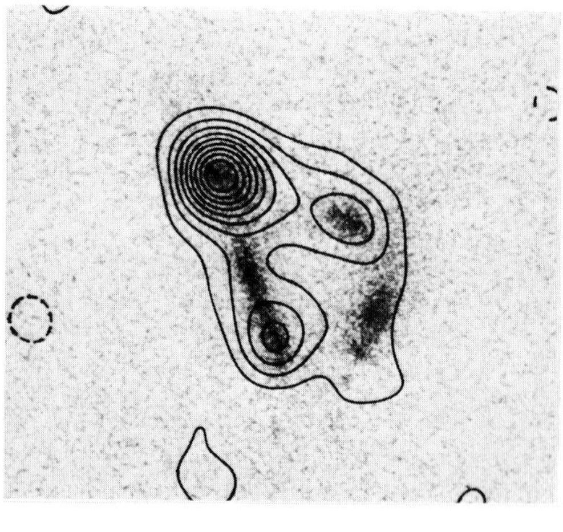

Fig. 2. 20-cm map, resolution 5"x 5", superposed on the optical image. Lowest and highest contours are 0.35 and 3.1 mJy/beam area.

References

Benvenuti, P., Casini, C., Heidmann, J. 1980. Proc. 2nd IUE Conference, 263.
Casini, C., Heidmann, J., Taringhi, M. 1979. Astron. Astrophys. 73, 216.

SS433 - OBSERVING EVOLUTION IN A PRECESSING, RELATIVISTIC JET

R. M. Hjellming
National Radio Astronomy Observatory, Socorro, NM, USA

K. J. Johnston
E. O. Hulburt Center for Space Research
Naval Research Laboratory, Washington, DC, USA

We commonly refer to the central "object" in extra-galactic radio sources as the "engine" that is the root cause of many radio source characteristics. We frequently ask, COULD "engines" at the cores of extra-galactic sources: (1) be compact objects with accretion disks; (2) eject well-collimated supersonic jets; (3) show relativistic effects in ejected material; (4) produce twin-jets; (5) produce one-sided jets; and (6) initiate highly polarized synchrotron radiation sources? SS433 is a binary star system with radio and optical jets that is relevant because it is a "little engine that could", and does, do all of these things. Further, we can observe changes in SS433 on time scales from hours to several months, and these data allow one to study evolution of jets in a more thorough fashion than is possible for extra-galactic sources.

The SS433 star system (V1343 Aql) has optical emission lines (Margon et. al. 1980) which change wavelength in a manner corresponding to a doppler shift range of 80,000 km/s with a periodicity of 164 days. These emission lines of H I and He I have been interpreted (Margon et. al. 1980) in terms of recombining material in twin-jet flows that have a velocity of 0.26c (c = the speed of light), a jet axis either 80 degrees or 20 degrees to the line of sight, and an ejection vector that rotates around the jet axis every 164 days at an angle of 20 degrees or 80 degrees. SS433 was independently found to be a radio source by Ryle et. al. (1979) and Seaquist et. al. (1979), and as can be seen from the large scale map of Geldzahler et. al. (1980), is a compact radio source inside the 1 degree by 2 degree "supernova remnant" W50. Johnston et. al. (1981) have shown that the relatively steady radio emission from SS433 is combined with flaring events that increase the flux density of the source by up to a factor of two with event time scales of 1-5 days. The radio emission is synchrotron emission from highly relativistic electrons because of the omnipresent non-thermal spectral index of 0.6 and the observation of linear polarization of up to 20% in the jets (Johnston et. al. 1981 and Hjellming and Johnston 1981a). The SS433 radio source was first found to have structure on angular scales of 0.1"-5" by Gilmore and Seaquist (1980). SS433 also shows structures on

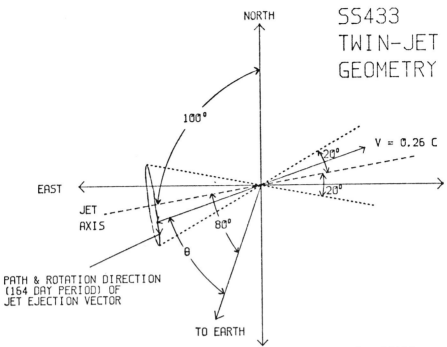

Figure 1. Geometry of the twin-jet ejection vectors for SS433 as determined from radio and optical data.

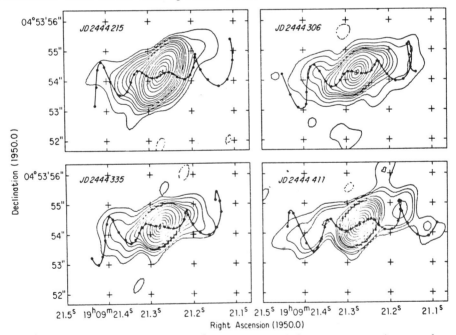

Figure 2. Four epochs of VLA 6 cm maps, with proper motion corkscrews, for SS433. Contours are roughly logarithmic ranging from -0.5% to 90%.

VLBI size scales of 0.005"-0.3" (Spencer 1979, Schilizzi et. al. 1982, Niell et. al. 1982, Walker et. al. 1981).

Hjellming and Johnston (1981a, 1981b) have shown that the SS433 star system is ejecting radio emitting material in the form of a corkscrew, exactly as one would expect from the twin-jet doppler shift model for the optical emission lines. The parameters of the model that fit both radio and optical data are shown in Figure 1. All of the parameters that could not be determined from the optical data are determined from high resolution VLA radio maps, such as the 6 cm maps shown in Figure 2. In Figure 2 the point source (0.4" X 0.7" HPBW) is located on the position of the star and the remaining structures are due to the ejected material which is found only on the corkscrews superimposed on the maps with filled circles at 20 day intervals in ejection time. As discussed by Hjellming et. al. (1981b), the relativistic time delay effects that can be seen in the corkscrews are used to make an absolute determination of the velocity, which is 0.26c to within 10-20% error. This and the observed proper motion of 3" per year, interpreted with the geometry of Figure 1, show the distance to be 5.5 kpc. This makes W50 violate the surface brightness vs. size relation for supernova remnants by about a factor of two, increasing the probability that W50 has either been caused by or heavily effected by SS433. The radio data also show that the "eastern" jet is the mainly blue-shifted jet in the foreground and the "western" jet is mainly red-shifted and predominately on the far side; it also shows that the sense of rotation is clockwise (left-handed) with respect to the eastern jet axis as shown in Figure 1. These results are borne out by improved 6 cm (0.4" x 0.4" HPBW) and 2 cm (0.13" x 0.13" HPBW) VLA maps made by Hjellming and Johnston (1981c) at 6 epochs between JD 2444579 and JD 2444717. All the extended structures are the accumulated effect of radio source ejection during the roughly 300 days prior to the time the source is observed.

Although to first order the evolution of the radio structures of SS433 is due mainly to the proper motion corkscrews of the twin-jet model, the high resolution VLA radio maps also contain information about the evolution of the radio emitting material as it moves out from the central source. This can be analyzed in two ways. Using maps such as those shown in Figure 2, one can plot the variation of intensity along each corkscrew as a function of time of ejection from the center. In some maps this results in a simple exponential decay with a time constant of 30-40 days. Hjellming and Johnston (1981c) have observed one case corresponding to an ejection at about JD 2444519 that eventually produced a stronger than normal double source with components moving away from the center at a rate of 0.25" per month. When the intensity of each component of this double is plotted as a function of time, a simple exponential decay is found with a time constant of 86 days. When one also considers the 1-5 day decay times for events observed by Johnston et. al. (1981), it is clear that one of the major variable parameters of the radio emitting plasma produced in the central regions of SS433 is the decay time, which is clearly exponential in character for the material with decay times greater than about 20 days - the only material that can contribute significantly to the extended structure in VLA radio maps. There are two obvious ways to obtain simple exponential decays for an evolving radio source: ionization losses due to relativistic electrons interacting with a co-moving thermal plasma, in which case this plasma typically has several electrons per cubic cm; or "catastrophic" losses due to relativistic electrons leaving the radio emitting region, possibly by diffusion across the boundaries of the radio emitting corkscrew. In the latter case the observed decay time would correspond to the diffusion time across the corkscrew.

Although the point source in VLA radio maps of SS433 is unpolarized with typical upper limits of the order of 1%, the extended structure is linearly polarized up to 20%. The linearly polarized intensity maps of Hjellming and Johnston (1981a) are shown in Figure 3 with the proper motion corkscrews of Figure 2 superimposed, and a small plus at the position of the star. Comparing Figures 2 and 3 it is clear that occasional portions of the radio emitting corkscrew are highly polarized and that both geometry and the averaging effects of the synthesized beam strongly affect the observed polarization. The resolution of the VLA maps is not sufficient to clearly show the polarization structure along the corkscrews. In Figure 4 selected linear polarized intensity contours are shown together with linear polarization vectors. In these and other maps it is very noteworthy that whenever a straight section of the corkscrew is seen over a relatively large region, the polarization vectors are roughly perpendicular to the corkscrew. Since the radio emission is optically thin, this means the magnetic field lines are aligned with the corkscrews. This is another indicator that the ejection of radio emitting plasma from the central regions produces contiguous and merged material, since the field lines would not be expected to follow the corkscrew under other circumstances.

If one assumes equipartition of magnetic and relativistic electron energies, one derives magnetic field strengths in the range 0.001-0.1 gauss, and total energies in particles and fields of the order of 10^{42}-10^{43} ergs. However, the short time scale of SS433 phenomena may mean the equipartition assumptions are not valid. If one assumes an electron concentration of several particles per cubic cm, as found from de-polarization across the source at 20 cm, the mass ejection from the central region is roughly 10^{-7} solar masses per year and the kinetic energy "luminosity" is 10^{38}-10^{39} ergs per second.

Because of the extensive amount of optical information about SS433 (Margon 1980) the root causes of the radio and optical jets are almost certainly high velocity (0.26c) well-collimated (roughly 4 degree opening angle) flows perpendicular to a thick accretion disk which is precessing with a period of 164 days, due to the influences of the companion star which is the source of infalling matter at a rate causing super-critical accretion. A schematic diagram illustrating this situation is shown in Figure 5, where the geometry is deliberately

SS433 – OBSERVING EVOLUTION IN A PRECESSING, RELATIVISTIC JET

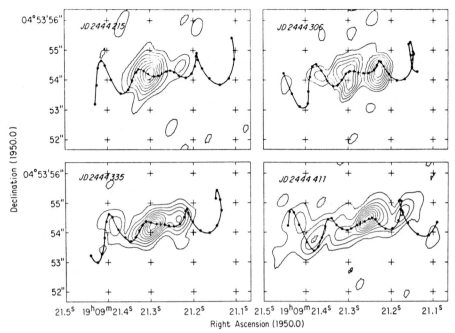

Figure 3. Four epochs of VLA 6 cm linearly polarized intensity maps, with proper motion corkscrews, for SS433. Contour intervals are linear with intervals of 10% starting at 90%.

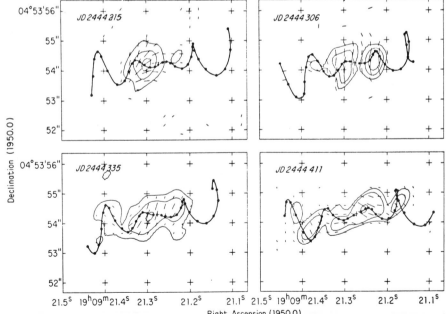

Figure 4. Polarization vectors are shown with selected linearly polarized intensity contours, and proper motion corkscrews, for four epochs of VLA 6 cm maps of SS433.

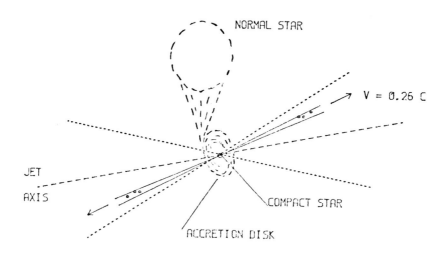

Figure 5. Schematic diagram relating the jet geometry (as in Figure 1) and the SS433 binary system. It also shows the probable origin of the jets in flows perpendicular to a super-critical, thick accretion disk around a neutron star or black hole.

drawn to correspond to the geometry diagram in Figure 1. The compact object which is the accretion target can be either a black hole or a neutron star. In Figure 5 the small circles in the flows schematically illustrate the recombining material producing the observed optical emission lines. The optically emitting regions must arise within a few light days of the compact object although the extended radio structures are observable only in the range 10-400 light days from the center. Both radio and optical emission regions are minor effects occurring in the flows, and SS433 is predominantly producing collimated kinetic energy.

VLBI observations of SS433, such as those presented by Schilizzi et. al. 1981 and Niell et. al. 1981 at this Symposium, provide essential information about the central "engine" for the production of radio emitting material. Because it is almost a certainty that the relativistic particles and magnetic fields are produced in the flows perpendicular to the accretion disk of SS433, the regions from the center to distances of 10-20 light days are where this production occurs, and this is the size scale accessible only to VLBI observations. Similarly, only VLBI observations will have sufficient resolution to show the detailed polarization and magnetic field structure of the SS433 corkscrew.

Because of the extent and type of information that is obtainable for the SS433 radio jets, and the related knowledge of the central "engine", SS433 is certainly the most effective "Rosetta Stone" that one has at the moment for understanding at least some of the physics of supersonic, radio-emitting jets. In this case, we know that a compact object, with an accretion disk and VERY supersonic outflows, is the basic cause of the jets. If these physical phenomena scale to larger masses, and Rees (1982) has argued that they do, the SS433 phenomena will be very important in understanding the physics of extra-galactic jets.

The National Radio Astronomy Observatory is operated by Associated Universities, Inc., under contract with the U.S. National Science Foundation.

REFERENCES

Geldzahler, B., Pauls, T., and Salter, C.:1980, Astron.Astrophys., 84, pp.237-244.
Gilmore, W. and Seaquist E.R.: 1980, Astron.J., 85, pp.1486-1495.
Hjellming, R.M. and Johnston, K.J.:1981a, Nature, 290, pp.100-107.
Hjellming, R.M. and Johnston, K.J.:1981b, Astrophys.J.(Letters), 246, pp.L141-L145.
Hjellming, R.M. and Johnston, K.J.:1981c, in preparation.
Johnston, K.J., Santini, N.J., Spencer, J.H., Klepczynski, W.J., Kaplan, G.H., Josties, F.J., Angerhofer, P.E., Florkowski, D.R., and Matsakis, D.N.:1981, Astron.J. (in press).
Margon, B.:1980, Sci.Amer., 243, pp.54-65.
Margon, B., Grandi, S.A., and Downes, R.A.:1980, Astrophys.J., 241, pp.306-315.
Niell, A.E., Lockhart, T.G., Preston, R.A., and Backer, D.C.: 1982, this volume, pp. 207-208.
Rees, M.:1982, this volume, pp. 211-222.
Ryle, M., Caswell, J.L., Hine, G., and Shakeshaft, J.:1979, Nature, 276, pp.571-575.
Seaquist, E.R., Garrison, R.E., Gregory, P.C., Taylor, A.R., and Crane, P.C.:1979, Astron.J., 84, pp.1037-1041.
Schilizzi et. al. 1982, this volume, pp. 205-206.
Spencer, R.E.:1979, Nature, 282, pp.483-484.
Walker, et. al.:1981, Astrophys.J. (in press).

DISCUSSION

SCHILIZZI: Are the moving blobs in SS433 entities like "plasmons"?

HJELLMING: I would believe that the radio-emitting portions of the "flows" from SS433 have detailed structure, and hence with sufficient resolution as with VLBI observations, you would, and do, see such details. However, there are probably local effects in what are otherwise continuous hydrodynamical flows.

GOSS: Westerbork Synthesis Radio Telescope observations in the HI 21 cm absorption line have been made of SS433. The derived limits for the distance are 3.7 to 4.7 kpc (Van Gorkom, Goss, Seaquist and Gilmore, MNRAS, in press, 1981).

HJELLMING: Although the 10% error estimate on our distance determination of 5.5 kpc is not significantly outside the 4.7 kpc limit, I would argue that a correction of 0.5-1.0 kpc is needed for the third (observed) arm of the Schmidt model you assumed.

SHERWOOD: I am happy to see the new distance. As Ann Downes et al. 1979 showed that W50 was being depolarized by the HII region, S74, I argued (Montreal IAU) that SS433 must be as far away as S74. I derived a distance of 5.4 kpc for S74 (Pub. Royal Obs. Edinburgh 9, 85, 1974).

HJELLMING: An interesting coincidence, perhaps meaning W50 and S74 are in the same spiral arm.

THE COMPACT RADIO STRUCTURE OF SS433.

R.T. Schilizzi[1], I. Fejes[2,3], J.D. Romney[2], G.K. Miley[4], R.E. Spencer[5] and K.J. Johnston[6].

[1]Netherlands Foundation for Radio Astronomy, Dwingeloo, NL.
[2]Max Planck Institut für Radioastronomie, Bonn, FRG.
[3]On leave from Satellite Geodetic Observatory, Penc, Hungary.
[4]Leiden Observatory, Leiden, NL.
[5]Nuffield Radio Astronomy Lab., Jodrell Bank, Cheshire, UK.
[6]Naval Research Laboratory, Washington, USA.

A campaign of VLBI observations of SS433 using the European network was begun in January 1980 and since October 1980 has continued at intervals of about two months except for the last two observations which were 6 days apart, see Table 1. The results of the first two epochs have been published in Nature (ref. 1), and those of the following six epochs are being prepared for publication.

Table 1: Observations of SS433.

Epoch (JD)	Date	Telescopes*	Wavelength (cm)
2444248	Jan. 80	E, W, K	6
393	June 80	E, W, C	6
517	Oct. 80	E, W	6
589	Dec. 80	E, W, J	21
651	Feb. 81	O, E, J	18
705	Apr. 81	O, E, W, J	6
750	May 81	O, E, W	6
756	June 81	O, E, W	6

* C: Chilbolton, E: Effelsberg, J: Jodrell Bank, K: Knockin (UK), O: Onsala, W: Westerbork.

Some conclusions from this work are:
(a) The trajectories (ref. 2, and independently by Fejes) of the three-dimensional kinematic model based on optical emission line data (ref. 3) are in reasonable agreement with the observed radio structures. The curvature is well-fitted, but the features (blobs) in the structure are more symmetrically spaced about the apparent centre than expected in the model from the finite travel time of the radiation from the back to the front of the source (see Fig. 1). However there is obviously uncertainty as to the exact centre of the structure.

(b) There is no evidence of the previously reported lag of the

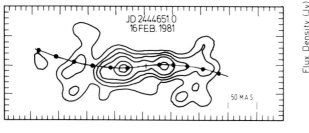 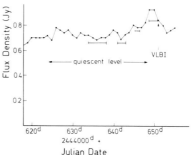

Figure 1: The 18 cm compact structure of SS433. North is up, east to the left. The restoring beam was 24 m.a.s. Also drawn is the expected trajectory from the kinematic model, with dots every 4 days.

Figure 2: The flux density of SS433 at 11 cm as a function of Julian Date, from measurements with the Green Bank interferometer. Horizontal bars indicate apparent ejection times of blobs in the structure.

radio structure behind the trajectory predicted from the kinematic model. The earlier reports (ref. 1,5) were presumably influenced by insufficient spatial resolution and limitations in the modelfitting of the structure. At a number of epochs, some features in the structures lie ahead of or behind their predicted positions on one side of the centre but not the other (e.g. Fig. 1). This may be evidence that the "jitter" discussed by Margon (ref. 4) in the pointing or phasing of the beams in the optical, is also affecting the radio.

(c) There is no strong correlation between the occurrence of a peak in radio flux density and the interpolated birth date of a pair of radio emitting blobs. Figure 2 shows the flux density variations at $\lambda 11$cm in the period leading up to and including the February observation.

(d) The dominant energy loss mechanism in the structure on the tens of m.a.s. scale is clearly not synchrotron radiation. For example, from the measured sizes and flux densities of the components at separations of $\sim$40 m.a.s. in the February 1981 data, and assuming a turnover frequency of 300 MHz (ref. 6), synchrotron lifetimes of $\sim$10 years are derived. If the turnover at 300 MHz is caused by some other mechanism than synchrotron self absorption then the synchrotron lifetimes are increased still further (ref. 7). Presumably the dominant loss mechanism is adiabatic expansion but this has yet to be convincingly demonstrated.

REFERENCES
1. Schilizzi, Miley, Romney and Spencer (1981) Nature 290, 318.
2. Hjellming and Johnston (1981) Ap. J. 246, L 141.
3. Abell and Margon (1979) Nature 279, 701.
4. Margon (1981) Univ. of Washington, preprint.
5. Niell, Lockhart and Preston (1981) to be published in Ap. J.
6. Seaquist, Gilmore, Nelson, Payten and Slee (1980) Ap. J. 241, L 77.
7. Hjellming and Johnston (1981) Nature 290, 100.

SS433: PERIODIC CHANGES IN THE RADIO STRUCTURE OF SCALE SIZE 10^{16} cm

A. E. Niell, T. G. Lockhart, R. A. Preston
Jet Propulsion Laboratory, Pasadena, California

D. C. Backer
University of California, Berkeley, California

Radio observations have clearly demonstrated that the kinematic twin-jet model (Milgrom 1979; Abell and Margon 1979) is the correct description of the general behavior of SS433 and have determined the orientation parameters that could not be obtained from the optical observations (Gilmore and Seaquist 1980; Hjellming and Johnston 1981 (HJ); Niell, Lockhart, and Preston 1981 (NLP)). Figure 1 shows the observed position angle of the radio jet at a distance of approximately 0".15 from the core during the period 1979 May to 1981 May as determined from our VLBI measurements at 2.3 GHz. The mean position angle is $98°\pm2°$ for a $20°$ half-angle cone of precession about an inclination of $79°$ to the line of sight. The phase is consistent with the expected propagation time out to 0".15 (10^{16} cm) at a speed of 0.26c for a distance to SS433 of 5 kiloparsecs (HJ; NLP).

From observations at three epochs using the antennas at Hat Creek (UC Berkeley), Big Pine (Caltech), and Goldstone (NASA), California at 2.3 GHz we have made hybrid VLBI maps of SS433. The angular resolution is about two days of travel time along the jets. For all three epochs the core is extended along the direction expected from the optical ephemeris and in the direction of the Eastern "blue" jet (HJ) and probably represents the beginning of another outburst. The maps (Figure 2) show that outside the core the jets are dominated by knots with brightness temperatures greater than 10^6K. (See also Schilizzi et al. in this volume). In February the two brightest knots were of comparable strength and age (approximately 8 days) suggesting that some ejections give rise to radio emission along both jets, even though the core appears extended to the East. In no case is the structure resolved perpendicular to the apparent direction of motion, thus setting an upper limit of 10^{15} cm on the transverse width of the knots.

Finally, we speculate that the differences in position between the predicted trajectories (given by the solid lines in Figure 2) and the locations of the observed knots are real and represent fluctuations in the direction of ejection of the source of the knots about the mean

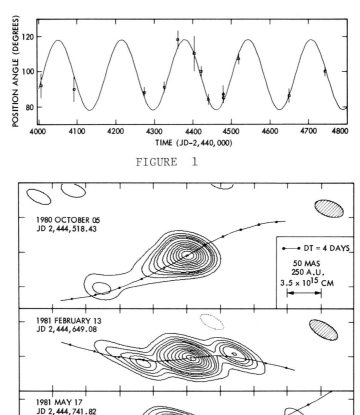

FIGURE 1

FIGURE 2

ephemeris. The observed angular fluctuations would produce deviations of the optical emission line radial velocities from the mean ephemeris which are comparable to those observed (Margon, Grandi, and Downs 1980).

REFERENCES:

Abell, G. O., and Margon, B.: 1979 Nature 279, 701.
Gilmore, W., and Seaquist, E. R.: 1980, A. J. 85, 1486.
Hjellming, R. M., and Johnston, K. J.: 1981, Ap. J. (Letters), 246, L141.
Margon, B., Grandi, S. A., and Downes, R. A. 1980, Ap. J. 241, 306.
Milgrom, M.: 1979, Astr. Ap. 76, 13.
Niell, A. E., Lockhart, T. G., and Preston, R. A.: 1981 Ap. J. (Letters) 250, (in press).
Schilizzi, R. T., Miley, G. K., Romney, J. D., and Spencer, R. E.: 1981, Nature 290, 318.

RADIATIVE ACCELERATION OF ASTROPHYSICAL JETS: LINE-LOCKING IN SS 433

Paul R. Shapiro
Dept. of Astronomy, University of Texas at Austin

Mordecai Milgrom
Dept. of Nuclear Physics, Weizmann Institute, Rehovot, Israel

Martin J. Rees
Institute of Astronomy, Cambridge, U. K.

ABSTRACT

Observations of SS433 are consistent with the view that the Doppler-shifted line emission originates in a pair of oppositely-directed, precessing jets in which a gas outflow is maintained at the remarkably time- and space-invariant speed of 0.26c. A radiative acceleration mechanism is described for the jets and a detailed, numerical, relativistic flow calculation presented which explain this terminal velocity as the result of "line-locking". The "line-locking" mechanism suggested here for SS433 may be important as well in extragalactic radio sources in which the radio luminosity is similarly weak compared with the kinetic energy and optical luminosities.

Measurements of the Doppler-shifted line emission from SS433 indicate that the outflow velocity has been constant to within roughly five per cent of 0.26c since the discovery of the moving lines. From the line widths, we know that the velocity is quite uniform in direction and magnitude across the jet (in the frame of the jet), as well. This has a very natural explanation (Milgrom 1979) if the following conditions are met: (1) A continuum radiation flux exists which is strong enough to accelerate gas away from the compact central object; (2) The dominant momentum transfer is through Lyman-line absorption by some hydrogenic ion; and (3) The continuum flux is sharply reduced at lab frame frequencies above the Lyman edge. In that case, the flow can accelerate only until the velocity is high enough to Doppler-shift the Lyman edge to the Ly-α frequency. This occurs when $\gamma(1-\beta) = 3/4$, or $v = 0.28c$. Thereafter, the flow speed is constant and described as "line-locked".

A detailed numerical calculation of this process has been performed which solves the relativistic flow equations, including matter-radiation coupling, radiative transfer, ionization balance, and the atomic level population equilibrium. We have assumed that the flow is steady-state, spherically symmetric, supersonic (ie. gas pressure is ignored), and isothermal, that special relativistic effects must be calculated exactly to all orders, but that gravity is Newtonian. Our first detailed results are of a pure hydrogen gas. We find that line-locking <u>can</u> be achieved and with rates of mass outflow within the range inferred for SS433 from observation, for a gas with $T \sim 10^4 K$, as long as the gas is highly clumped (e.g. $\propto$ (Mach number)^2). For example, if initial radius, luminosity, temperature, and central object mass are 10^{12}cm, 10^{38} erg s^{-1}, 10^4K, and 1 $M_\odot$, respectively, then a 10^{17}gm s^{-1} outflow will achieve line-locking by roughly 3×10^{12}cm. In general, the maximum outflow is limited by the condition $(1/2)\dot{M}\beta^2 c^2 \lesssim \beta L$.

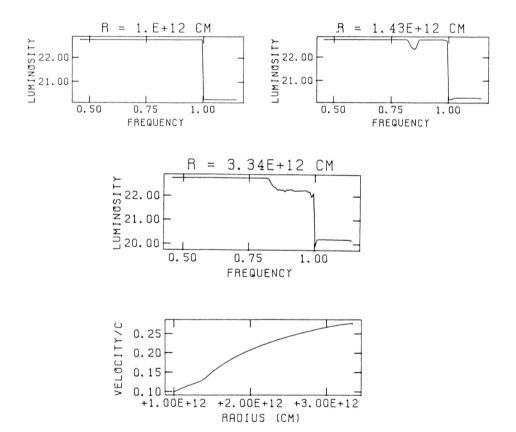

Figure 1. Results for illustrative case mentioned in text. Luminosity is $\log_{10}(dL/d\nu)$. Frequency is in units of $\nu_{Ly-edge}$.

REFERENCE

Milgrom, M. 1979, Astron. Astrophys., <u>78</u>, L17.

MECHANISMS FOR JETS

Martin J. Rees
Institute of Astronomy
Cambridge, England

ABSTRACT. The evidence is now compelling that "jets" delineate the channels along which power is supplied from galactic nuclei into extended radio sources - this accumulating evidence, reported by many speakers, has been one of the major themes of this conference. Jets (often apparently one-sided) have been discovered inside many symmetrical double sources. And M87, familiar optically as a "one-sided jet" for over 60 years, is now found to have weak double radio lobes. The VLA has resolved $\sim$ 70 jets in extended sources; there are now many instances where small jet-like structures are found on the VLBI scale; indirect arguments (some of which I'll mention) indicate that there is directed outflow on still smaller scales, and that the primary collimation may occur right down in the central "powerhouse" (scales $\leq 10^{15}$cm). The length-scales relevant to jet production and propagation thus span 9 orders of magnitude; the physical processes and conditions may vary widely over this vast range of scales. This paper will deal briefly with two aspects of jet physics: firstly, some direct inferences from radio maps; and, secondly, some possible mechanisms in galactic nuclei that could set up a collimated outflow.

1. PHYSICS OF RESOLVED RADIO JETS

Velocity

The jet velocity in the galactic object SS433 is well-determined (at $\sim$ 80,000 km s^{-1}), but we do not have equally firm estimates for any extragalactic jet. The only two cases where there is optical evidence are DA240 (Burbidge, Smith and Burbidge 1975), where the so-called "jet" may well be an unrelated foreground feature, and Coma A (Miley *et al.* 1981) where the relation between the moving material and the jet speed is somewhat unclear. For the great majority of jets, we must fall back on indirect arguments; but these yield velocity estimates ranging from $\sim$ 300 km s^{-1} right up to $\sim$ c.

Several of the brighter compact radio sources mapped by VLBI display a one-sided core-jet morphology, and in addition show apparent superluminal expansion. As discussed in Dr Cohen's paper, this indicates relativistic motion nearly along our line of sight; the apparent one-sidedness could then just stem from Doppler boosting of radiation from the approaching jet over that from the receding counter-jet. I shall not discuss detailed models for superluminal sources; however, it is hard to avoid the conclusion that relativistic outflow and "Doppler favouritism" are part of the story.

It is then a natural extension to interpret the one-sidedness of many extended jets (as described in Dr Bridle's contribution) in terms of Doppler favouritism, and to conclude that relativistic bulk velocities generated in the nucleus persist out to scales of $\gtrsim 10^5$ pc. Very fast jets have the further advantage that they minimise the mass flux required to provide the thrust and energy input into strong double sources such as Cygnus A.

But there are counter-arguments, suggesting that, at least in lower-powered sources, the flow along jets is much slower:

(i) Some one-sided jets show bends. If the bulk flow follows the jet, then the smoothness of the brightness contours around the bend indicates a subrelativistic jet velocity (Pottash and Wardle, 1979; Van Groningen, Miley and Norman, 1980)

(ii) In sources that show a high degree of reflection or inversion symmetry, one can sometimes give a (model-dependent) velocity estimate. For example, if the bending in 3C31 and 3C449 results from acceleration of the parent galaxy in its orbit about a close companion, then the jet velocities in these sources are in the range 300-500 km s^{-1} (Blandford and Icke, 1978).

(iii) In some jets - NGC6251 being the best example - there seems to be some internal Faraday depolarisation. This indicates the presence of thermal material in the jet, and yields an estimate of its density. According to Saunders *et al.* (1981), the energetics of the extended lobes then suggest a speed $v_j \simeq 1.5 \times 10^4$ km s^{-1}. The energy flux (for a given density in the jet) goes as v_j^3, so a relativistic velocity would demand 10^3 times more energy. The residual uncertainty in this argument comes from doubt about whether the apparent depolarization is really internal to the jet.

If the jets are not relativistic, Doppler shifts will be negligible, and some other explanation is needed for the one-sidedness often observed. Possibilities include:

Flip-flop behaviour. The one-sidedness may be intrinsic: there may be no beam at all on the other side. But then, to give rise to symmetric double radio lobes, beams must have squirted for comparable times in each direction - averaged over the source lifetime. Moreover, the last injection on the now-defunct side must have been recent enough to generate the highest energy electrons still radiating. But when jets are as long as in NGC 6251 the beam must squirt for $\geq 10^6$ $(v_j/c)^{-1}$ yrs between reversals. It is surprising that this interval should be so long. Only a small range of timescales can be squeezed between the two constraints. There is no obvious physical mechanism that might cause this "flip flop". One possibility might be that the massive power house has been displaced from a precisely central position and is oscillating in the potential well of the galaxy (or its stellar core); timescales of $10^6 - 10^7$ yrs could then arise actually.

Asymmetric internal dissipation. There are many strong classical doubles where the beams are not directly seen, but can be inferred to be active. In such sources, presumably the kinetic energy of the beam is transported out to the "hot spots" without there being too much internal dissipation or boundary friction along its path. It is unknown what determines the amount of such dissipation - and, in consequence, the amount of radiation from the beam. All that is required is the conversion of a few percent of the kinetic energy. Perhaps there is some kind of "turbulent transition", which happens to one jet but not to its counterpart on the other side. For instance, the shear across the jet may be larger on one side, due to different conditions near the nucleus, or to effects of the interstellar environment. If the probability of such a transition decreased with increasing jet power, then the apparent correlation of jet properties with radio luminosity could be understood. The fact that the compact jets in NGC 6251, M87 and NGC 315 are on the same side as the extended jets indicates that the asymmetry must be maintained over many scale lengths.

At present, the arguments support relativistic speeds in the superluminal sources, but are evenly balanced for sources like M87 and NGC 315. An interesting test would be to see if the extended lobes are in any way systematically different on the side where there is a jet; if so, this would tell against "Doppler favouritism" as an explanation for one-sided jets.

The Content of Jets

If extended jets are sub-relativistic, and the apparent depolarization is caused internally, the typical mean densities of thermal electrons are $\sim 10^{-2}$ cm^{-3}. Thus, the pressures and densities in extended jets resemble those encountered in our interstellar medium, and conditions could be just as complex. The gas may be concentrated in dense cool clouds where the field is stronger than average, and the magnetic field may have many reversals. When associated optical emission is seen, this could be thermal radiation from clouds in the jet (cf. SS433); alternatively, it could be synchrotron radiation if high-energy electrons are being accelerated in the jet via shock waves, reconnection etc. (cf. M87).

As Dr Bridle has explained, the collimation and confinement of large-scale jets does not seem to fit any simple picture. The cone angle (in, for instance, NGC 315) seems to widen, and then, further out, narrow down again; there may be a transition from pressure confinement to free expansion and back again; there are some jets where the external pressure cannot be high enough to provide confinement, but which cannot be free either. Moreover, the jet fluid is not flowing adiabatically: internal dissipation, or friction at the boundary, must be tapping some of the bulk energy and converting it into relativistic particles - otherwise the outer parts of jets would not be detected at all.

The extended jet in 4C 32.69 poses problems (Pottash and Wardle 1979). Its internal pressure - even the minimum based on assuming equipartition - is too high to be confined by a plausible external medium. On the other hand, there are problems with supposing that it is free: the thrust estimated on this assumption ($\gtrsim \theta^{-2}$ times the internal pressure) is far larger than could be opposed by the ram pressure of the intergalactic medium.

The serious problems with pressure confinement and with free jets motivate one to consider the properties of the magnetic field, and to ask whether magnetic stresses - squeezing by a field wound round the jet - can aid confinement.

Magnetic Confinement

The evolution of the jet's magnetic field not only controls the synchrotron emission and its polarization, but may also have a strong effect on the collimation, confinement and stability. In the absence of resistivity, the magnetic flux in the jet is conserved along the jet trajectory (unless field is entrained from the surroundings), although the field strength may be amplified by internal shear. If there is no velocity gradient across the jet, the longitudinal field, $B_{||}$, scales with the jet diameter as $B_{||} \propto d^{-2}$, while the transverse field $B_\perp$, scales as $B_\perp \propto d^{-1}$ (for $v_j \simeq$ constant).

However, a transverse velocity gradient of magnitude $\nabla v \gtrsim v_j/r$ acting over any scale within the jet, would create sufficient shear to make the longitudinal field dominate. Spread across the diameter of the jet, this much shear would correspond to a maximum velocity differential of only $\nabla v_j/v_j \gtrsim \theta/2$. If the shear reflects the overall velocity profile of the jet, as determined by a turbulent or viscous boundary layer, then the greatest amplification of parallel field might be expected in a sheath around the jet. (Other speakers have discussed the data on jet polarization.)

The field strengths inferred within both extended and compact radio jets are compatible with having been advected away from the neighbourhood of a central compact object according to the scaling law $B \propto r^{-1}$. If the jet carries a net current, then it can be magnetically self-confined under the tension associated with toroidal field lines (Benford, 1979; Chan and Henrinksen 1980). (The current involved would actually be comparable with the current postulated to flow through the

Crab pulsar.) In an axisymmetric jet, magnetic confinement can be achieved by re-arranging the density profile so that the toroidal field $B_\perp$ varies inversely with distance x from the axis. The magnetic stress then varies as x^{-2} and can be reduced to the value of the external pressure. The return current flows at larger radii still, and the stresses associated with it can be negligibly small. This scaling breaks down in the core of the jet, which is compressed to the point where the combination of particle pressure and magnetic pressure due to $B_{||}$ balances the confining stress associated with the surrounding toroidal field. In this way, the pressure in the core of the jet can be substantially larger than that in the external medium.

The field causing the confinement would be basically perpendicular to the jet direction; however, the hypothesis of magnetic confinement is reconcilable with polarization evidence that $B_{||}$ often dominates, because the synchrotron emission may come predominantly from the "core" rather than the confining sheath.

2. THE INNERMOST FEW PARSECS

If the VLBI "maps" of 3C273 and other compact sources ever became as detailed as those we already have from the VLA for large-scale jets, they would probably reveal the same type of complex and inhomogeneous structure. There must be "in situ" particle acceleration in the individual superluminal blobs; the external pressure has the value appropriate to a galactic nucleus (or quasar emission line region), but it may still not be high enough to confine the blobs. Over the next few years, study of variability in VLBI jet structure may yield clues to the physics of larger-scale jets.

Even the milli-arc-second (parsec-scale) jets may be secondary manifestations of a collimation process on a scale orders-of-magnitude smaller still. There are three reasons for attributing the basic collimation to sub-parsec dimensions:

(i) The apparent power output and rapid optical variability of sources such as AO 0235+164 and OJ 287, and the random behaviour of their optical polarization (Angel and Stockman, 1980), indicates relativistic beaming, and this would be on a scale smaller than that probed by VLBI.

(ii) The immediate environment of a massive collapsed object is the most propitious place for generating a high energy-per-particle.

(iii) The long-term stability of the jet axis can be ensured by the Lense-Thirring effect if the collimation occur close enough to a massive spinning relativistic object (Bardeen and Petterson, 1975; Rees, 1978).

Collimation via the well-known twin-exhaust mechanism (Blandford and Rees, 1974) may develop on any scale, and its qualitative features are scale-independent: collimation may occur at a few hundred Schwarzschild radii if the cloud is bound to the central black hole, at tens to hundreds of parsecs if it is bound to the nuclear star cluster, and on scales of kiloparsecs if the cloud is of galactic scale. On these larger scales, the nature of the galactic environment (e.g. elliptical or spiral) may affect what happens (cf. Sparke and Shu, 1980). Phenomena on all scales may play a role in moulding the shapes and determining the emissivity of the observed radio jets. But, for the reasons just mentioned, the primary collimation probably occurs around a massive compact object.

3. JET PRODUCTION NEAR BLACK HOLES

The cloud which confines and collimates the jets must be gravitationally bound in a potential well; but its pressure must be sufficient to prevent it from collapsing into the centre (or into a thin disc if it is rotating). Consequently, the value of (P/ρ) for the cloud material must be of the same order as the gravitational binding energy. For a cloud in a $\sim 1/r$ potential around a massive black hole, where P/ρ may exceed $m_e c^2$, it becomes implausible to suppose that the pressure comes from an electron-ion plasma with $T_i = T_e$: the electrons would then need to be relativistic, and their cooling (via synchrotron and compton processes) would be very rapid. There are two classes of model for a pressure-supported cloud in a relativistically deep potential well:

(i) The cloud may be supported primarily by radiation pressure. The gas temperature can then be lower by the same factor by which radiation pressure exceeds gas pressure. If the cloud is sufficiently dense and opaque, the radiation will acquire a black body spectrum. If radiation pressure provides the primary support, the leakage of energy must correspond to the "Eddington luminosity" $L_E = 4\pi GMcm_p/\sigma_T$ for the central mass.

or: (ii) The cloud may be supported by ion pressure, the electrons being cooled by radiative losses to $\lesssim 1$ Mev. This option is plausible only when the density is low, so that the electron-ion coupling time is long enough to prevent all the ion energy from being drained away during the inflow timescale.

When a rotating cloud of either of these types is established around a massive black hole, a distinctively relativistic feature of the gravitational potential well comes into play. There will be a "funnel" around the rotation axis, bounding a "region of non-stationarity", within which no combination of pressure gradients and centrifugal force can support a stationary axisymmetric flow. All material within the funnel must either have positive energy (in which

case it will escape) or else have so little specific angular momentum
that it falls freely into the hole. Along the walls of the funnel,
which are roughly paraboloidal in shape, the specific angular momentum
is nearly constant and equal to the minimum specific angular momentum
which can be swallowed by the hole.

Any cloud surrounding a black hole thus has a toroidal shape (with
a lower-density funnel "cored out" around the rotation axis). If
relativistic plasma is generated near the hole, it therefore may not
need to excavate an escape route via the twin-exhaust mechanism, since
there is a pre-existing channel (whose walls will however be modified
in shape by the pressure of the outflowing beam).

Detailed studies of tori supported by radiation pressure have been
carried out by the Warsaw group and their collaborators (Jaroszynski,
Abramowicz and Paczynski 1980; Abramowicz, Calvani and Nobile, 1980;
and references cited therein). The radiation emerges especially
intensely along the rotation axis. This is because centrifugal effects
greatly enhance the effective gravity on the walls of the 'funnel'
around this axis. Consequently, radiation pressure might preferentially
eject jets. However, even though, along the axis, the radiation flux
per unit solid angle can be $\sim$ 100 times the Eddington value (Sikora
1981), this is not an efficient mechanism for producing ultra-relativistic
jets. This is because the high ambient density of multiply-reflected
radiation within the funnel provides a Compton drag which prevents the
attainment of relativistic speeds until the material reaches the outer
radius of the torus. If the density of the jet becomes too high, its
optical depth exceeds unity; the acceleration must then be treated not
just in terms of individual test particles, but by regarding the jet
as a fluid within which radiation pressure provides a high (P/ρ).
Taking this into account, Sikora and Wilson (1981) argue that radiation
pressure acceleration can only be efficient in the "fluid" rather than
the "test particle" limit.

These radiation-supported tori, which require a "supercritical"
fuelling rate, may be appropriate models for the majority of quasars -
those with low polarization, weak radio emission, and an optical continuum
that appears predominantly thermal - but may be less relevant to those
active nuclei whose main output is conspicuously non-thermal (the
"blazars").

Ion-Supported Tori

Tori supported by radiation pressure will always emit thermal
radiation with luminosity $\sim L_E$. This will typically be in the optical
or ultraviolet band. But the observed luminosity of most active galactic
nuclei that display jets is <u>much less than</u> L_E for a $10^8 M_\odot$ black hole
($\sim 10^{46}$ erg s^{-1}). $10^8 M_\odot$ is probably the minimum mass which can have
produced the large internal energy contents of the extended radio lobes.
(This discrepancy is particularly acute in the case of M87 which is
argued (Young et al. 1978) to have a black hole of mass $M \simeq 3 \times 10^9 M_\odot$
and an observed nuclear luminosity thus only $\sim 3 \times 10^{-5} L_E$.)

At low mass accretion rates, spherical accretion has a low radiative efficiency, because the infalling material is unable to radiate its internal energy on the free-fall time scale. Disk accretion likewise will be inefficient at low accretion rates if the magnetic(?) viscosity is high enough: gas can still get rid of its angular momentum, and swirl inwards towards the hole, in a timescale shorter than the cooling time. A torus can then form which is supported by ions (whose temperature is ~ 100 Mev near the hole), but in which the electrons cool down below ~ 10 Mev. Unless collective effects couple the ions and electrons much more efficiently than Coulomb interactions do, the ions will be unable to cool on the inward drift timescale if $\dot{M}/\dot{M}_{crit} < 50$ $(v_{infall}/v_{free\ fall})^2$. Even though this torus may itself radiate very little, it can nevertheless anchor a magnetic field part of which threads the hole, thereby allowing electromagnetic torques to tap the hole's spin energy (in the manner outlined by Blandford and Thorne elsewhere in these proceedings). Poynting flux and/or ultrarelativistic particles would then be collimated by the torus into jets. This mechanism is analysed by Rees $et\ al.$ (1981), who propose that the primary power supply for all the most purely non-thermal active nuclei (e.g. M87, strong double sources, highly polarised "blazars", etc.) could be powered by spinning holes enveloped by ion-supported tori. A hole of $10^8 M_\odot$ could store several times 10^{61} ergs of spin energy; even a low value of $\dot{M}$ could maintain a magnetised torus which could gradually extract this energy.

Production of Beams with Lorentz Factor ≥ 5

The VLBI data imply that, at least in the superluminal sources, the bulk Lorentz factor of the beams, γ_b, is in the range 5 - 10. (As explained earlier, however, the evidence on beam speeds in other objects is ambiguous.) For an electron-proton plasma to attain this energy, each proton must acquire 5 - 10 Gev, which is ≥ 20 times the maximum mean energy per particle that can be made available in an accretion process. This is a general difficulty with purely gas-dynamical processes; it suggests that a different process must be invoked. Two possibilities are:

(i) Electromagnetic mechanisms may channel most of the energy of infalling matter or of a spinning hole into a small fraction of the particles, or into a Poynting flux which can accelerate high energy particles in a beam.

or (ii) The beams may be composed of $e^+ - e^-$ plasma rather than containing ions; $\gamma_b \simeq 5$ can then be attained with only ~ 5 Mev rather than ~ 5 Gev per electron. These pairs may be produced by pulsar-type vacuum breakdown processes, or by photon-photon interaction in a compact source of X-rays and γ-rays. Radiation pressure is of course more efficient for an $e^+ - e^-$ plasma than for an electron-ion plasma. Note however that if powerful jets of $e^+ - e^-$ originate at small radii they will annihilate unless γ_b is initially large.

4. INSTABILITIES?

Even if equilibrium flow patterns exist that can give rise to jets, they may be subject to serious instabilities. Given the difficulty of predicting the stability of terrestrial and laboratory fluid flows, one should be cautious about attaching too much weight to stability calculations, when they are applied to the flow of magnetized (possibly relativistic) plasma flowing throuhg a medium of uncertain properties. Kelvin-Helmholtz instabilities can occur anywhere along the beam's path; but even more serious is the possibility that the setting-up of the collimated flow may be completely prevented by Kelvin-Helmholtz and Rayleigh-Taylor instabilities, in the nozzle region.

Recent calculations by Norman *et al*. (1981) find that the twin-exhaust pattern seems stable only if the width of the nozzle is comparable with the scale height – i.e. for a limited range of energy fluxes (the external pressure being given). At too high a flux the flow pattern is disrupted by violent Kelvin-Helmholtz instabilities, while at too low a flux the channel quasi-periodically pinches off, due to Rayleigh-Taylor instability. These calculations are the best we yet have, but they are still based on assuming simple equations of state for each fluid. Furthermore, they use a 2-D code rather than full 3-D hydrodynamics. Thus the "instabilities" are due to ring-shaped protuberances around the flow boundary. Smarr reports elsewhere in these proceedings some calculations of jet propagation and stability carried out using an improved code.

The simulations by Norman, Smarr and their collaborators are an impressive portent of how numerical techniques will soon permit real "experimental" study of the stability of flow patterns. This numerical approach is likely to be more fruitful than linear stability analyses. However, in comparing these simulations with the observations, it is important not to forget that the only datum we have is the distribution of <u>radio surface brightness</u>. Any apparent "blobbiness" in this, implies inhomogeneity in the quantity B^2 x (path length through source) x (density of relevant relativistic electrons). Unless the magnetic field and the relativistic particles are dynamically dominant, such features could merely indicate regions where particle acceleration is concentrated or the field is specially strong, rather than being substantial features in the overall flow.

5. SCALING LAWS: MINI- AND NANO-QUASARS

The radio structure of Sco X1 looks like a miniature version of an extragalactic double source; SS433 involves collimated jets; and jet-like structures are found in some regions of star formation. This prompts the question of whether these resemblances are merely superficial, or whether similar mechanisms for jet production can indeed operate on vastly different scales.

The flow pattern around (or onto) a compact object is basically controlled by two parameters: the ratio L/L_E (which fixes the relative dynamical importance of radiation pressure and gravity) and the ratio $t_{cool}/t_{dynamical}$ (which fixes the temperature when a stationary flow pattern is set up). Suppose we have an object with given values of M and L, where the flow pattern is axisymmetric and characterised by a velocity $\underline{v}(r,\phi)$, ϕ being the angle made with the symmetry axis; we can then inquire about the properties of a scaled-down object with mass $M' = xM$ ($x \ll 1$). If we scale $L' = xL$, then the miniature version has the same value of L/L_E. The gravitational radius r_g scales as M. If the flow pattern is similar in the miniature version (which is equivalent to requiring similar viscosity parameters) then $\underline{v}'(xr,\phi) = \underline{v}(r,\phi)$, and the dynamical timescales then scale as $t'_{dyn} = xt_{dyn}$. If the objects were fuelled by accretion, and the efficiencies were the same in both cases, then $\dot{M}' = x\dot{M}$. The characteristic densities then scale as $\rho'(xr,\phi) = x^{-1}\rho(r,\phi)$. Now in general $t_{cool} \propto \rho^{-1} \times$ (function of T): the ρ^{-1} dependence applies not only to two-body cooling processes, but also to cyclotron-synchrotron cooling if B^2 scales with ρ^{-1}. We then have $t'_{cool} = xt_{cool}$ – in other words, the ratio t_{cool}/t_{dyn} is the same in the "miniature" flow pattern, given that L is scaled with M.

This means that flow patterns with a given value of L/L_E and $\dot{M}/\dot{M}_{crit}$, around black holes of very different mass, would be very similar. (The precise scaling breaks down when optical depth effects are important, but the above argument does have great generality). The apparent analogy between stellar-scale phenomena and active galactic nuclei may indeed reflect an underlying physical similarity. The relevant parameter is $\dot{M}/\dot{M}_{crit}$, and we can make the following schematic comparisons.

	$\dot{M}/\dot{M}_{crit} \gtrsim 1$	$\dot{M}/\dot{M}_{crit} \ll 1$
$M = 10^9 M_\odot$	Low-polarization "thermal" quasars	"Blazars", M87 (main power perhaps extracted electromagnetically from hole, not deriving from accretion)
$M = 10^6 M_\odot$	Seyfert nuclei	Galactic centre
$M = (1-10) M_\odot$	SS433	Sco X1 "radio stars"? γ-ray sources?

So there may be miniquasars (Seyferts) and nano-quasars (SS433, Sco X1 etc.); the production of jets may be a generic feature of the flow pattern around collapsed objects on all scales.

6. CONCLUSIONS

In summary, I have outlined how collimated jets can be set up close to a central black hole. However, even if the primary collimation is indeed produced in this way, the jets may experience many vicissitudes (dissipation, reconvergence, etc.) before attaining the much larger dimensions where they are observed in the radio band. The large-scale radio structures are influenced by the galactic and extragalactic environment. The energy densities are low, the speeds are probably not relativistic; but even though the physics is not extreme, a detailed understanding of large-scale source morphology may be as challenging and difficult as computational meteorology. It may turn out that the primary energy production is easier to understand: even though it may entail more "extreme" conditions - relativistic flows, black holes, etc. - the crucial processes may be quite symmetric and standardised, and thus more amenable to serious modelling.

I have benefited from helpful discussions about radio sources with many colleagues, and am particularly indebted to my collaborators, Mitch Begelman, Roger Blandford and Sterl Phinney.

REFERENCES

Abramowicz, M.A., Calvani, M., and Nobile, L.: 1980, Ap.J. 242, p.772.
Angel, J.R.P., and Stockman, H.S.: 1980, Ann.Rev.Astr.Astrophys. 18, p,321.
Bardeen, J.M., and Petterson, J.A.: 1975, Ap.J.(Lett), 195,L65.
Benford, G.: 1979, M.N.R.A.S., 183, p.29.
Blandford, R.D., and Icke, V.: 1978, M.N.R.A.S., 185, p.527.
Blandford, R.D., and Rees, M.J.: 1974, M.N.R.A.S., 169, p.395.
Burbidge, E.M., Smith, H.E., and Burbidge, G.R.: 1975, Ap.J.(Lett), 199, L137.
Chan, K.L., and Henriksen, R.N.: 1980, Ap.J., 241, p.534.
Jaroszynski, M., Abramowicz, M.A., and Paczynski, B.: 1980, Acta Astron. 30, p.1.
Miley, G.K., Heckman, T.M., Butcher, H.R., and Van Breugel, W.J.M.: 1981, Ap.J. (Lett), 247, L5.
Norman, M.L., Smarr, L., Wilson, J.R., and Smith, M.D.: 1981, Ap.J., 247, p.52.
Pottash, R.I., and Wardle, J.F.C.: 1979, Astron.J., 84, p.707.
Rees, M.J.: 1978, Nature, 275, p.516.
Rees, M.J., Begelman, M.C., Blandford, R.D., and Phinney, E.S.: 1981, Nature (in press).
Saunders, R. *et al*.: 1981, M.N.R.A.S. (in press)
Sikora, M.: 1981, M.N.R.A.S., 196, p.297.
Sikora, M., and Wilson, D.B.: 1981, M.N.R.A.S.(in press).
Van Groningen, E., Miley, G.K., and Norman, C.: 1980, Astron.Astrophys., 90, L7.
Young, P.L. *et al*.: 1978, Ap.J., 221, p.721.

DISCUSSION

LAING: There is evidence that Faraday rotation and depolarization may be caused by gas in front of the sources. This has also been suggested for the sources in which diffuse optical line emission is seen. Estimates of densities and velocities of jets which depend on polarization measurements should therefore be treated with great caution.

REES: It is certainly important to decide whether the depolarization in (e.g.) NGC 6251 comes from (a) small-scale structure in a foreground Faraday screen; (b) entrained gas in a sheath around the jet; or (c) gas pervading the body of the jet. If (a) and (b) could both be excluded, such evidence would point towards <u>sub-relativistic</u> speeds for extended jets.

BURBIDGE: I gather that you believe that relativistic particles can be generated at high efficiency near a supermassive object. Even so the total energies are assumed to be the equipartition values. This implies certain magnetic field strengths. If you do not have a natural way to get them the total energies may be higher. Also, in places where you require reacceleration, e.g., the jet in M87, can you really get this to occur? What I am really hinting at is that the energies required are much greater than equipartition values unless high efficiencies are present at all stages and equipartition arises naturally. Do you believe that these latter arguments are plausible?

REES: Our best estimates of the overall energy requirements come from the extended radio components. One knows in some cases that the jets cannot be too far from equipartition, because there is now evidence on the external confining gas pressure; similarly, one knows that the proton energy cannot overwhelm the electron pressure (c.f., the Crab Nebula, where similar arguments apply). It is true that there is no general reason for expecting equipartition. However, when the flow pattern involves systematic shear, the magnetic stresses tend to build up until they react back on the flows (i.e., the magnetic energy becomes competitive with kinetic energy). The other part of your question refers to the efficiency with which relativistic electrons can be generated. We know that they occur, with an efficiency of at least a few percent, in supernova remnants (probably via shock fronts), even though the velocities involved are only ~ 0.01 c. It is plausible to expect higher efficiencies when the velocities involved are larger (e.g., for "in situ" production in radio lobes, the "knots' of the M87 jet, and the "blobs" in superluminal sources). Near a black hole--where the bulk velocities are $\sim$ c, and powerful electromagnetic effects can be drawn upon--one would expect that the wavelength energy (up to $\sim$ 30% of rest mass) would go <u>mainly</u> into relativistic plasma. This plasma can radiate near the hole; alternatively, its relativistic internal motions may be converted, via adiabatic expansion, into bulk relativistic outflow, and efficiently transformed back into random relativistic motion at remote locations (source components, etc.). For these reasons (and also because beaming can be invoked to bring down the overall luminosity of extreme objects such as AO 0235+164), I honestly do not believe that the "energetics" of active nuclei raise fundamental problems. If the class of model I've discussed is basically wrong, the flows can only be detected when proper detailed calculations have been carried out.

VISCOUS DISSIPATION IN JETS

Mitchell C. Begelman
Astronomy Department, University of California, Berkeley
and
Institute of Astronomy, Cambridge

ABSTRACT The slow decline of surface brightness along many large-scale jets indicates that their internal pressures are determined largely by dissipation. If dissipation arises from a viscous interaction between a jet and its environment, then the observed degree of collimation enables one to constrain the nature of the viscous stress. Simple phenomenological models of the stress account for the frequently observed "gaps" and provide a means of slowing down jets without their becoming decollimated.

   The variation of radio surface brightness along multikiloparsec-scale jets is often flatter than one would expect from a simple magnetohydrodynamic model (Bridle, this volume). Corresponding values of the minimum pressure are generally comparable with or larger than estimates or upper limits for the ambient pressure; hence it is likely that relativistic electrons and magnetic fields are near equipartition, and contribute a substantial fraction of the total pressure. If one treats the equipartition pressure as proportional to the total pressure and the radio contours as indicative of the flow pattern, then one concludes that the flow is "subadiabatic", i.e., the gas within the jet becomes hotter as it expands (Begelman et al. 1982). The likeliest source of energy for this heating is the kinetic energy of the jet. This energy may be tapped by viscous stresses which carry some of the jet's momentum into the surrounding medium. Unfortunately, as with accretion disks we have little understanding of the viscosity mechanism, except to recognize that the dominant processes probably involve fluid turbulence or magnetic fields rather than "molecular" interactions. Nevertheless, there are some qualitative constraints which the dissipation mechanism must satisfy in order to be compatible with observations.

   Consider a pressure-confined fluid jet, subject to a viscous stress S and heat flux H. Schematically, the steady-state equations of motion and thermodynamics are

$$\rho v \cdot \nabla v = -\nabla p - \nabla \cdot S \qquad (1)$$

$$\frac{1}{\gamma-1} p v \cdot \nabla \ln(p/\rho^\gamma) = -\nabla \cdot H - \frac{\partial v}{\partial x} S \qquad (2)$$

where γ is the adiabatic index and $\partial v/\partial x$ represents the shear. H probably scales $\sim Sv$; the absence of limb-brightening may be evidence for the internal redistribution of heat, corresponding to a Prandtl number of order unity. Under this assumption, the requirement that dissipative heating be effective over several pressure scale heights h yet not destroy the jet implies that S must have the scaling

$$S \sim \frac{d}{h} p \qquad (3)$$

where d is the half-thickness of the jet. If S becomes much larger than this, then p/ρ^γ will exponentiate and the jet will heat and decollimate dramatically. Is S is much smaller, then the heating will be too weak to alter the adiabatic relation between p and ρ. Substituting (3) into (1) and setting $\nabla \cdot S \sim S/d$ we find that the pressure and stress terms are comparable, hence the stress has an important effect on the thrust of the jet only if the jet is marginally sonic or subsonic.

A physically suggestive way to express the scaling relation (3) is to write S in the form

$$S \sim \alpha p \qquad (4)$$

where α is a dimensionless quantity which need not be constant. Given a specific form for α, one can determine whether the jet approaches a dissipative configuration

$$\frac{d}{h} \sim \alpha \qquad (5)$$

and whether this configuration will be maintained over large distances. For simplicity, suppose that α is a constant $\ll 1$. This model is simply the "α-model" of accretion disk theory (Shakura and Sunyaev 1973). If a jet starts out from the nucleus of a galaxy with an opening angle $\lesssim 1$ radian, the effects of dissipation will not be important until the jet has been collimated to an opening angle of order 2α. In this initial zone, adiabatic losses outweigh the heating due to viscous dissipation, and if the latter is responsible for most of the particle acceleration then we should not expect to observe much radio emission from this region. We might therefore associate these regions with the "gaps" of several kiloparsecs often seen in weaker sources.

Once dissipation becomes important, the exponential sensitivity of p/ρ^γ to S in equation (2) forces the jet to track relation (3). For constant α and a power-law run of p, the jet evolves with a constant opening angle $\sim 2\alpha$. At first, the jet remains supersonic. Thrust and velocity are only weakly affected by the dissipation, but the Mach number M decreases with distance r, $\propto r^{-1} p^{-1/2}$, due to a steady increase in the internal sound speed. When $M \lesssim 1$, the thrust is affected by both the stress and the pressure gradient, but in opposite senses. Further evolution depends on the details of the dissipation process, particularly on the rate of entrainment. The Mach number may continue to decrease, or may hover near unity if the pressure forces come into balance with the viscous forces. In either case, the jet will continue to track condition (3). If the jet does not lose much heat to the ambient medium, the energy flux, $\propto pvd^2$, is approximately conserved, hence the velocity decreases $\propto p^{-1}r^{-2}$. Variations in Mach number and mass discharge are then related through the scalings $\dot{M} \propto M^2/v^2 \propto M^2 p^2 r^4$. Note that if M is constant or decreasing, there must be substantial entrainment along the jet.

Thus, a subsonic, dissipative jet can decrease its velocity and increase its mass discharge by orders of magnitude, without being decollimated. It can do so because the external pressure gradient continually transforms heat back into kinetic energy through adiabatic expansion along the jet. The principal requirements are that the stress be always smaller than the pressure, and have a feedback property which enables the jet to stably track condition (3). Other stress prescriptions besides the α-model possess this property. For example, models with $S/p \propto M^\alpha$ will track condition (3) if α is not too negative. The precise stability criterion depends on the run of p as well as initial conditions and the equation of state in the absence of dissipation. Qualitatively, the requirement is that the importance of stress relative to pressure should increase with increasing Mach number.

Despite the absence of sound theoretical grounding, simple phenomenological stress prescriptions may help to unify several observational features of jets. First, a jet evolving dissipatively in a power-law pressure distribution should have an opening angle $\sim 2\alpha$, independent of the power-law index. If the viscosity arises from turbulence and/or tangled magnetic fields, then α should differ little from source to source (Eardley and Lightman 1975, Lynden-Bell and Pringle 1974) and may help to explain the rough consistency in jet opening angles. Second, the model provides a means for smoothly decelerating and adding mass to jets without disrupting or decollimating them. This feature may aid in understanding the correlation between jet visibility and the power of the associated nuclear and extended sources (Bridle, this volume). Third, the rather sharp onset of strong dissipation predicted by phenomenological stress models can provide a natural interpretation of the observed "gaps".

REFERENCES

Begelman,M.C., Blandford,R.D. and Rees,M.J. 1982. Rev.Mod.Phys., in preparation.
Eardley,D.M. and Lightman,A.P. 1975. Ap.J., 200, pp. 187-203.
Lynden-Bell,D. and Pringle,J.E. 1974. M.N.R.A.S., 168, pp. 603-637.
Shakura,N.I. and Sunyaev, R.A. 1973. Astron.Astrophys., 24, pp. 337-355.

SIMPLE FORMULA FOR RADIO JET SURFACE BRIGHTNESS

R.N. HENRIKSEN
Institute for Plasma Research
Stanford University

Henriksen, Bridle and Chan (1981; HBC) have proposed that the energy for the synchrotron emission of radio jets is derived ultimately from the turbulent hydrodynamic eddy cascade between the Taylor wave number, k_T, and the Kolmogorov wave number. This cascade is established during the development of a turbulent jet by the shearing action of large-scale vortical entrainment of ambient material (Brown & Roshko, 1974).

Their formula for the surface brightness, $B(\nu)$, of an optically thin radio jet is

$$B(\nu) = \sqrt{1 - \bar{\omega}^2/R^2} \; e(\nu) \; A \; (Vj/R)^2 \; (dR/dz)^3 \tag{1}$$

where $R(z)$ is the 'radius' of the jet, Vj is the jet velocity, A is the mass flux in the jet (a function of R when the ambient density profile is sufficiently flat), $\bar{\omega}$ is the cylindrical distance from the jet axis (z) of the measurement, and $e(\nu)$ allows for the spectral distribution of the power in the cascade as

$$e(\nu) = (1 - \alpha) / (4\pi \nu_{max}) \; (\nu/\nu_{max})^{-\alpha} \tag{2}$$

Either the observed spectral index may be used for α, or the theoretical value of 0.75 (HBC and below) may be used.

The formula needs only $Vj(z)$ in order that the 'shape' $B_\nu(z)$ may be compared to the observation. For NGC 315 this is taken from the CH model fit to $R(z)$ published by Bridle, Chan and Henriksen (1981). The fit to the brightness data of Willis et al. (1981) is shown in Fig.1. The smooth curve is equation 1. The two sections correspond to the two regimes of expansion in NGC 315.

The self-consistent spectral index, α, is a matter for careful calculation (Eilek and Henriksen, 1982). However, HBC have shown that $\alpha = 0.75$, if their 'Lighthill' rate of resonant Alfven wave emission just balances the synchrotron losses in each energy interval. Moreover, the 'Lighthill' wave driving rate given by HBC as

$$I_a(k) \simeq \rho \left((\Delta v)^3 / R\right) \left(\Delta v / v_a\right)^{3/2} k_T^{1/2} k^{-3/2}, \tag{3}$$

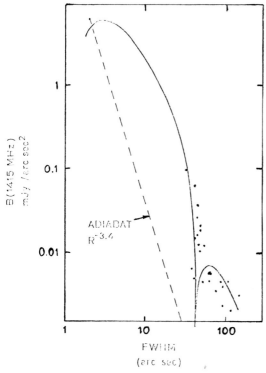

FWHM
(arc sec)

(with $\Delta v \simeq Vj(dR/dz)$, v_a the mean Alfven speed) may be shown (e.g. Eilek, 1979) to yield a steady MHD wave spectrum with energy density $Wa \propto k^{-3}$, provided that the particle spectrum is $dN \propto E^{-2.5} dE$. But when $Wa \propto k^{-3}$, Lacombe (1979) has shown that there is a $E^{-2.5}$ self-similar solution for the particle distribution with wave acceleration and synchrotron losses, of the form

$$dN = 2N_T \pi^{-1/2} x^{-2.5} e^{-1/x} Gt \, dE/c \quad (4)$$

where $x \equiv Gt E/c$; provided that the ratio of the acceleration time to the loss time is $t_A/t_S = 4.5$. Thus our scheme is self-consistent (ignoring back reaction on the hydrodynamic turbulence) if some reason may be found for the special value of t_A/t_S.

In fact, we have from HBC and Lacombe (1979) that when $Wa \propto k^{-3}$ then (γ_1 is the low energy cut-off of the relativistic particles)

$$\left. \begin{array}{c} t_A = 1/Gp \, , \quad t_S = 1/Sp \\ G \simeq (4\pi/3) \, (c/eB^3) \, (\Omega_e k_T/c\gamma_1)^{1/2} \cdot \rho \cdot \left((\Delta v)^3/R\right) (\Delta v/v_a)^{3/2} \\ S = (4/9) \, B^2 e^4/(M_e^4 c^6) \end{array} \right\} \quad (5)$$

Typical values for the main jet in NCG 315 show that indeed $t_A/t_S = O(1)$. Because t_A/t_S varies with energy when W_a does not vary as k^{-3}, no other spectral shape is stable in time. In fact there is a tendency for t_A/t_S to return to its self-similar value (Eilek and Henriksen, 1982), because of the dependence of t_A on the spectral shape.

REFERENCES
Bridle, A. H., Chan, K. L., Heinriksen, R. N. 1981, J.R.A.S.C., 75.
Brown, G. L., and Roshko, A. 1974, J. Fluid Mech., 64, 775.
Eilek, Jean A. 1979, Ap. J. 230, 373.
Eilek, Jean A., and Henriksen, R. N. 1982, this volume, p. 233.
Henriksen, R. N., Bridle, A. H., Chan, K. L. 1981, Ap. J. submitted and SUIPR report #844, Inst. for Plasma Research, Stanford University.
Lacombe, Catherine. 1979, Astronomy and Astrophysics 71. 169.
Willis, A. G., et al. 1981, Astronomy and Astrophysics, submitted.

INSTABILITIES IN PRESSURE CONFINED BEAMS AND MORPHOLOGY OF EXTENDED RADIO SOURCES

A.Ferrari, S.Massaglia, E.Trussoni and L.Zaninetti
Istituto di Cosmo-geofisica del C.N.R., Torino, Italy
Istituto di Fisica Generale dell'Universita', Torino, Italy

Several authors have suggested that radio jet morphologies resolved in extragalactic sources are the effects of large-scale Kelvin-Helmholtz instabilities in high-speed, pressure-confined fluid beams ejected from parent active galactic nuclei (Ferrari et al. 1978,1979,1981; Hardee 1979;Benford et al. 1980). In particular results from studies for cylindrical geometries indicate how to connect the "wiggles" (observed in 3C449, NGC 6251, M87 and Cen A) with helical perturbations and the "knots" (observed in NGC 315, M87, Cen Aetc.) with pinching modes. Correspondingly small scale MHD perturbations, generated by the same instability or nonlinear cascade processes, are efficient in accelerating relativistic electrons via stochastic scatterings (Lacombe 1977; Ferrari et al. 1979). This picture may satisfy both the requirements for in situ re-acceleration and the intrinsic correlation between morphology and emission.

While most of the theoretical results so far obtained have been worked out in the linear regime, application to observed morphologies requires a discussion of the level at which the instability saturates before disrupting the beams. For this we may consider the main dissipation mechanisms in collisionless plasmas: a) interactions of unstable modes with both thermal and suprathermal components of the plasma distribution function; b) nonlinear mode-mode interactions; c) formation of shocks. In the following we assume, consistently with instability calculations, that the fastest growing modes have wavenumber $k \sim 1/a$ where a is the beam radius, and their typical frquency is $\omega \sim kv_{beam}$. In the linear regime (small amplitude perturbations) instability saturates if

$$\dot{E}_{abs} > \dot{E}_{KH} \sim \frac{\delta B^2/4\pi}{\tau_{KH}} \qquad (1)$$

where $\dot{E}_{abs}$ = energy absorption rate of interaction of MHD modes with plasma particles, $\dot{E}_{KH}$ = instability energy release rate and $\tau_{KH} = (kv_s \alpha)^{-1}$ = linear time scale of instability; numerical calculations show $\alpha \simeq 0.1 \div 1$. For supersonic beams the most effective absorption process is direct Fermi- acceleration of suprathermal particles (namely synchrotron emitting electrons);in fact for flows with Mach numbers

$$M \gtrsim 4 \cdot 10^2 \frac{a}{v_{08}}, \qquad M = v_{beam}/v_s \qquad (2)$$

($v_{08} = v_{beam}/(10^8 \text{ cm/s})$) Eq (1) can be satisfied for $\delta B^2 \ll B^2$. For lower Mach numbers saturation may be provided (still in linear regime) by mode-mode interactions; both mode coalescence ($\omega_1 + \omega_2 \to \omega_3$) and decay ($\omega_1 \to \omega_2 + \omega_3$) drag energy away from the fastest growing mode ($ka \sim 1$). However decay would create long wavelength modes incompatible with geometrical conditions; at the same time for not-too-high magnetic fields

($v_A \lesssim v_s$) coalescence is generally more efficient. This transfers the instability energy to short wavelengths via a cascade process and long wavelength modes cannot grow above $\delta B^2 < B^2$. Namely Eq. (1) can be fulfilled as long as

$$v_A \gtrsim 2 v_s a \qquad (2)$$

Short wavelengths may then easily dissipate their energy either via synchrotron electron acceleration or via nonlinear Landau damping on thermal particles.

Finally for transonic flows, $M \gtrsim 1$, and for $v_A \lesssim v_s$, MHD modes cannot be stabilized in the linear regime. However the perturbation amplitude cannot exceed $\sim \lambda/2M$ ($\lambda = 2\pi/k$) both for pinching and helical modes, above which limit a shock is expected to form. Such a shock dissipates the instability energy both to thermal and suprathermal particles; acceleration of relativistic electrons prevails for densities of the order of the typical estimates in radio jets.

The occurrence of different saturation mechanisms allows correlating observed morphology with beam parameters. Namely: a) low-luminosity beams (also "invisible" beams) are expected to be characterized by highly supersonic flow speeds; b) high-brightness and well-defined beams are formed in supersonic flows in the presence of relatively high (longitudinal) magnetic fields; c) bright and high structured beams correspond to low Mach number flows modulated by shocks; d) subsonic flows are disrupted by instability close to the parent galactic nucleus.

We finally remark that the well-known jets in M87 and Cen A are not necessarily to be included in this scheme, as they appear to be not in pressure equilibrium with the external medium. Most likely their morphological structures must be ascribed to instabilities in magnetically confined beams (see Ferrari and Trussoni 1981, Coppi and Ferrari 1981 for a discussion of kink resistive instabilities).

References

Benford,G.,Ferrari,A.,and Trussoni,E.:1980, Astrophys.J.241,pp.98-110
Ferrari,A.,Trussoni,E.,and Zaninetti,L.:1978, Astron.Astrophys.64, pp.43-52
Ferrari,A.,Trussoni,E.,and Zaninetti,L.:1979, Astron. Astrophys. 79, pp.190-196
Ferrari,A.,Trussoni,E.,and Zaninetti,L.:1981,Mont.Not.Royal astr.Soc.,196, in press
Ferrari,A.,Trussoni,E.:1981,Proc.of the 15th ESLAB Symp. on X-Ray Astr.,Amsterdam.
Ferrari,A.,Massaglia,S.,Trussoni,E.:1981,Mont.Not.Royal astr.Soc,in press
Hardee,P.H.:1979,Astrophys.J.234,pp.47-55
Lacombe.C.:1977,Astron.Astrophys. 54,pp.1-16

CONNECTIONS BETWEEN TURBULENCE AND JET MORPHOLOGY

Gregory Benford
Physics Department, University of California, Irvine
Irvine, California 92717

Radio jet morphologies may be caused by large-scale Kelvin-Helmholtz (KH) instabilities. If high-speed, pressure-confined fluid beams lie at the core of these jets, they are susceptible to KH modes, as studied by a number of authors.[1-5] The "wiggles" seen, for example, in 3C449, NGC 6251, M87 and Cen A, may arise from helical instabilities.[3,4] "Knots" may be radial oscillations (not necessarily unstable) forming sausage-like regions of compression, as in NGC 315, M87, Cen A etc. There seems no reason why such macroscopic modes should not appear. The smaller scale waves, however, cannot be easily resolved by radio astronomy, and can have profound effects, such as particle reacceleration. It would be rather more satisfying if there were observable large-scale implications of the microturbulence.

As discussed in Ref. 2, we expect morphology of scale $ka \sim 1$ (with k the wavenumber and a the beam radius) when the Alfven speed $v_A < v_s \emptyset$, where v_s is the sound velocity and $0.1 < \emptyset < 1$. When $v_A > v_s \emptyset$, coalescence of modes allows a cascade of turbulence from high k to low k, so that $ka \gg 1$ morphology appears, and cyclotron resonance is the primary damping agent. The crucial problem is how to generate large scale turbulence and convey the stored energy to reaccelerated particles, without simultaneously heating the jet so that it expands drastically. I shall assume that the cascade process of Ref. 2 is efficient enough, and allows estimations of the time scale for energy transfer. The magnetic perturbations created at $k_0 = 1/a$ are weak in the sense that $[\delta B(k_0)] < B_0$, the ambient field. They are built up by the linear instability. In a time

$$t_c = \frac{a}{2\pi v_A (2-\nu)} \frac{[1 - (k_0/k)^{2-\nu}]}{[(\delta B/(k_0)/B_0]^2}$$

energy moves from k_0 to k. This time scale is dependent on the power law spectrum of the (assumed) k-spectrum, $F(k) \alpha k^{-\nu}$. Typically, $\nu \approx 1.5$. Note that t_c is insensitive to ν and k_0/k if $k_0/k \ll 1$. The central point of this paper is that this time scale may dictate some macroscopic signatures of the cascade.

Gaps. Consider a jet leaving an active nucleus. Its relativistic electrons soon exhaust their available synchrotron energy. They move to low-pitch-angle distributions and cannot radiate again until they are scattered into large pitch angles by the magnetic turbulence. This turbulence will not appear until the modes at k_o cascade to a scale $k^* > 10^6 k_o$. This will be true <u>if</u> the beam is generated without microturbulence already present at scales between k_o and k^* -- i.e., a "smooth start." Assuming this, the jet will be invisible for a distance $L \approx v_b t_c \approx 10a$ to $100a$. In principle this relation can be bounded by observations, and compared with the gaps observed in some cources, such as 3C388.

Expansion in Steps. Some jets display a step-like expansion as they move along their axis; for example, NGC 315. This can be explained by several gradients in the external pressure, but it could as well be an effect of increasing internal pressure, generated by downward cascade of magnetic energy. A possible sequence begins with a relatively "smooth" start for a beam, followed by a buildup in particle scattering after a time t_c. The beam expands under this new particle pressure, since all electrons and protons can be scattered by the waves. However, increasing radius increases t_c, slowing the transfer of energy into sidewise pressure. Thus the beam stops its expansion until new KH turbulence develops and then cascades down to $k \sim k^*$. Then expansion begins again. In some circumstances the pattern can have an on-off appearance, leading to several "steps" along z, with distance between the expansions, D, given approximately by

$$D = Ma \left(\frac{da}{\pi a} \right) \left(\frac{B_o}{\delta B} \right)^2_{k_o} \sim Ma$$

with M the Mach number; $2 \leq M \leq 20$ from theoretical work[1] so the steps are separated by distances considerably exceeding a.

Conclusion. The daunting details of turbulence theory in jets demand a formidable array of calculations to fit observations. The cascade model must be worked out to provide efficiencies, luminosities, etc. for particular sources. However, the simple physical arguments and scalings that I have made here may be testable without recourse to mammoth calculations. Systematic success would then allow us to distinguish cases in which the external medium does not dominate the gross morphology of jets.

References
1. A. Ferrari, E. Trussoni, L. Zaninetti, Astr. Ap. 64, 43 (1978); Astr. Ap. 79, 190 (1979); Mon. Not. R.A.S. 193 469 (1980).
2. G. Benford, A. Ferrari, E. Trussoni, Ap. J. 241, 98 (1980).
3. G. Benford, Ap. J. 247, 458 (1981).
4. P. E. Hardee, Ap. J. 234, 47 (1979).
5. G. Benford and G. Cavallo, Astron. and Astrophys. 93, 171 (1981).

PARTICLE ACCELERATION IN RADIO SOURCES WITH INTERNAL TURBULENCE

J.A. Eilek
Physics Department, New Mexico Tech
Socorro, N.M. 87801 U.S.A.

R.N. Henriksen
Institute for Plasma Research, Stanford University
Stanford, CA 94305 U.S.A.

Many extended radio sources seem to need in situ regeneration of the relativistic electrons. MHD turbulence generated by surface instabilities has been suggested as the reacceleration mechanism. However, Eilek (1981) has shown that short wavelength MHD waves, which are the most effective particle accelerators, are strongly damped in radio sources. This results in the turbulent region being confined to a thin layer on the edge of the source, so that particles accelerated here must propagate into the radio source if this reacceleration mechanism is to account for the internal synchrotron luminosity. Most likely the particles diffuse across tangled field lines; using numerical modelling of the turbulence, Eilek showed that the particles propagate only a small distance (10 pc - 1 kpc) in from the edge. This predicts that larger sources should appear limb brightened; but such limb brightening is rare. If short wavelength MHD waves are indeed the source of reacceleration, they must be generated internally.

In this paper we propose an attractive alternative: that the flowing plasma displays vortical hydrodynamic turbulence, and that this turbulence drives MHD waves throughout a large portion of the source. Fluid turbulence will generate MHD waves, in a process akin to the Lighthill radiation of sound waves (Kato, 1968). Henriksen, Bridle and Chan (1981) have shown that this process can account for the surface brightness distribution of some radio jets, if all of the Lighthill energy goes directly into the radiating relativistic particles. But a more detailed investigation of the microphysics is needed. Are the strength and spectrum of the MHD waves generated in this process sufficient to reaccelerate the particles in the face of synchrotron and expansion losses? What effect does this reacceleration have on the particle spectrum? In this paper we discuss these questions generally; a fuller presentation and numerical modelling will appear elsewhere (Eilek and Henriksen, 1982).

The energy balance at wave number k is

$$\frac{dW(k)}{dt} = P(k) - \sum_i \gamma_i(k)W(k) \qquad (1)$$

If $W(k)$ is the energy density in MHD waves, $P(k)$ is the driving function, and $\gamma_i(k)$ is the damping rate due to the i'th dissipation process. When the Lighthill driving is energetically small, the fluid turbulence is Kolmogorov and the driving function for Alfven waves can be shown to obey $P(k) \propto k^{-3/2}$. For times short compared to the particle response time, we set $dW(k)/dt = 0$ and solve (1) for $W(k)$. With Alfven waves, the dominant damping is usually cyclotron resonant acceleration of the relativistic electrons. If the electron momentum distribution is $f(p) \propto p^{-s}$, Kolmogorov/Lighthill driving results in an Alfven wave spectrum,

$$W_A(k) \propto k^{-(s - 3/2)}. \qquad (2)$$

Once $W(k)$ is known quantitatively, the particle acceleration rate can be found by integrating over the wavenumber spectrum. Eilek and Henriksen (1982) show that self consistent models of radio sources can be found with enough energy in the high-k waves to offset particle energy losses.

But what of the particle distribution? The acceleration time

$$t_A(p) \propto p^{-(s - 7/2)}; \qquad (3)$$

but the synchrotron lifetime

$$t_{sy}(p) \propto p^{-1}. \qquad (4)$$

Thus, an arbitrary electron distribution will tend to evolve towards a power law with $s = 4.5$. The synchrotron spectral index predicted is then $\alpha = (s - 3)/2 \sim 0.7$, which agrees with the observed trend in radio sources.

REFERENCES

Eilek, J.A., 1981, to appear in Ap. J.
Eilek, J.A. and Henriksen, R.N., 1982, in preparation.
Henriksen, R.N., Bridle, A.D., and Chan, K.L., 1981, submitted to Ap. J.
Kato, S., 1968, Publ. Ast. Soc., Japan, 20, 59.

JETS FROM DISCS AND DOUGHNUTS

P.M. Allan
Sterrewacht Leiden
Wassenaarseweg 78
Leiden
Netherlands

Various authors have suggested that there is a close connection between jets in radio galaxies and the precessing beams of SS433, and that the underlying mechanism for forming the jets is the same on both scales (Rees, 1981). We examine a possible model for generating gas jets close to the central object in either of these two cases.

Elsewhere (Allan, 1981), we have proposed that in cases where viscosity is small, accreted gas may form a doughnut around a compact object instead of the usual accretion disc. We have calculated the flow of gas off the surface of such a doughnut under the influence of the radiation from the doughnut and the gravitational attraction of the compact object, and the general result is that orbits are bent in towards the rotation axis, which causes a standing shock around the axis. The flow of gas along the axis and inside the shock is then calculated in a manner similar to the flow in an accretion column (Bondi & Hoyle, 1944).

The present situation is rather more complicated than the usual accretion column due to the fact that the velocity of the incoming gas parallel to the z (rotation) axis, $V_{//}$, and the accretion rate per unit lenght, A, are not constant, but can be represented by the formulae

$$V_{//} = V_o (1 - (z_o/z)^n)$$
$$A = A_o/(1 + (z/z_o)^2)$$

where V_o, A_o, z_o and n are constants. The equations of conservation of mass and momentum can be combined to give the equation describing the flow of gas in the jet as

$$u \frac{du}{dz} + \frac{GM}{z^2} - \frac{KL}{4\pi GMc} (\frac{r^2}{r^2+z^2})^{3/2} \frac{z}{r} = \frac{z_c+z_o}{z+z_o} \frac{u(V_{//} - u)}{z - z_c}$$

where u is the gas velocity, M is the mass of the compact object, L is the luminosity of the doughnut, K is the opacity and r is the orbital distance of the doughnut from the compact object. $z=z_c$ is a critical point in the flow and in fact it must be a stagnation point. Furthermore if the gas in the column is to be accelerated by the gas flowing in from outside then $z_o < z_c$. For reasonable values of z_c, the velocity increases to a limiting value of V_o, which is of order $(GM/r)^{1/2}$, which is typically c/3.

This model of gas jets may also be relevant to flow off accretion discs if most of the outflowing gas originates at the inner edge of the disc. If this is the case then this model should have some general validity in axisymmetric accretion systems and can provide a single explanation for jets in SS433 and radio galaxies.

REFERENCES

Allan, P.M.: 1981, Mon. Not.R. astr. Soc., in press.
Bondi, H. & Hoyle, F.: 1944, Mon. Not.R. astr. Soc. 104, p. 273.
Rees, M.J.: 1981, in 'Proceedings of the XVth ESLAB Symposium on X-Ray Astronomy, Space Sci. Rev. 30, p. 87

VORTEX ACCRETION FUNNEL / RELATIVISTIC BEAM MODELS OF DOUBLE RADIO SOURCE

H. A. Scott and R. V. E. Lovelace
Cornell University

A consistent model of extra-galactic double radio sources must evidently involve all aspects of the source from the underlying power source to the production of the radio lobes. Here, we give an overview of our work on the different aspects of a self-consistent model which includes gravitational accretion of fluid with angular momentum as the power source, the production of hydrodynamic or relativistic particle-beam jets, and the formation of expanding radio components.

Axisymmetric accretion flows of perfect fluids with angular momentum have recently been studied (Scott and Lovelace 1981a). We find that an empty vortex funnel about the symmetry axis may occur as determined by the conservation of energy and angular momentum. The funnel becomes cylindrical far from the accreting object with a width varying linearly with the specific angular momentum of the fluid.

Vortex funnels, as well as hydrodynamic jets have been found to occur in viscous incompressible flows with angular momentum (Lovelace, et al. 1981). In this case, dimensional analysis implies a self-similar form for the axisymmetric flows so that the Navier-Stokes equations reduce to a set of non-linear ordinary differential equations. These equations have been solved analytically for flows with constant specific angular momentum and numerically for more general flows. Vortex funnels and jets occur for a wide range of parameters. The jets require a source of momentum at the origin along the $\pm z$ axes, which is possibly due to radiation pressure (Davidson and McCray 1980) or to electromagnetic effects (Lovelace 1976, Lovelace, et al. 1979).

Close to a central black hole, the accreting fluid forms an accretion disc in the equatorial plane. The rotating disc and an axial magnetic field, which is "amplified" by the accretion, generate an electric field which acts to produce oppositely propagating beams of ultra-relativistic particles, and, through a cascade process, oppositely propagating electron-positron beams or jets (Lovelace, et al. 1979). The vortex funnel of the accretion flow, which traps an axial magnetic field, provides a path of egress from the central source along which particle beams can propagate freely, i.e., ballistically.

The relativistic particles stream outward, in the ± z directions, along magnetic field lines which may initially be closed. When the energy density in particles exceeds that in the magnetic field, the particles stretch the field lines, giving rise to a highly elongated field configuration as indicated in Figure 1. The beam front, where the magnetic field reverses directions, advances with a speed which can be derived from the conservation of momentum of the relativistic particles magnetic field, and ambient medium (Scott and Lovelace 1981b). Relativistically expanding radio components on the scale of parsecs result for plausible values of the parameters.

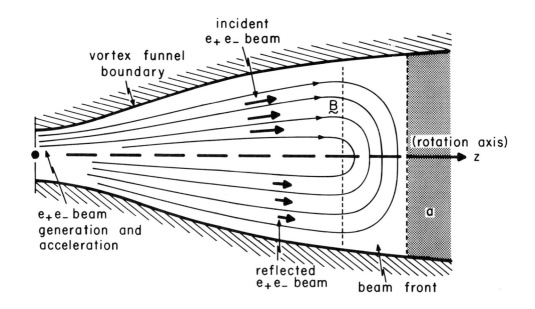

Figure 1. Schematic drawing of beam, vortex funnel geometry. The region (a) may be occupied by an ambient medium.

REFERENCES

Davidson, K. and McCray, R.: 1980, Ap. J. 241, pp. 1082-1089.
Lovelace, R.V.E. : 1976, Nature 262, pp. 649-652.
Lovelace, R.V.E., MacAuslan, J. and Burns, M.:1979, in J. Arons, C. McKee and C. Max (eds.), "Particle Acceleration Mechanisms in Astrophysics" AIP Conf. Proc. 56, pp. 399-415.
Lovelace, R.V.E., MacAuslan, J., Ruchti, C. and Scott, H.A.:1981, in preparation.
Scott, H.A. and Lovelace, R.V.E.:1981a, Ap. J., in press.
Scott, H.A. and Lovelace, R.V.E.:1981b, in preparation.

INFRARED OBSERVATIONS OF RADIO GALAXIES

George H. Rieke
University of Arizona
Tucson, Arizona 85719

I. INTRODUCTION

For technical reasons, infrared studies of active galaxies have lagged far behind optical and radio ones. This is unfortunate, since entirely new aspects of these sources are often revealed in the infrared. The extreme efficiency of dust at degrading ultraviolet photons into cool thermal emission frequently makes the luminosity of an extragalactic source inaccessible to optical and radio astronomers. At the same time, the effects of dust on optical emission line ratios and continuum shapes can be profound. The complete identification of samples of radio sources will require infrared observations to supplement the optical techniques now generally employed, and the extreme properties of the sources bright in the infrared can provide new insights to conditions in extragalactic nonthermal sources. To illustrate these points, I will discuss three cases: 1.) galaxies undergoing a powerful burst of star formation, 2.) intermediate type Seyfert galaxies, and 3.) an extreme infrared identification of an extragalactic radio source.

II. STARBURSTS

Early observations of the nuclei of apparently normal spiral galaxies revealed unexpectedly large excesses at 10 and 20μm (Kleinmann and Low 1970), suggesting that the luminosities of these regions were substantially underestimated in optical studies. This suggestion has been confirmed in more systematic infrared surveys (Rieke and Lebofsky 1978) and by extending the wavelength coverage into the far infrared, where the bulk of the luminosity is emitted (Telesco and Harper 1980; Harper, private communication).

There are a variety of indications that the observed luminosities are thermally reradiated by dust heated by hot stars produced in recent bursts of rapid star formation. Absorption and emission features associated with interstellar dust are prominent in the infrared spectra

(Willner, Soifer, and Russell 1977; Lebofsky and Rieke 1979). The infrared sources have an extent of a few hundreds of parsecs (e.g., Rieke and Low 1975; Rieke 1976; Becklin et al. 1980; Rieke et al. 1980; Telesco 1981; Telesco and Gatley 1981), similar to the size of the complex of enormous HII regions around our Galactic Center. In some cases, the infrared emission originates in a ring of HII regions centered on the galactic nucleus (Telesco 1980; Telesco and Gatley 1981), strengthening the analogy with the Galactic Center. There are indications of a correlation of infrared luminosity with the mass in interstellar clouds (Rickard, Harvey, and Thronson 1980). There is a correlation of infrared luminosity with nonthermal radio luminosity (van der Kruit 1971; Rieke 1978), as might be expected if the radio emission is produced in supernova explosions of the young massive stars.

A quantitative test of the starburst hypothesis requires a comprehensive set of radio, infrared, and optical observations. Where such data are available, it has been possible to construct consistent models that account for the radio, infrared, optical, and x-ray properties of the galaxy, within the limitations of our current knowledge of the evolution of stars and of supernova remnants. These models are most complete for NGC 253 and M82 (Rieke et al. 1980). If current estimates of the masses of these galactic nuclei are correct, the models require that a very large percentage of the available interstellar material be converted into massive stars, with a corresponding suppression of the formation of solar-mass stars; star formation in these regions must proceed substantially differently from how it does in the solar neighborhood.

Despite the plausibility of these models, there has been little direct evidence for starbursts. Direct detection of the predicted population of luminous stars has been frustrated in the optical by extinction and in the infrared by technical difficulties. Recently, we (Lebofsky and Rieke, in preparation) have obtained improved infrared spectra of M82 and other starburst galaxies which show increased strength of the 12CO and 13CO stellar absorption bands near 2.4µm, as would be expected if a significant proportion of the near infrared fluxes from these galaxies is produced by supergiants. This observation provides a direct confirmation of the starburst hypothesis and should allow a more accurate study of the star formation process in galactic nuclei.

III. DUST IN SEYFERT GALAXIES

It is generally agreed that dust influences the properties of type 2 Seyfert galaxies by heavily reddening their emission lines and continua and by shifting the bulk of their luminosity into the infrared. As a result, it may be difficult to study the underlying luminosity sources in these objects. There is even a possibility that some starburst galaxies have been mis-classified as type 2 Seyferts. For example, both NGC 253 and M82 have very bright compact nuclei (obscured toward our direction because both galaxies are nearly edge-on); they both produce luminosities of the same order as typical type 2 Seyferts; and their emission lines show that the gas clouds around their nuclei undergo substantial noncircular motions. Viewed from a distance of 20 Mpc and more nearly face-on, it might be difficult indeed to distinguish these galaxies from type 2 Seyfert galaxies.

Spectroscopic observers, particularly Osterbrock and co-workers, have called attention to intermediate-type Seyfert galaxies with permitted emission lines that have narrow cores and broad wings. As shown by their violent optical variability, many of these objects are powered by compact nonthermal sources in their nuclei. However, coordinated photometry, polarimetry, and spectroscopy of NGC 4151, the prototype of these sources, indicates that its output between 1 and 3μm is dominated by reradiation by hot dust near its nucleus, and the emission between 3 and 100μm arises from cool dust in the region producing the narrow emission line components (Rieke and Lebofsky 1981). The cool dust significantly reddens the emission lines and probably also the nonthermal continuum. The amount of cool dust required in this model is not unreasonably large. In fact, it is consistent with the general observation that below $\sim$ 900 K, about 1% of the mass in interstellar gas will condense into dust, where the mass in gas is estimated from recombination theory and the strength of the narrow hydrogen line components. Therefore, comparable effects due to reddening and reradiation should be found in other intermediate type Seyfert galaxies.

The cool dust might be detected through 1.) a steep rise of the infrared continuum from 2 to 10μm; 2.) reddening of the hydrogen recombination lines; and 3.) reddening of the optical continuum.

To apply the first of these tests, I have supplemented the data of Rieke (1978) with that of McAlary, McLaren, and Crabtree (1979) to estimate the steepness of the infrared continuum for 23 type 1 or intermediate type Seyfert galaxies. The identification of intermediate type galaxies is very dependent on the resolution and signal to noise of the optical spectrophotometry; within this sample, Osterbrock (1977) identified Mrk 6, 79, 315 and NGC 3227, 4151, and 5548 as intermediate. A higher resolution spectrum of NGC 7469 (Heckman et al. 1981) shows it also to have narrow line cores. IC 4329A, not observed by Osterbrock, appears to have narrow line components (Wilson and Penston 1979; Pastoriza 1979). I have excluded Mrk 231 becaause of its many peculiarities. So far as is known, the other members of the sample are pure type 1s. The average infrared slope, $-\alpha$, for the intermediate type galaxies is 1.57+0.17 and for the type 1s is 0.88+0.08. The errors here are the standard deviation of the mean of the slopes. Thus, the average slope for the intermediate types is more than three standard deviations steeper than for the type 1s. In fact, with the exception of NGC 5548 (type 1.5) and Mrk 279 (type 1), all of the intermediate type galaxies have slopes steeper than any of the type 1s.

To apply the second test, I excluded Mrk 231 because of its peculiarities and IC 4329A because its edge-on aspect will produce additional extinction not associated with the nucleus. I also excluded the type 1.8 and 1.9 Seyferts (Osterbrock 1981), in case their exceedingly steep Balmer decrements have some origin other than reddening. The remaining intermediate type galaxies with available spectrophotometry (primarily by Osterbrock 1977 and Koski 1978) have an average $\langle H\gamma/H\beta \rangle = 0.33 \pm 0.022$, while the type 1s have $\langle H\gamma/H\beta \rangle = 0.44 \pm 0.026$. The intermediate galaxies have steeper decrements by more than 3 standard deviations

To apply the third test, Mrk 231 and IC 4329A were again excluded. It was assumed that U-B and B-V give an estimate of the slope of the optical continuum; the photoelectric photometry was taken from the compilatiion of Weedman (1977). The average B-V and U-B for the intermediate types were 0.695 ± 0.053 and -0.41 ± 0.084 respectively; for the type 1s they were 0.53 ± 0.042 and -0.63 ± 0.044. Each color is more than two standard deviations redder for the intermediate types.

Therefore, all three tests suggest that there is significant reddening in intermediate type Seyfert galaxies. In fact, the correlations between infrared slope and Balmer decrement and continuum slope that led me to conclude that dust plays an important role in type 1 Seyfert galaxies (Rieke 1978) existed from the inclusion of the intermediate type galaxies in the sample.

To explore this hypothesis more quantitatively, I have applied the model developed for NGC 4151 (Rieke and Lebofsky 1981) to a number of intermediate type galaxies. If $M_d/M_g = g$, and the parameters for NGC 4151 are unprimed while those for the other galaxy are primed, equation (1) of Rieke and Lebofsky (1981) yields

$$g'/g = (S'/S)(d\, I\, \theta^3 / d'\, I'\, \theta'^3)^{.5} \qquad (1)$$

where S is the flux at the peak of the thermally reradiated spectrum, d is the distance to the galaxy, I is the intensity of the narrow component of $H\beta$, and θ is the angular diameter of the reradiating region. However, θ is proportional to $S^{.5}$, and we assume that d is proportional to redshift, z. Therefore,

$$g'/g = (S\, z\, I / S'\, z'\, I')^{.5} \qquad (2)$$

Table 1 shows the estimated values of g' for intermediate type galaxies. With the exception of Mrk 372, the calculated values all lie near the expected value of 1%. The steep Balmer decrements for this galaxy and also for Mrk 6, 315, and NGC 3227 indicate significantly stronger reddening than for NGC 4151. If the estimates of I/I' are corrected for the additional reddening in these cases, the final estimates of M_d/M_g are obtained. In all cases, the value is in satisfactory agreement with 1%.

Therefore, it seems likely that dust plays a very important role in intermediate type Seyfert galaxies. Whether the effects of dust are entirely sufficient to account for the extreme properties of type 1.8 and 1.9 galaxies, however, needs additional investigation since these sources are relatively little studied in the infrared.

Table 1

Name	type	S/S'	z/z'	I/I'	g'	$\frac{M_d}{M_\odot}/\frac{M_g}{M_\odot}$
Mrk. 6	1.5	9	0.19	1.6	0.01	0.003
79	1.2	8	0.15	10	0.02	0.02
315	1.5	20	0.083	26	0.04	0.02
372	1.8	100	0.11	67	0.16	0.05
NGC 3227	1.2	5	1.0	16	0.05	0.03
4151	1.5	1.00	1.00	1.00	0.006	0.006
5548	1.5	7	0.20	7	0.02	0.02
7469	1.2	2.3	0.20	3.5	0.008	0.008

The pure type 1 galaxies require relatively little gas in their narrow line regions to produce the observed forbidden emission lines, thus accounting for the lack of cool dust. However, a very small amount of hot dust within or near their broad line regions would still dominate their outputs in the near infrared (in the case of NGC 4151, $\sim 0.1\ M_\odot$ of hot dust was sufficient). Since the infrared frequently contributes only a small fraction of the total luminosity of these sources, the expected degree of reddening can be small. This model can be tested by accurate measurements of the shape of the optical-infrared spectrum during variations. For example, observations of III Zw 2 suggest that its infrared excess component cuts off steeply near 1 μm as would be expected if it arose through thermal reradiation by hot dust (Lebofsky and Rieke 1980).

IV. NONTHERMAL SOURCES

Not all extragalactic infrared sources radiate thermally! A particularly interesting class of nonthermal source is the objects with very steep infrared spectra that are found at the positions of exceptionally faint optical identifications of radio sources (Rieke, Lebofsky, and Kinman 1979). These sources are discriminated against in optical surveys because so little of their total luminosity emerges in the visible or blue (Rieke and Lebofsky 1980). A number of the sources have power law slopes as steep as ν^{-3}. The state-of-the-art limit for source detections at 2μm is about magnitude 19. A source of this brightness and with a power law slope of -3 would be 23rd magnitude at 9000 Å and 25th at 6000 Å. Thus, given current technology, the completest possible identifications of extragalactic source samples will require infrared observations.

These steep spectrum sources provide unique opportunities to study the processes producing the optical-IR emission of QSOs. Some of these possibilities have been explored by Bregman et al. (1981) for the source 1413+135. They found this source to be strongly variable, highly polarized, and free of emission lines, therefore establishing a close relation to BL Lac-type objects. The source lies within a distant galaxy ($z = 0.26$); because of the steepness of the nonthermal spectrum, the galaxy can be studied in the optical without worry about contamination from its luminous nonthermal nucleus.

The above-mentioned characteristics indicate that the optical-infrared flux is produced by synchrotron radiation in a very compact region. The steep synchrotron spectrum does not connect smoothly to the x-ray flux; together with the compactness of the source deduced from its variability, this characteristic implies that a second mechanism, Compton scattering, produces the x-ray flux. Application of conventional synchrotron-self-Compton models to the source then indicates that it has a large magnetic field (~ 10 gauss) and exhibits bulk relativistic motion toward the observer (Bregman et al. 1981).

The steepening of the spectrum near 5µm approaches the maximum abruptness that can be achieved by a synchrotron-emitting source. The relativistic electron spectrum must terminate sharply at a maximum energy corresponding to a Lorentz factor of ~ 300, and the magnetic field must be highly uniform in the emitting region. This extreme behavior is not unique to this source; early in its outburst in 1975, AO 0235+164 also exhibited a strongly curved spectrum with a dramatic increase in slope near 3µm and a cut-off toward shorter wavelengths nearly as sharp as is theoretically possible for synchrotron emission (Rieke et al. 1976).

If the source parameters derived for 1413+135 are typical of BL Lac type objects, a number of additional conclusions can be reached. For example, some of these objects show rotation of the polarization position angle between the optical and infrared (Rieke et al. 1977). A possible explanation was Faraday rotation in the relativistic plasma. However, with allowance for the dependence of relativistic electron density on magnetic field strength, it can be shown from the discussion of Pacholczyk (1974) that the degree of rotation goes inversely as the magnetic field strength to the power $(\gamma - 1)/2$ if the index in the power law energetic electron spectrum is γ. Thus, the large magnetic field strength in 1413+135 indicates that Faraday rotation will play a negligible role. An alternate possibility is that the position angle rotation arises when outbursts of relativistic electrons are directed into regions of differing magnetic field orientation (Impey, Brand, and Tapia 1981; Moore et al. 1981). Further support for this hypothesis is that the sources where rotation has been observed (AO 0235+164, 0735+178, OI 090.4, BL Lac; Impey, Brand, and Tapia, 1981; Rieke et al. 1977; unpublished observations) have exceptionally strong position angle variability (see, e.g., Angel and Stockman 1980), and relatively strong intensity variability. In addition, the spectra of these sources are frequently variable during outbursts (Rieke et al. 1976; O'dell et al.

1978). Thus, the wavelength dependence of polarization can arise in a natural way when a new injection of electrons with a different energy spectrum and polarization position angle adds its contribution to the pre-existing spectrum of the source.

It is a pleasure to acknowledge the assistance of M. J. Lebofsky in preparing this discussion. This work was supported by the National Science Foundation.

References

Angel, J. R. P., and Stockman, H. S. 1980, Ann. Rev. Ast. and Astrophys. 18, 321.
Becklin, E. E., Gatley, I., Matthews, K., Neugebauer, G., Sellgren, K., Werner, M. W., and Wynn-Williams, C. G. 1980, Ap. J., 236, 441.
Bregman, J. N., Lebofsky, M. J., Aller, H. D., Rieke, G. H., Aller, M. F., Hodge, P., Glassgold, A. E., and Huggins, P. J. 1981, Nature, in press.
Heckman, T. M., Miley, G. K., van Breugel, W. J. M., and Butcher, H. R. 1981, Ap. J., 247, 403.
Impey, C. D., Brand, P. W. J. L., and Tapia, S. 1981, preprint.
Kleinmann, D. E., and Low, F. J. 1970, Ap. J. (Letters),, 159, L165.
Koski, A. T. 1978, Ap. J., 223, 56.
Lebofsky, M. J., and Rieke, G. H. 1979, Ap. J., 229, 111.
Lebofsky, M. J., and Rieke, G. H. 1980, Nature, 285, 335.
McAlary, C. W., McLaren, R. A., and Crabtree, D. R. 1979, Ap. J., 234, 471.
Moore, R. L., et al. 1981, preprint.
O'Dell, S. L., Puschell, J. J., Stein, W. A., and Warner, J. W. 1978, Ap. J. Suppl., 38, 267.
Osterbrock, D. E. 1977, Ap. J., 215, 733.
Osterbrock, D. E. 1981, Ap. J., in press.
Pacholczyk, A. G. 1974, "Planets, Stars, and Nebulae Studied with Photopolarimetry," ed: Gehrels (Univ. of Ariz., Tucson, Ariz.), p. 1030.
Pastoriza, M. G. 1979, Ap. J., 234, 837.
Rickard, L. J., Harvey, P. M., and Thronson, H. 1980, IAU Symp. No. 96.
Rieke, G. H. 1976, Ap. J. (Letters), 206, L15.
Rieke, G. H. 1978, Ap. J., 226, 550.
Rieke, G. H., and Lebofsky, M. J. 1978, Ap. J. (Letters), 220, L37.
Rieke, G. H., and Lebofsky, M. J. 1980, IAU Symp. 92, p. 263.
Rieke, G. H., and Lebofsky, M. J. 1981, Ap. J., in press.
Rieke, G. H., and Low, F. J. 1975, Ap. J., 197, 17.
Rieke, G. H., Lebofsky, M. J., and Kinman, T. D. 1979, Ap. J. (Letters), 232, L151.
Rieke, G. H., Lebofsky, M. J., Kemp. J. C., Coyne, G. V., and Tapia, S. 1977, Ap. J. (Letters), 218, L37.
Rieke, G. H., Lebofsky, M. J., Thompson, R. I., Low, F. J., and Tokunaga, A. T. 1980, Ap. J., 238, 24.
Rieke, G. H., Grasdalen, G. L., Kinman, T. D., Hintzen, P., Wills, B. J., and Wills, D. 1976, Nature, 260, 754.

Telesco, C. M. 1981, preprint.
Telesco, C. M. and Gatley, I. 1981, preprint.
Telesco, C. M., and Harper, D. A. 1980, Ap. J., 235, 392.
van der Kruit, P. C. 1971, Astron. Astrophys., 15, 110.
Weedman, D. W. 1977, Ann. Rev. Ast. and Astrophys., 15, 69.
Willner, S. P., Soifer, B. T., and Russell, R. W. 1977, Ap. J. (Letters), 217, L121.

THE NATURE OF THE ENERGY SOURCE IN RADIO GALAXIES AND
ACTIVE GALACTIC NUCLEI

F. Pacini, Arcetri Astrophysical Observatory and
   University of Florence, Italy
M. Salvati, Institute of Space Astrophysics CNR
   CNR Frascati, Italy

THE ENERGY REQUIREMENTS

For more than 20 years it has been known that extragalactic radio sources contain up to 10^{60}–10^{62} ergs in the form of relativistic electrons and magnetic fields. One arrives at these figures if one assumes that the radio emission is due to the synchrotron process and the source contains an equal amount of energy in electrons and fields (Burbidge 1956). Any deviation from the postulated equipartition increases the energy required to account for the observed luminosities. Some authors believe that the real demands on the energy source may be still higher because of the probable presence of high energy protons. The ratio E_p/E_e is determined by the way in which particles gain and lose energy, and it is impossible to estimate it a priori. Observationally one has two conflicting lines of evidence: (a) in galactic cosmic rays one measures $(E_p/E_e) \simeq 10^2$; (b) in the Crab Nebula one infers $(E_p/E_e) \lesssim 1$ (otherwise the dynamical pressure of the proton gas would cause a nebular expansion much faster than observed).

A totally different way to estimate the energy requirements in active galactic nuclei consists in multiplying the luminosity across the electromagnetic spectrum (up to, say, 10^{47} erg s^{-1}) by the average lifetime deduced from statistical arguments (10^8 years?). Again one obtains yields reaching 10^{62} erg. We note that this figure is the equivalent of about 10^8 $M_\odot$ c^2.

THE SOURCE OF ENERGY

Many processes, both conventional and exotic, have been invoked in order to account for the source of energy. Only the conventional models have been worked in some detail while the exotic ones (quarks' fusion, white holes, etc.) have remained in the stage of proposals. At present, most people agree that gravity is the most likely cause of the energy supply. Indeed, its efficiency can be larger than that of nuclear fusion (roughly 40% against 1%), and objects are actually seen where gravitational energy is released in the form of high-energy particles and photons (e.g., supernovae, pulsars, strong binary X-ray sources).

There are two different classes of models based upon the conversion of gravitational energy.

(a) <u>Very dense clusters of 1-10 $M_\odot$ stars</u>. The stars are assumed to have frequent collisions deep in the galactic nucleus, or to ignite a chain of SN explosions, or perhaps to be already in the form of pulsars. The rationale of this approach is to gather in a small volume of space a collection of stellar-size gravitational engines, each capable of $\lesssim 10^{53}$ erg, and to add their individual outputs.

(b) <u>One (or, at most, a few) supermassive objects</u>. Their mass would have to be of the order of 10^8-10^9 $M_\odot$ and their radius close to the gravitational radius $R_s \simeq 10^{13}$ $(M/10^8$ $M_\odot)$ cm in order to meet the energy requirements.

In our opinion the observational evidence collected in the last several years leaves little room for models of class (a). Superimposed on the average emission, there are indeed sudden outbursts of luminosity at various wavelengths. These outbursts often involve more than 10^{53} erg, occasionally much more (Angel and Stockman 1980; Jones and Burbidge 1973; Stein <u>et al</u>. 1976). Their energetics cannot be accommodated in the framework of stellar collisions or individual SN explosions, and some kind of collective behavior should be postulated, which would change class a models into class b. On the other hand, in the case of supermassive objects with a binding energy $\simeq 10^{62}$ erg, it is hardly difficult for the ingenuity of theorists to invent mechanisms which can release suddenly the required amount of energy (e.g., instabilities in the object's magnetosphere).

The nature of the central supermassive body is unknown, but at present many researchers favor the idea that it might be a black hole, following the original suggestion by Salpeter (1964; for more recent contributions, see Rees 1980). Others have argued that the central body could be a magnetoid (Ginzburg and Ozernoy 1977) or perhaps a rotationally supported spinar (Cavaliere <u>et al</u>. 1969; Morrison 1969; Woltjer 1971). From many points of view the observational consequences of these configurations are the same and there are no critical arguments in favor of any of them. It is important to stress that all dynamical evidence for a central density enhancement in some galactic nuclei (e.g., M87) cannot distinguish black holes from other configurations, provided these are compact but not necessarily collapsed anywhere near the gravitational radius. However, <u>if</u> the inevitable fate of all mass condensations above a certain threshold is the collapse into a black hole, then black holes should be more common in galactic nuclei than their relatively short-lived precursors.

THE ENERGY SOURCE IN RADIO GALAXIES AND ACTIVE GALACTIC NUCLEI

In our opinion the nature of the central object is unclear at present and may remain unclear for some time. As we shall see in the following, it is still profitable to discuss the processes which extract energy from that object without worrying too much about its precise physical state.

ACCRETION VS ELECTRODYNAMIC MODELS

Once again, two kinds of mechanisms have been proposed for the basic engine:

(a) <u>Accretion</u>, i.e., the direct release of gravitational energy by matter (gas and/or disrupted stars) falling into the central body. The latter only acts as the source of the gravitational field and would remain invisible if no fuel were available.

(b) <u>Electrodynamic models</u>, i.e., the release of energy from a region of size R permeated with strong electromagnetic fields. This can take either the form of a genuine Poynting flux B^2R^2c in pulsar-type situations, or represent the rate $d/dt\,(B^2R^3)$ at which e.m. fields are burned-out in flare-like phenomena. In either case gravity has done its work at an earlier time by converting potential energy into fast rotation and/or strong fields.

It is well known that both (a) and (b) can work in real life: Accretion is the basic element for the activity of binary X-ray sources, while electrodynamic models are involved in the relationship between rotating neutron stars and SN remnants. In some cases the distinction between the two mechanisms could become rather artificial: For instance, on-going accretion could increase the strength of the magnetic field and deform its geometry, thus leading to reconnection and flaring.

The numbers required for accretion or for electrodynamic models are generally reasonable. Specifically, if L is the energy yield to be provided, accretion requires

$$\frac{\dot M}{M_\odot\,\mathrm{yr}^{-1}} = \frac{L}{10^{47}\,\mathrm{erg\,s}^{-1}} \qquad (1)$$

In the case of electrodynamic models

$$\frac{B}{100\,\mathrm{G}} = \left(\frac{L}{10^{47}\,\mathrm{erg\,s}^{-1}}\right)^{1/2} \frac{10^{16}\,\mathrm{cm}}{R} \qquad (2)$$

Here we shall not enter into the details of accretion. We stress, however, that the standard approach followed in connection with thermal X-ray sources cannot be adopted without major modifications. As is well known, the usual thermal disks would radiate mostly in the UV range because of their low temperature

$$T = 5 \; 10^7 \left(\frac{M_\odot}{M}\right)^{1/4} \; °K. \tag{3}$$

One is therefore led to consider alternative ways to heat the infalling plasma and/or the outflowing radiation, such as the possible occurrence of flares dissipating magnetic energy (e.g., Maraschi et al. 1979) and the comptonisation of the low-energy photons. Even so, in a thermal scenario it appears rather difficult to explain the emission of hard X-rays and of gamma rays (up to 100 MeV) as observed from 3C 273 (Swanenburg et al. 1978). The latter would require a gas at 10^{12} °K over a region of size 10^{17}-10^{18} cm, very far from the central body (Cavallo and Rees 1978). It is much more likely that in most cases there is a combination of thermal and nonthermal phenomena. Something similar occurs also in the galactic source Sco X-1 where the X-ray emission extends into the hard region and where there is definite evidence for an associated nonthermal radio emission.

In the following, we shall consider the basic features of the electrodynamic models and in particular those aspects which are geometry-independent.

THE ELECTRODYNAMIC MODELS

Various authors have considered AGN models leading to electrodynamic outflows. Some of them involve the large scale equivalent of pulsars (Cavaliere et al. 1969; Morrison 1969; Woltjer 1971); others deal with electro-optical cascades near black holes (Blandford and Znajek 1977; Thorne and Blandford, this conference); still others can be applied both to spinars and black holes (see, e.g., Lovelace 1976). The common feature is that a large region of space is envisaged where the particles can be accelerated in situ by electromagnetic fields anchored to a central supermassive object. In this way one overcomes two major well-known difficulties with the theory of extreme nonthermal environments:

(a) The very short lifetime of the electrons against radiation losses. In this case the lifetime has only a formal meaning because the particles are re-accelerated continuously by the same fields in which they radiate.

(b) The low density of thermal gas compatible with the polarization measurements, e.g., in BL Lac sources (Blandford and Rees 1978). This demands a continuous stirring of the same particles, rather than the steady injection of new high-energy electrons which would then decay into thermal ones.

In a previous paper (Pacini and Salvati 1978), we have shown that the physical conditions in the active region can be reasonably constrained by means of general considerations, without resorting to detailed modeling. Indeed, the power radiated across the e.m. spectrum should match the energy carried by the electrodynamic outflow (see eq. 2) if the efficiency of the acceleration process is close to 1. The further conversion from particle energy to photon energy is always very efficient due to the short radiated lifetime. Also, one naturally expects emission in (at least) two different frequency ranges, one associated with the synchrtron process

$$\nu_s = 10^6 \, B \, \gamma^2 \, \text{Hz} \tag{4}$$

and the other corresponding to the simultaneous self-Compton process

$$\nu_{ic} = \gamma^2 \, \nu_s \tag{5}$$

The typical parameters derived from the above equations--if one assumes $L \simeq 10^{47}$ erg s^{-1} and $R \simeq 10^{16}$ cm--are $B \simeq 10^2$ G, $\gamma \simeq 10^2$-10^3, in good agreement with recent experimental results (Bregman et al. 1981 a,b).

It is also worthwhile to compare the theoretical expectations with the observational evidence concerning the broadband distribution of the emitted radiation. The optical emission should be accompanied by inverse Compton X-rays of comparable intensity: This is because--as a consequence of eq. 2--the synchrotron photon energy density is of the same order as the energy density of the primary e.m. field. The prediction is consistent with the general observational trend (Tananbaum et al. 1979; Setti and Woltjer 1979). The inverse Compton emission would be less important than the synchrotron emission only in case of strongly anisotropic sources. In our opinion, the measured ratio $L_x/L_{opt} \simeq 1$ represents one of the most important clues as to the merits of electrodynamic models. Higher order inverse Compton emission could have a comparable intensity, but it is more likely that the output is severely affected by the reduced Klein-Nishina cross-section and by internal reabsorption processes (Cavallo and Rees (1979).

In addition to the global condition that the energy flow should be conserved, one can also consider the possibility that in the active region the energy attained by the individual particles is limited by the radiative reaction. If so, one should write

$$e \, v \, E = \frac{B^2}{4\pi} \, c \, \sigma_T \, \gamma^2 \tag{6}$$

where $E = B$ and v is the particle drive velocity along E (the instantaneous velocity is of course $= c$).

Equation 6 entails a characteristic frequency $\nu_{max} = (e^2/mc^2)(c/\sigma_T)$ $\simeq$ 50 MeV, provided that v is set equal to c on purely dimensional grounds. It is worthwhile to note that this frequency represents a natural upper limit to the spectrum emitted via synchrotron radiation in any loss-limited model. It is, however, unlikely that most particles will reach the limiting energy, and this should lead to the observed much softer spectrum.

CONCLUSIONS

Although progress in our understanding of the prime mover in active galactic nuclei has been slower than one may have hoped, we have now reasonable evidence that the energy source is associated with a supermassive configuration which has undergone gravitational collapse down to the black hole stage or its proximity. The choice between accretion and electrodynamic models cannot yet be made, but models of the latter kind seem able to account for the general energy budget and the gross spectral distribution. The parameters required for the strength of the e.m. field (such as, for instance, $B \simeq 10^2-10^3$ G over distances $10^{15}-10^{16}$ cm) are the same as those inferred from recent observations (Bregman et al. 1981 a,b). Many details of the acceleration process, radiation transport, formation of jets and connection with the large-scale activity must still be worked out and at present their analysis is at a preliminary stage.

This work has been supported partly by the National Research Council (CNR) and partly by NATO Grant 1946. One of us (F.P.) is indebted to the Institute of Astronomy, Cambridge, UK for hospitality during the writing of this paper.

REFERENCES

Angel, J.R.P., and Stockman, H.S. 1980, Ann. Rev. Astron. Astrophys. 18, 321.
Blandford, R. D., and Rees, M. J. 1978, Pittsburgh Conference on BL Lac Objects, ed. A. M. Wolfe, p. 328.
Blandford, R. D., and Znajek, R. L. 1977, M.N.R.A.S. 179, 433.
Bregman, J. N., et al. 1981a, preprint.
Bregman, J. N., et al. 1981b, preprint.
Burbidge, G. R. 1956, Astrophys. J. 124, 416.
Cavaliere, A., Pacini, F., and Setti, G. 1969, Astrophys. Lett. 4, 103.
Cavallo, G., and Rees, M. J. 1978, M.N.R.A.S. 183, 359.
Ginzburg, V. L., and Ozernoy, L. M. 1977, Astrophys. Space Sci. 50, 23.
Jones, T. W., and Burbidge, G. R. 1973, Astrophys. J. 186, 791.
Lovelace, R.V.E. 1976, Nature 262, 649.
Maraschi, L., Perola, G. C., Reina, C., and Treves, A. 1979, Astrophys. J. 230, 243.
Morrison, P. 1969, Astrophys. J. 157, L73.
Pacini, F., and Salvati, M. 1978, Astrophys. J. 225, L99.

Rees, M. J. 1980, Proceedings of the 1980 Texas Symposium on
    Relativistic Astrophysics, Baltimore, in press.
Salpeter, E. E. 1964, Astrophys. J. 140, 796.
Setti, G., and Woltjer, L. 1979, Astron. Astrophys. 76, L1.
Stein, W. A., O'Dell, S. L., and Strittmatter, P. A. 1976, Ann. Rev.
    Astron. Astrophys. 14, 173.
Swanenburg, B. N., et al. 1978, Nature 275, 298.
Tananbaum, H., et al. 1979, Astrophys. J. 234, L9.
Woltjer, L. 1971, Nuclei of Galaxies, ed. O'Connell, Pontificiae
    Academiae Scripta Varia 35, 477.

DISCUSSION

REES: To be stable against bifiburcation, fragmentation, etc., a
spinar must be pressure-supported along the rotation axis. This
pressure will come predominantly from radiation. Therefore, any spinar
must emit thermal radiation at the Eddington limit. This rules out
spinars as the power source in most double radio sources (when a large
mass is required on energy grounds, but high nuclear thermal luminosity
is not seen). Spinars are perhaps--if they ever form--no more than
short-lived processors of massive black holes.

BLACK HOLES AND THE ORIGIN OF RADIO SOURCES

Kip S. Thorne and Roger D. Blandford
California Institute of Technology

ABSTRACT

Powerful, extragalactic radio sources might be fuelled by energy release near a massive black hole. In this article we describe some relativistic effects which may be relevant to this process. We use Newtonian language so far as possible and illustrate the effects with "simple" analogies. Specifically, we describe the gravitational field near a black hole, Lens-Thirring and geodetic precession, electromagnetic energy extraction of the spin energy of a black hole and the structure of accretion tori around a black hole.

1. INTRODUCTION

Ever since 1964 massive black holes have been recognized as possible prime movers for quasars and extragalactic radio sources. With the subsequent discovery of rapid X-ray and optical variability, small scale radio jets and apparent superluminal expansion, this notion has become more appealing. However, we still lack direct observational evidence for the existence of massive black holes; and the chain of theoretical argument linking them to the observations reported at this symposium is consequently rather long. Some of these observed properties, in particular the alignment and apparent re-alignment of radio jets, may be explained by crucially general relativistic effects. It is our purpose here to describe some of these effects in a manner that is accessible to an astronomer unfamiliar with the full relativistic formalism.

2. THE GRAVITATIONAL FIELD OF A BLACK HOLE

We shall adopt the "3+1" point of view on relativistic gravity (e.g. Thorne and Macdonald 1981). In this view, which is mathematically equivalent to the usual 4-dimensional one, spacetime is split up into a curved 3-dimensional space with metric $g_{jk}(j=1,2,3)$ and a universal time t which differs from "proper time" by a gravitational redshift factor.

Because space is curved, as we move from one equatorial circle surrounding a black hole to another, the circumference increases less rapidly than Euclid would have demanded: d(circumference)/d(radius) < 2π. If we extract the equatorial plane from around the hole and embed it in Euclidean 3-space, then we obtain a trumpet-horn surface. For a nonrotating hole (and far outside a rotating hole) the equation of this surface in cylindrical coordinates ($\tilde{\omega}$, z) is $(z/m) = 4(\tilde{\omega}/2m - 1)^{\frac{1}{2}}$ where $m = GM/c^2$.

The gravitational field around the hole can be described using an electromagnetic analogy. We can decompose it into a gravitoelectric field $\vec{g}$ and a gravitomagnetic field $\vec{H}$. In the limit of weak (Newtonian) gravity, $\vec{g}$ reduces to the usual gravitational acceleration and $\vec{H}$ disappears. For systems with weak gravity and low velocity (v << c), the Einstein equations for $\vec{g}$ and $\vec{H}$ become almost identical to Maxwell's equations, and the geodesic equation for the motion of a freely falling particle becomes equivalent to the Lorentz force law:

$$\nabla \cdot \vec{g} = -4\pi G\rho, \quad \nabla \cdot \vec{H} = 0, \quad \nabla \times \vec{g} = 0, \quad \nabla \times \vec{H} = 4[-4\pi G\rho\vec{v}/c + 1/c(\partial\vec{g}/\partial t)]$$

$$(d\vec{v}/dt) = \vec{g} + (\vec{v}/c) \times \vec{H} \tag{1}$$

See, e.g., Braginsky et al. (1977), where the equations are written to second order in the source velocity rather than just first order as here. $\vec{g}$ and $\vec{H}$ can be derived from scalar and vector potentials: $\vec{g} = -\nabla\varphi$, $\vec{H} = \nabla \times \vec{\gamma}$. ($\varphi$ and $\vec{\gamma}$ can be related to the time-time and time-space components of the metric tensor by $\varphi = \frac{1}{2}(g_{00} + c^2) \gamma_j = cg_{0j}$.) Note the minus signs in the "Poisson" and "Ampere" equations expressing the attractive character of gravity and the extra factor of 4 in the equation for $\nabla \times \vec{H}$ which is related to the spin-2 nature of the gravitational field in contrast to the spin-1 character of the electromagnetic field.

From our electromagnetic experience, we can infer immediately that, at a distance r >> m where the space is nearly flat, a spinning hole will be surrounded by a radial gravitoelectric field $\vec{g} = -(GM/r^2) \hat{r}$, and a dipolar gravitomagnetic field $\vec{H} = (2G/c)[\vec{S} - 3(\vec{S} \cdot \hat{r}) \hat{r}]/r^3$. The gravitoelectric monopole moment is the hole mass M. Using the gyromagnetic ratio familiar from classical mechanics, the gravitomagnetic dipole moment is $\frac{1}{2}G\vec{S}/c$ where $\vec{S}$ is the hole's spin angular momentum. The gravitomagnetic field is then -4 times the usual dipolar field. Note that $\vec{g}$ and $\vec{H}$ communicate information to a distant observer about the two parameters M and $\vec{S}$ which characterize completely an uncharged spinning black hole.

3. RELATIVISTIC PRECESSION AND JET ALIGNMENT

A. Gravitomagnetic Precession and the Bardeen-Petterson Effect

Consider a ring of gas in orbit about a hole with spin angular momentum $\vec{S}$. Let the specific orbital angular momentum $\vec{\ell}$ be slightly misaligned with $\vec{S}$. A torque $\vec{r} \times (\vec{v} \times \vec{H})/c$ per unit mass will act upon the ring. Using the formula for the dipolar gravitomagnetic field and averaging over azimuth gives a mean torque per unit mass of $2(\vec{S} \times \vec{l})/r^3$. The ring will therefore undergo gravitomagnetic or Lens-Thirring precession with angular frequency $2\vec{S}/r^3$. Adjacent rings precess at different rates, and friction between them drives the inner rings into the equatorial plane (Bardeen and Petterson 1975; Rees 1978). The disk should become equatorial when the inflow time becomes comparable to the precession time; i.e. at a radius $r_{BP} \sim (S/S_{max})^{2/3} (v_r/v_K)^{-2/3}$ m where $S_{max} = m^2 c^3/G$ is the maximum spin angular momentum that the hole can have, v_r is the inward drift speed and v_K is the Keplerian velocity. According to this mechanism, any model that produces a jet near a black hole must align it parallel to $\vec{S}$ regardless of the initial orbital angular momentum of the accreted gas. A corollary is that the hole can change its spin direction substantially only when it has increased its mass by a fractional amount $\sim (S/S_{max})(m/r_{BP})^{\frac{1}{2}}$ (Rees 1978).

B. Spin-Orbit Coupling, Space-Curvature Precession and Jet Precession

Consider a gyroscope moving in a circular orbit with velocity $\vec{v}$ through the gravitoelectric field $\vec{g}$ of a massive body. This motion induces a gravitomagnetic field $\vec{H}' = -(\vec{v}/c) \times \vec{g}$ in the frame of the gyroscope, which in turn drives it to precess with $\vec{\Omega}_{SO} = -\vec{H}'/2c = (m/2r)\vec{\omega}$ where $\vec{\omega}$ is the orbital angular velocity. This precession is the analog of electromagnetic "spin-orbit coupling" in a hydrogen atom. By contrast with the hydrogen atom there is no "Thomas precession" to be added to $\vec{\Omega}_{SO}$ because the gyroscope is "freely falling" rather than accelerated relative to local inertial frames. There is however an additional contribution to the acceleration, with no electromagnetic analog, attributable to the space curvature. In Fig. 2a, we see an embedding diagram for the curved space around a massive body. The gyroscope is only sensitive to the local geometry so we may replace the trumpet horn by a cone tangent at the radius of the orbit. We can imagine constructing this cone from a flat piece of paper on which the gyroscope maintains a fixed direction (Fig. 2b). In one orbit the gyroscope moves from A to A' and therefore precesses through an angle $\psi = 2\pi[1 - \sqrt{1 - 2m/r}] \simeq 2\pi m/r$. The

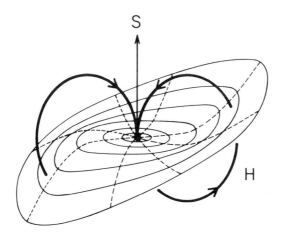

Fig. 1. The Bardeen-Petterson Effect

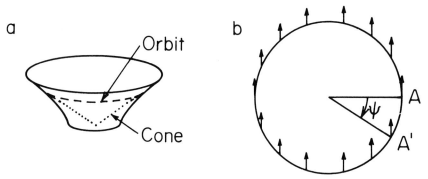

Fig. 2. Precession of a gyroscope induced by space-curvature.

space curvature precession rate is then seen to be $\vec{\Omega}_{SC} = 2\vec{\Omega}_{SO}$. The two components of precession together bear the name geodetic precession and their total precession rate is $\vec{\Omega}_{geo} = 3/2(m/r)\vec{\omega}$.

For a gyroscope the gravitomagnetic precession is smaller than the geodetic precession. However for a disk or ring, only the gravitomagnetic precession leads to an observable effect.

If our gyroscope is a spinning black hole in an orbit with radius a about another black hole of mass M, then the spin and any jet aligned with it will precess geodetically with period $2\pi/\Omega_{geo} \sim 10^4 (a/0.01\ pc)^{5/2} (M/10^8\ M_\odot)^{-3/2}$ yr. Geodetic precession may be responsible for the inversion symmetric radio sources (Begelman, Blandford and Rees 1980).

4. ELECTRODYNAMICAL EXTRACTION OF ROTATIONAL ENERGY

A. Energy Extraction from a Black Hole

The mass of a spinning black hole (as measured by its gravitational force on a distant test particle) need not increase with time. Up to 29 per cent of the mass may be extracted as usable energy by classical processes. We may think of this removable mass as "gravitomagnetic" or "rotational" energy stored outside but near the hole's horizon. A particle on a suitable orbit outside a spinning hole can exert slow-down torques on the hole (Penrose 1969). However, if this energy is to be usable, this particle must be coupled dynamically to the external world. One way of achieving this is to replace the particle by an electromagnetic field.

B. Electromagnetic Energy Extraction

Consider a rotating black hole surrounded by a magnetized accretion disk (Fig. 3). Although the magnetic field threading the disk may be rather chaotic, the field lines threading the hole and unconstrained by gas pressure should be rather well ordered: if made chaotic, they will

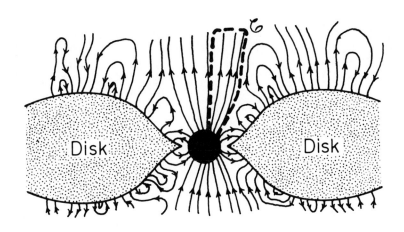

Fig. 3. Electromagnetic Extraction of Rotational Energy of a Black Hole

rearrange themselves on a timescale $\sim m/c$ so as to minimize the electromagnetic energy near the horizon while holding fixed the flux through the horizon (Macdonald and Thorne 1981). These magnetic field lines are held on the hole by Maxwell pressure from surrounding field lines, which in turn are anchored in the disk by currents. If the disk were suddenly removed, the field would convert itself into radiation and fly away in a time $\sim m/c$.

Consider a closed curve $\mathcal{C}$ at rest in the curved space. The general relativistic version of Faraday's law of induction gives for the EMF around the curve

$$\oint_{\mathcal{C}} \vec{E} \cdot \vec{d\ell} = -\frac{1}{c}\frac{d}{dt}\Phi - \oint_{\mathcal{C}} (\vec{\gamma} \times \vec{B}) \cdot \vec{d\ell} \qquad (3)$$

where $\vec{\gamma}$ is the gravitomagnetic potential and Φ is the magnetic flux bounded by $\mathcal{C}$ (Macdonald and Thorne 1981). The fields $\vec{E}$ and $\vec{B}$ should be regarded as defined by the Lorentz force law for a charge q, $d\vec{p}/dt = q(\vec{E} + \vec{v} \times \vec{B}/c)$, where t is the universal time of §2.

Now choose for the curve $\mathcal{C}$ the dashed line in Fig. 4, which connects the horizon of the hole with the "acceleration region" where charged particles can cross the field lines. This curve is a path along which current can flow. (Current which enters the horizon can be regarded as flowing along the horizon until it exits again; see Znajek 1978.) If the magnetic field is stationary and axisymmetric, the EMF is produced solely by the gravitomagnetic potential and is of magnitude EMF $\sim (S/S_{max})$ Bm $\sim 10^{20} (S/S_{max})(B/10^4 \text{ G})(M/10^8 \text{ M}_\odot)$ V. In effect the B field becomes a DC transmission line for transporting power in the form of a Poynting flux from the hole to the acceleration region. (It is, however, necessary that charges of both sign be supplied to the magnetosphere. This can be effected by cross field diffusion or more probably by a discharge process involving the production of electrons and positrons by γ-rays.) The field lines will be approximately equipotential, but both the hole and the acceleration region will possess resistances R $\sim 30\Omega$ (Znajek 1978; Damour 1978). Since the load and "battery" are impedance matched, the power transmitted to the load is

$$L \sim (\text{EMF})^2/4R \sim 10^{45}(S/S_{max})^2 (B/10^4 \text{ G})^2 (M/10^8 \text{ M}_\odot)^2 \text{ ergs}^{-1} \quad . \quad (4)$$

This power will probably be deposited ultimately in a flux of high energy particles. This constitutes a possible origin for the relativistic plasma responsible for radio jets and for other non-thermal activity associated with active galactic nuclei. Final collimation of the jets would probably occur at large distances from the hole either by external pressure or toroidal magnetic fields.

This mechanism of energy extraction by magnetic torques can also operate in an accretion disk (e.g. Blandford and Payne 1981). However, here the electromagnetic energy extraction occurs by a slightly different mechanism. Far enough from the hole, $\vec{\gamma}$ is negligible and the EMF around a stationary circuit is zero. However the gas in the disk should have a very low resistivity and thus the electric field in the instantaneous rest frame of the gas is zero. In an inertial frame there will be a potential difference $\sim \int (\vec{v}_K \times \vec{B}) \cdot d\vec{\ell}$ between field lines. This same potential difference is available in the magnetosphere for the heating and bulk acceleration of plasma. Note that a magnetic torque extracts angular momentum as well as energy from the disk, and under extreme conditions can replace the viscous torque invoked in standard accretion disk theory.

5. ACCRETION TORI

As described by Dr. Rees in these proceedings, the accretion disk surrounding a black hole may thicken close to the hole either as a result of radiation pressure when the accretion rate is supercritical, or as a result of ion pressure when it is subcritical. The gas flow is then usefully idealized as a torus with a black hole at its center. The shape of this torus is governed by the balance between gravitational and centrifugal forces, just like the shape of a whirlpool.

At the inner, equatorial edge of the disk, if Newtonian gravity were valid, force balance would say $\ell^2/r_I^3 = GM/r_I^2 +$ (inward pressure force); or $r_I \lesssim \ell^2/GM$ where ℓ is the specific angular momentum and r_I is the inner radius. Because centrifugal forces always overwhelm gravity at $r < r_I$, matter can never fall out of the Newtonian disk and into the Newtonian "black hole" at $r = 0$.

General relativity changes this in a crucial way: the gravitational force becomes infinite at a finite radius, the black hole's horizon. As a result there is a region of non-stationarity close to the hole, where centrifugal forces can no longer counterbalance gravity; and if the disk encroaches upon this region, matter will fall out of it and into the hole.

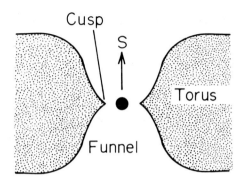

Fig. 4. Accretion torus around black hole and funnel.

This effect of relativity can be mocked up in Newtonian theory by replacing the Newtonian gravitational potential by $\phi = -mc^2/(r - 2m)$ (Paczynski and Wiita 1980). Consider the centrifugal plus gravitational energy, $\varepsilon = \ell^2/2r^2 + \phi$, for motion of a gas particle in the equatorial plane. This "effective potential for radial motion" has a minimum only if $\ell > 3\sqrt{3/2}$ mc. For larger values of ℓ, gas will trickle out of the disk and into the hole if the minimum radius of the torus is at the maximum of the potential. This maximum corresponds to a cusp in the surface of the torus. However, if $\ell > 4$mc, the energy at the cusp becomes positive and gas can escape to infinity. There is a region of non-stationarity (2m < r < 4m in the equatorial plane) within which gas must either fall into the hole or escape to infinity. For a large torus of negligible binding energy and constant specific angular momentum, force balance guarantees that $\varepsilon = \ell^2/2\tilde{\omega}^2 + \phi = $ constant $\simeq 0$ over the torus' surface, which implies that the surface has the shape of a paraboloid of revolution out of radii $\tilde{\omega} \gg m$: $z = (\tilde{\omega} - 4m)(\tilde{\omega} + 4m)/8m$ (Fig. 4). The resulting funnel may collimate any wind that blows off the inner part of the disk (e.g. Lynden-Bell 1978). However, for a radiation-driven wind, radiation drag makes it hard to achieve the high terminal Lorentz factors demanded by observations of superluminal radio sources.

6. CONCLUSION

Black holes appear to be capable of accounting for the alignment, precession, energizing and collimation of radio jets by the mechanisms outlined above. It is quite possible that virtually all of the relativistic particle energy of most of the powerful radio sources that we see be derived electromagnetically from the hole's spin--either a spin existing from birth or one built up in an early phase of high mass accretion. Such electromagnetic energy extraction would likely be accompanied by a subcritical mass accretion rate, as the modest optical outputs of powerful radio sources suggest. Supercritical accretion produces a high-luminosity, radiation-supported torus, which may explain optically brilliant but radio-quiet quasars in the case of high-mass holes associated with elliptical galaxies, and may explain Seyfert nuclei in the case of modest-mass holes in spiral galaxies.

Of course, many quite different mechanisms have been proposed for the explanation of quasars and radio sources. It is not unreasonable to expect that this diversity of theory finds a counterpart in a heterogeneity of the real objects. We are certainly a long way from a direct proof of the presence of a black hole in any extragalactic radio source, and we must once again turn to our observational colleagues for further enlightenment.

REFERENCES

Bardeen, J.M., Petterson, J.A.:1975, *Astrophys. J. Letters* 195, pp. 65-67.
Begelman, M.C., Blandford, R.D. and Rees, M.J.:1980, *Nature* 287, pp. 307-309.
Blandford, R.D. and Payne, D.G.:1981, *Monthly Notices Roy. Astron. Soc.* (submitted).
Braginsky, V.B., Caves, C.M. and Thorne, K.S.:1977, *Phys. Rev.* D15, pp. 2047-2068.
Damour, T.:1978, *Phys. Rev.* D18, pp. 3598-3604.
Lynden-Bell, D.:1978, *Phys. Scripta* 17, pp. 185-191.
Macdonald, D. and Thorne, K.S.:1981, *Monthly Notices Roy. Astron. Soc.* (in press).
Paczynski, B. and Wiita, P.J.:1980, *Astron. Astrophys.* 88, pp. 23-31.
Penrose, R.:1969, *Nuovo Cimento* 1, pp. 252-276.
Rees, M.J.:1978, *Nature* 275, pp. 516-517.
Thorne, K.S. and Macdonald, D.:1981, *Monthly Notices Roy. Astron. Soc.* (in press)
Znajek, R.L.:1978, *Monthly Notices Roy. Astron. Soc.* 185, pp. 833-840.

ACKNOWLEDGMENT

The authors thank the National Science Foundation (AST 79-22012 and AST 80-17752) for financial support.

SUPERCRITICAL ACCRETION AND ITS POSSIBLE RELATION TO
QUASARS AND RADIO SOURCES

David L. Meier
Jet Propulsion Laboratory, Pasadena, California

This paper reviews some of the properties of supercritical accretion which make it a possible model for the quasar optical central source and for producing the jet in radio sources. Some of the problems with this scenario can be remedied with alternative, but related, models.

The model supposes that material is accreting at a rate $\dot{M}$ onto a black hole of mass $M_8 = M/10^8$ $M_\odot$ with energy conversion efficiency ε and is producing luminosity $L = \varepsilon \dot{M} c^2$. "Standard" accretion theory is assumed: 1) a substantial fraction of L is converted into heat and radiated thermally and 2) the ratio of the free-fall time to the heating time $\alpha < 1$ (Shakura and Sunyaev 1973, hereinafter SS). (If magnetic dynamos and reconnection participate in the heating, then a limit on the field strength can be found by noting that the ratio of magnetic to total energy density $\alpha_m < \alpha$ [SS].) In order for ε to be a substantial fraction, the infall velocity must be small ($\sim \alpha$ times the free-fall speed) as would occur in a disk or in a spherical distribution of occasionally-colliding clouds orbiting the hole. Supercritical accretion occurs when the outward radiation pressure at a radius r due to electron scattering ($\kappa_{es} L/4\pi r^2 c$) exceeds that of gravity (GM/r^2), i.e. when $\dot{M}$ exceeds $\dot{M}_{cr} \equiv 4\pi GM/\kappa_{es} c \varepsilon = 0.22\ \varepsilon^{-1}\ M_8\ M_\odot yr^{-1}$, where $\kappa_{es} = \sigma_T m_p^{-1}$ is the electron scattering opacity. Some or most of the accreting matter is expected to be driven away from the black hole when $\dot{m} \equiv \dot{M}/\dot{M}_{cr} > 1$.

The new result we note is that standard accretion theory cannot reproduce the optical spectrum of a quasar <u>unless</u> the accretion is nearly critical or supercritical. It is well-known that quasars have "nonthermal", steep optical spectra, often suggested to be synchrotron emission, with a low energy cut-off of 10^{14} Hz (3μ) or less. This cannot occur unless the emitting region is optically thin to both free-free and synchrotron self-absorption down to this frequency. Figures 1 and 2, however, show that optical thinness of the accreting material occurs only for a limited range of α, $\dot{m}$, and M. (Calculations are taken from SS for $\dot{m} < 1$ and Meier 1981 for $\dot{m} > 1$. For $\dot{m} \ll 1$ optical thickness to free-free is a density effect and for $\dot{m} \gg 1$ it is due to the large <u>geometrical</u> thickness of the disk. Optically thinness to synchrotron occurs only when the disk is geometrically large, i.e. $\dot{m}$ or M large.) We conclude that necessary (but not sufficient) conditions to fit quasar spectra are rapid heating ($\alpha \sim 1$), nearly critical accretion ($\dot{m} > 0.1$), and a supermassive black hole ($M > 10^{5-6} M_\odot$).

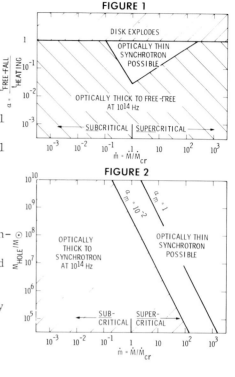

Other properties make supercritical accretion an interesting model (see Meier 1981 for details). If the optical emission is synchrotron, the lack of polarization must be explained (see, e.g., R. Moore in this symposium). The moderate optical depths to electron scattering expected here ($\tau_{es} \sim 1-10$) provide a natural de-polarization mechanism. In addition, if $\alpha \sim 1$, X-rays are produced and, if $\dot{m} > 1$ also, the ejected wind has velocities up to 0.3c, similar to quasar broad absorption and emission line widths. Finally, the geometrically thick supercritical disk has evacuated funnels along the rotation axis and hence the symmetry for collimating jets (see Abramowicz and Piran 1980, Begelman and Meier 1981, and reviews in this volume by Thorne, Rees, and Shklovsky).

One problem with this model is that constant velocity outflow does not fit the quasar absorption or emission line profiles. However, the model may provide the wind needed in the Scott, Christiansen, Weymann, and Schiano (1981) ablated cloud model for the broad absorption profiles. Another problem is that low luminosity ($< 10^{43}$ erg s^{-1}) objects with no strong optical component (e.g., Cen A, M 87) still seem to be ejecting jets. This implies that either jets can be ejected even when the accretion is subcritical or that their hole masses are below $10^5 M_\odot$. Disks thickened by gas, rather than radiation, pressure are a possible mechanism for producing subcritical jets (see Rees, this volume).

REFERENCES

Abramowicz, M. A. and Piran, T.: 1980, Astrophys. J. Letters, 241, p.L7.
Begelman, M. C. and Meier, D. L.: 1981, Astrophys. J., in press.
Meier, D. L.: 1981, Astrophys. J., in press.
Scott, J. S., Christiansen, W. A., Weymann, R. J. and Schiano, A.: 1981, in preparation.
Shakura, N. I. and Sunyaev, R. A.: 1973, Astron. Astrophys., 24, p.337 (SS).

The research described in this paper was carried out primarily at JPL and was sponsored by NASA contract NAS 7-100.

GALACTIC CENTERS AND TWIN-JETS

Wolfgang Kundt
Institut für Astrophysik der Universität Bonn
Auf dem Hügel 71
53 Bonn 1
West Germany

Abstract. A schematic model is suggested to describe the various activities in the centers of massive galaxies. Basic assumptions are that (i) a uniform model can account for all the observed phenomena, such as quasars, blazars, radio galaxies, Seyferts, and the centers of normal galaxies (with disks), (ii) activity in galactic centers is repetitive, and (iii) in-situ-acceleration of highly relativistic electrons (other than by adiabatic compression) has an insignificant efficiency, i.e. is ignorable.

The central engine. At this meeting, Pacini has summarized the spinar model; see also [Morrison (1969)]. Its difficulties are that (i) after one active cycle, it collapses to a black hole, i.e. it cannot describe repetitive behaviour, (ii) the frozen-in transverse magnetic dipole moment would give rise to a distinctly periodic non-thermal output on the spin time scale, of order months for the strong sources, which is not observed, and (iii) it is not clear how a spinar could give rise to the (much more extended) narrow line emission regions.

Thorne, Blandford, and Rees have discussed features of the black hole model. Its difficulties are that (i) the unresolved central mass of nearby galaxies is typically of order several 10^6 $M_\odot$ [Bisnovatyi-Kogan & Blinnikov (1976), Duncan & Wheeler (1980)] and may be smaller than several 10^3 $M_\odot$ in the case of our own Galaxy [Parijskij (1981)] whereas the central black hole of a formerly active galaxy would be expected to have grown well beyond 10^9 $M_\odot$ by now, (ii) a black hole cannot anchor a transverse magnetic moment, and (iii) the large observed mass outflow rates from galactic centers, of order $M_\odot$/yr, would ask for much larger (supercritical) inflow rates into the black hole, and hence for excessive fuelling rates.

For these reasons, I have suggested the massive core of a supermassive magnetized disk (=SMD) as the central engine [Kundt (1979), and fig. 1]: Continual mass spiral-in into the extremely fast-spinning core, through the $\lesssim 10^5$ K warm disk [Sorrell (1980)], is balanced by an unsteady massive wind which supplies the material seen in the broad and - further out - narrow emission and absorption lines. This thermal component of the wind is driven by hydrogen burning; it has a small filling factor, between 10^{-5} and 10^{-10}. Most of the radiated power $\dot{E}$ is drawn from rotation via low frequency magnetic multipole radiation, with $\dot{E}=10^{45.5}\eta\, u_{-3}^{4}$ erg/s, where $\eta \lesssim 1$ is an efficiency factor for poloidal flux generation, $u:=R_S/R$ is the inverse radius in units of the Schwarzschild radius, and $u_{-3}:=u/10^{-3}$. This power is independent of the mass in the central core; the latter determines the spin period, and refilling timescale, and may fall between 10^2 and 10^7 $M_\odot$. Discharges outside the atmosphere give rise to pair production. As in pulsars [Kundt (1981), Kundt & Krotscheck (1980)], the $e^{\pm}$-plasma is post-accelerated by the forming (strong) low frequency waves, turns the thermal component into magnetized filaments in pressure equilibrium, and is eventually focussed into two antipodal beams by magnetic pinching.

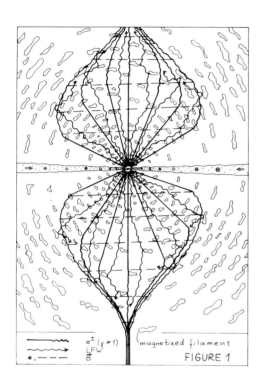

FIGURE 1

Fuelling. Geometrically, there are two preferred ways of fuelling the central engine: by spherical accretion or by accretion from a disk. A time-integrated energy output of 10^{63} erg, at a conversion efficiency ε of some 10^{-3} of the rest energy, corresponds to an average fuel demand of some 56 $M_\odot$/yr. This calls for an efficient and steady angular momentum loss mechanism. A massive gas disk stirred by large-scale turbulence can convect $\dot{M} \lesssim c_{sound} v_{turb}^2/G = 10 c_\varepsilon v_7^2$ ($M_\odot$/yr) to its center [c.f. Bailey & Clube (1979)]. Galactic-scale dust lanes have been detected in some 40 active elliptical galaxies, roughly at right angles to the radio axis; they may well originate from intergalactic accretion. Their huge mass and spin would explain the long-term steadiness, and direction stability of jets in extragalactic radio sources. Besides, accretion from a flat disk can greatly exceed the Eddington rate whereas spherical plasma accretion cannot.

Wind. Quasar emission lines reveal chemical compositions not very different from local galactic ones, i.e. we observe processed material. This suggests that we see the ashes rather than the fuel, as they are blown out from the burning core. The differences among the various types of extragalactic point sources may be largely due to different proportions of the thermal and relativistic component: a dense thermal shell gives rise to a compact, radio- and X-ray-quiet source whereas a more expanded (filamentary) shell becomes radio- and X-ray transparent and has a more uniform expansion velocity and larger volume, hence narrower and more intense emission lines.

Jet Formation. Focussing of the jets requires pressures $\dot{E}/Ac$ some 10^{10} times higher than galactic, (A=nozzle area). If these are provided by extrinsic plasma walls, as in various versions of the twin-exhaust model [c.f. Smith, Smarr, Norman & Wilson (1981)], there are the following difficulties: (i) the walls of the nozzles are in metastable dynamic equilibrium, hence must be continually replenished, (ii) the walls are hot ($T \lesssim 10^{10}$ K) and massive, hence are liable to be detected in emission, (iii) the walls could be detected through depolarization of the central radio source, (iv) the walls should filter out the (relativistic) jet component from the thermal component - which latter is observed e.g. through the forbidden lines. For these reasons, I prefer a model in which the expanding $e^{\pm}$-plasma is focussed by its frozen-in toroidal magnetic flux, i.e. by a magnetic pinch. Note that the extrinsic model needs a supermassive center to keep the scale height small whereas the intrinsic model does not.

Jet speed. Several estimates of the bulk speed of matter in the jets have been summarized by M. Rees at this meeting. These estimates are quite different in magnitude, depending on the specific assumptions made. In [Kundt & Gopal-Krishna (1980, 1981)] we have argued that the bulk Lorentz factor of the jet material is large, typically $\gamma \gtrsim 10^2$, because (i) the outward speed of the hot-spots in the outer lobes ($\gtrsim 0.1c$) must be subsonic w.r.t. the shocked material, (ii) the largest radio lobes would otherwise be intolerably old, (iii) beam bending via ram pressure poses problems for non-relativistic speeds in at least some of the sources, (iv) the frequent 1-sidedness of radio through X-ray jets, on scales between one pc and hundreds of Kpc, finds a simple explanation as forward-peaked synchrotron radiation caused by interactions with obstacles (such as channel walls), c.f. figure 2 and next section, (v) superluminal expansions are hard to explain without relativistic bulk motion, (iv) low frequency variability, absence of intergalactic scintillation, and absence of large inverse Compton contributions to the spectrum all indicate that relativistic beaming is important, (vii) the absence of boron lines from quasar spectra speaks against a significant fraction of ions in the emission volume [Baldwin et al (1977)], (viii) independent arguments have been given in [Kundt (1981)] that the corkscrew-shaped jets in the SS 433 system are beams of extremely relativistic $e^{\pm}$-plasma twisted because of a precessing injection direction.

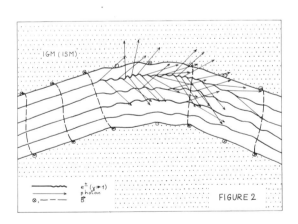

FIGURE 2

One-sidedness. Different explanations have been given for the frequent 1-sidedness of jet structures, among them: (i) isotropic radiation from expelled plasma clouds boosted in the direction of motion, (ii) alternating jet production, so that the observed 1-sidedness is intrinsic, and (iii) relativistic beaming caused by obstacles [Kundt & Gopal-Krishna (1981)]. A difficulty of interpretation (i) is that it implies implausibly small inclination angles. Difficulties of interpretation (ii) are that there are 2-sided jets (like in NGC 1265), that 1-sided structure is seen on largely different length scales (like in Cen A), and that for 1-sided emission to be sustained, the reflection symmetry of the feeding disk would have to be violated for many dynamic time scales of the central engine. Interpretation (iii), on the other hand, would explain why sometimes hot-spots are seen far removed from the outer edge of a lobe: they may be "knees", i.e. curved channel segments pointing towards us. In straight sources (e.g. Cyg A), most of the jet power is dumped in the heads of the outer lobes whereas in multiply bent sources (in particular head-tail sources), most of the power is dissipated by wall interaction. The degree of jet bending, in turn, depends on the ratio of jet power to relative velocities of the active galaxy w.r.t. its ambient medium.

References

Bailey, M.E. & Clube, S.V.M., 1978: Nature 274, 37
Bisnovatyi-Kogan, G.S. & Blinnikov, S.I., 1976: Sovj.Astron. 20, 275
Duncan, M.J. & Wheeler, J.Cr., 1980: Astrophys.J. 237, L27
Kundt, W., 1979: Astrophys. & Space Sci. 62, 335
Kundt, W., 1981a: IAU 95 'Pulsars', eds. W. Sieber and R. Wielebinski, Reidel, Dordrecht, p. 57
Kundt, W., 1981b: Rome meeting on SS 433, to appear in Vistas in Astronomy
Kundt, W. & Krotscheck, E., 1980: Astron. & Astrophys. 83, 1
Kundt, W. & Gopal-Krishna, 1980: Nature 288, 149
Kundt, W. & Gopal-Krishna, 1981: Astrophys. & Space Sci. 75, 257
Morrison, Ph., 1969: Ap.J. 157, L73
Parijskij, Y.N., 1981: private communication
Smith, M.D., Smarr, L., Norman, M.L. & Wilson, J.R., 1981: Bubbles, jets and clouds in active galactic nuclei, preprint Urbana, Illinois
Sorrell, W.H., 1981: Nature 291, 394

X-RAY AND OPTICAL OBSERVATIONS OF QUASARS

Harvey Tananbaum and Herman L. Marshall
Harvard-Smithsonian Center for Astrophysics

1. INTRODUCTION

The first Einstein observations established that quasars, as a class, are luminous X-ray emitters (Tananbaum et al. 1979). Since then, one of our major programs has been to carry out further X-ray and optical observations to improve our estimate of the contribution of quasars to the 2 keV extragalactic isotropic X-ray background and also to reconcile number counts of discrete X-ray sources with reported optical number counts of quasars (cf. Zamorani et al. 1981 and references therein). This paper is a preliminary report on our observations of the 1.7 square degree field studied originally by Braccesi and his colleagues (Formiggini et al. 1980 and Braccesi et al. 1980). A more detailed presentation of our results is in preparation (Marshall et al., 1981).

The method of selecting or counting quasars which has been used commonly is a search for "stellar" optical images with an ultraviolet excess (UVX) as determined from 2 or 3 exposures with different filters. Since emission lines present in the U-band can contribute significantly to the ultraviolet excess, this method becomes unreliable (or incomplete) for objects with redshifts greater than 2.2, when prominent lines (including Ly-α and CIV) have been redshifted out of the U-band and into the B-band.

Figure 1 is a partial compilation of counts of UVX objects obtained primarily by this method. At the brighter magnitudes - a preliminary Green and Schmidt (1978) 16^m survey and the Braccesi et al. (1970) 18.3 survey - the quasar counts are determined only after spectroscopic follow up observations have confirmed the presence of redshifted broad emission lines. This is necessary due to the contaminating presence of large numbers of stars with UV excess at relatively bright magnitudes. At magnitudes fainter than $18^m_\cdot3$ much less spectroscopic work has been done. Here the number counts show considerable scatter, some of which may be due to statistical fluctuations, but most of which is probably due to systematic effects. The solid line represents the expected source counts based upon a model which utilizes pure density evolution $\rho(z) = \rho(0)(1+z)^8$ with appropriate cutoffs at high redshifts. The steep power law with slope 2.16 fit to the source counts by Braccesi et al. (1980) requires density evolution at least as great as $(1+z)^8$. However, we note that Zamorani et al. (1981) found that such a steep power law for the optical source counts could not extend much above 20^m without encountering problems with limits set by the extragalactic 2 keV background and by the numbers of discrete X-ray sources actually observed in Einstein deep surveys. (See also Setti and Woltjer 1979, and Cavaliere et al. 1980.)

The support for a steep power law fit to the quasar counts at faint magnitudes was further reduced by the report of Bonoli et al. (1980) that a significant fraction of the UV excess objects fainter than ~$19^m\!.5$ in the 1.7 square degree Braccesi field show "extended" images, and therefore may not meet the operational definition commonly used to select quasars. The lowered estimates then generated for the source counts at $19^m\!.5$ and 20^m are also shown in Figure 1. A turnover in the source counts consistent with these data of Bonoli et al. and with Kron's (1980) limit based on a color-color analysis of stellar objects brighter than ~23.5 can be produced by models which invoke pure luminosity evolution. The dashed line in Figure 1 corresponds to such a model, with $L(z) = L(0)e^{\frac{8z}{1+z}}$.

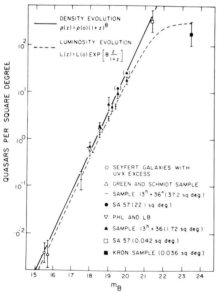

Figure 1: Optical source counts of quasar candidates selected by UV excess method. The logarithm of the number of quasar candidates per square degree brighter than a given magnitude is plotted versus magnitude.

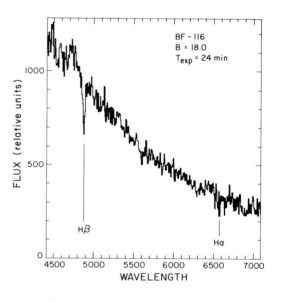

Figure 2: MMT Spectrum of BF-116 showing Hβ and Hα absorption lines.

2. OPTICAL OBSERVATIONS

The scatter of the data in Figure 1 suggests that spectroscopic observations are needed to confirm quasar candidates at relatively faint magnitudes. We have begun such a program by using the Multiple-Mirror Telescope (MMT) slit spectrograph to observe UV excess quasar candidates selected by Braccesi and his colleagues in the 1.7 square degree survey region. We ultimately will obtain a complete quasar sample to photoelectric B magnitude = 19.8; at present we have a complete spectroscopic survey containing 16 candidate objects with $B \leq 19.2^m$ (corresponding to Braccesi et al. (1980) photographic magnitude $b \simeq 19.35$).

Figure 2 shows an MMT spectrum obtained for one of the quasar candidates. The data are conspicuous by the absence of quasar emission lines; the spectrum shows $H\beta$ and $H\alpha$ absorption lines characteristic of a star. The continuum slope shown is not real, since we have not yet normalized the data for the detector response function.

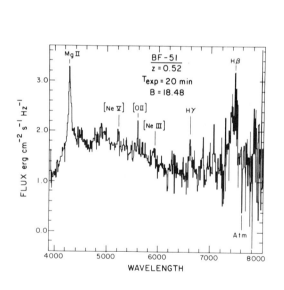
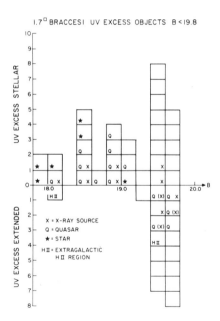

Figure 3: MMT spectrum of BF-51 showing broad, permitted emission lines from $H\gamma$, $H\beta$, and MgII as well as the more narrow forbidden lines of NeIII, OII, and NeV.

Figure 4: Histogram versus blue magnitude for UVX "stellar" objects in the top half and for UVX "extended" objects in the bottom half.

Figure 3 shows another MMT spectrum, a 20 minute exposure for the 18.5^m quasar candidate BF-51. This time we see broad, permitted emission lines from Hβ, Hγ, and MgII as well as the more narrow forbidden lines of NeIII, OII, and NeV. This rather typical quasar spectrum is fit with a redshift of 0.52.

Of the 16 objects brighter than $19\overset{m}{.}2$, we have found 6 stars and 10 quasars. The colors of the objects are such that simple modifications of the color selection criterion could not have excluded the stars and retained the quasars. The number of stars found (6) is somewhat higher than the 2 predicted from the 10-15% contamination rate estimated by Braccesi et al. (1980) for this magnitude range. Figure 4 summarizes the observations with a histogram versus blue magnitude for UVX "stellar" objects shown in the top half and for UVX "extended" objects discussed by Bonoli et al. shown in the bottom half. Objects we find spectroscopically to be stars are so indicated, while quasars are denoted by the letter "Q". For the complete sample brighter than $19\overset{m}{.}2$, we see that the contamination by stars is a problem at the brighter end. "X"´s are used to denote objects detected in our IPC X-ray observations; the figure shows that 5 of the 10 quasars in our complete sample from 18.0 to $19\overset{m}{.}2$ have been detected as X-ray sources, as have a number of fainter UV excess objects. The "Q" notation is also attached to 5 of the extended UV excess objects; the significance of this result is discussed below.

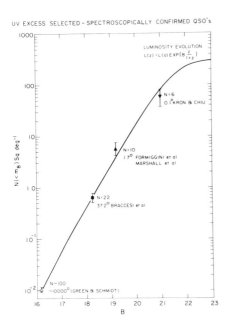

Figure 5: Number-magnitude relation for quasars selected by UVX criterion and verified spectroscopically.

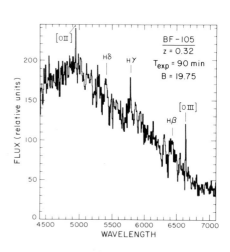

Figure 6: MMT spectrum of BF-105 showing broad emission lines from Hδ, Hγ, and Hβ and forbidden lines from OII and OIII.

Figure 5 shows the available number count data for quasars which have been verified by spectroscopic observations of candidates selected by the UV excess criterion. Data points indicate the ~100 quasars brighter than $16^m.2$ observed in ~10,000 square degrees in the Green and Schmidt (1981) survey, the 22 quasars brighter than $18^m.3$ in the 37.2 square degree Braccesi sample, the 10 quasars brighter than $19^m.2$ from the present observations of the 1.7 square degrees Braccesi field, and the 6 spectroscopically confirmed quasars brigther than 21^m in 0.1 square degrees studied by Kron and Chiu (1981). The solid line represents the quasar counts calculated from pure luminosity evolution proportional to $e^{\frac{8z}{1+z}}$, taken directly from Figure 1 without an attempt to fit or normalize to the current data. Since the model calculation does not take into account the $z \simeq 2.2$ cutoff in the UV excess selection method, the observed number of quasars so selected should be ~25% below the model curve at 21^m and ~60% below the model curve at 22^m. It should be clear that spectroscopic observations for complete samples (using all selection methods including objective prism grating surveys and X-ray surveys, as well as UV excess) in the region from 19^m to 22^m are required for determining accurate quasar counts and for evaluating various models for evolution.

The importance and the feasibility of making spectroscopic observations for faint objects is illustrated by our MMT data for BF-105 shown in Figure 6. This $19^m.75$ object was observed for 90 minutes, because it was a prime candidate for identification with one of our IPC X-ray sources, even though Bonoli et al. had described it as extended and therefore not a quasar. The spectrum is very similar to that of the quasar BF-51. Broad emission lines from $H\beta$, $H\gamma$, and $H\delta$ are indicated as are forbidden lines from OII and OIII. The redshift determined for this object is 0.32. In addition, the tentative identification of BF-105 with an X-ray source has been confirmed to an accuracy of better than 5" by a follow up observation with the Einstein HRI. Kron (1981) has carried out spectroscopic observations of several other UV excess objects described as "extended" by Bonoli et al., and finds spectra characteristic of quasars. Three of these objects are also tentatively identified with IPC X-ray sources. Veron and Veron (1981) have obtained a 3.6m plate in excellent seeing for ~1/3 of the 1.7 square degree field. They find that all of the UV excess "extended" objects in their field appear stellar and attribute the result of Bonoli et al. to the graininess of the plates used. Thus, the higher quality imaging data, the spectroscopic data, and the X-ray data demonstrate that most of these objects are in fact quasars.

3. X-RAY OBSERVATIONS

In July 1980, we used the Einstein Observatory to obtain 3 deep IPC exposures, with durations of ~40,000 seconds, to cover the 1.7 square degree Braccesi field. At least 40 X-ray sources can be detected in the field. Nine X-ray sources are identified with confirmed quasars - UV excess objects with measured redshifts. In addition, 3 UV excess objects are coincident within 5 arc seconds with HRI X-ray sources. These 3 objects are almost certainly quasars, but this still must be confirmed spectroscopically.

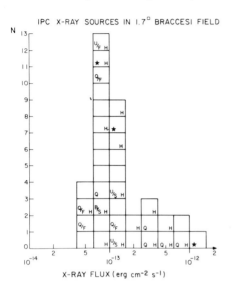

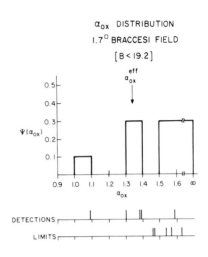

Figure 7: Histogram of 36 IPC sources than 4×10^{-14} erg s^{-1} cm^{-2} versus flux; the histogram is incomplete at the lowest X-ray fluxes.

Figure 8: Lower portion: Values of α_{ox} for X-ray detected quasars and lower limits on α_{ox} for quasars not yet detected in X-rays.
Top portion: Maximum likelihood estimate for the differential distribution $\psi(\alpha_{ox})$.

The Einstein IPC observations for the 1.7 square degree field are summarized in Figure 7. Here we show a histogram of 36 IPC sources brighter than 4×10^{-14} erg s^{-1}cm^{-2} versus flux; the histogram is incomplete at the lowest X-ray fluxes. The 9 spectroscopically confirmed quasars are indicated by "Q" and "Q/F" for "stellar" and former "extended" objects respectively. Three X-ray sources identified with stars brighter than $10^m.5$ are also indicated. An "H" labels sources that have been more accurately located in HRI follow up observations that covered portions of the 1.7 square degree field. The 3 UV excess objects confirmed as X-ray emitters via the HRI observations are noted by "U/S" and "U/F". At least 7 IPC sources have accurate HRI locations but no UV excess quasar candidates brighter than $19^m.8$ at present. Identification of these sources, as well as others detected by the IPC, is one of the objectives of our ongoing program. We expect that many of these sources will be identified with UV excess quasars fainter than 19.8, and also that some will be identified with non UV excess quasars at redshifts greater than 2.2. One such candidate is indicated by "B/S" in this figure; it is a stellar object of magnitude 19.6, with blue color but without UV excess (u-b>0), and it is coincident within 5" with the HRI location for the X-ray source.

4. CONTRIBUTION TO THE X-RAY BACKGROUND

Tananbaum et al. (1979) and Zamorani et al. (1981) have used the parameter α_{ox} plus optical source counts to estimate the contribution of quasars to the 2 keV X-ray background. α_{ox} is the power law energy index between the monochromatic optical luminosity at 2500 Å in the source frame and the monochromatic X-ray luminosity at 2 keV in the source frame. Zamorani et al. found that α_{ox} is different for radio loud and radio quiet quasars and also that it varies with redshift and/or optical luminosity. Since they did not have X-ray observations of a complete sample they computed a weighted average for α_{ox} of 1.45, to be used in estimating the contribution of quasars to the X-ray background.

We now have X-ray data for a small, but complete sample of optically selected, spectroscopically confirmed quasars in the range $18.0-19^m.2$. The lower portion of Figure 8 shows the values of α_{ox} determined for the 5 quasars detected in our IPC observations of this sample. Lower bounds on α_{ox} are shown for the 5 quasars not yet detected in X-rays. We apply the non-parametric maximum likelihood method developed by Avni et al. (1980) for using both the detections and bounds to determine the best estimate of the binned differential distribution $\psi(\alpha_{ox})$, which is shown in the top portion of Figure 8. From this distribution we determine the nominal value and uncertainty for $\alpha_{ox}^{effective}$, which corresponds to the average ratio of X-ray luminosity to optical luminosity for our sample. The data in the figure give $\alpha_{ox}^{effective}$ = 1.34, while a slightly more accurate calculation using unbinned values of α_{ox} gives $\alpha_{ox}^{effective}$ = 1.37. The 1σ uncertainty in $\alpha_{ox}^{effective}$ is ±0.10. These values for α_{ox} are consistent with those previously determined for quasars at brighter optical magnitudes.

Using $\alpha_{ox}^{effective}$ =1.37 and our observed distribution of optical source counts from $18.0-19^m.2$, we determine that these objects contribute ~17% of the 2 keV X-ray background. With X-ray data for a complete sample of optically selected quasars, a more direct calculation of the contribution to the background can be made by applying the method of Avni et al. directly to the observed X-ray fluxes and upper limits. In this way we find that the quasars from $18.0-19^m.2$ comprise 15% of the 2 keV extragalactic background, with a ±1σ range of 9% to 23%.

With number counts consistent with those predicted by the luminosity evolution model proportional to $e^{\frac{8z}{1+z}}$ and $\alpha_{ox}^{effective}$ = 1.45, Zamorani (1981) has computed that quasars brighter than 20^m contribute 18% of the 2 keV background; quasars brighter than 21^m, 40%; and quasars brighter than 22^m, 60%. We find $\alpha_{ox}^{effective}$ = 1.37 ± 0.10 for our small, complete sample, and we note that the "extended" objects reported by Bonoli et al. (1980) are in reality quasars that contribute to the X-ray background. These results suggest that the Zamorani numbers may well be conservative estimates, and reinforce our earlier conclusion that quasars make a very significant contribution to the 2 keV extragalactic background.

We thank Yoram Avni and Gianni Zamorani for many helpful discussions and comments, as well as assistance with much of the computation. This research was supported by NASA Contract NAS8-30751.

REFERENCES

Avni, Y., Soltan, A., Tananbaum, H., and Zamorani, G. 1980, Ap.J., 238, 800.
Bonoli, F., Braccesi, A., Marano, B., Merighi, R., and Zitelli, V. 1980, Astron. Astrophys., 90, L10.
Braccesi, A., Formiggini, L., and Gandolfi, E. 1970, Astron. Astrophys., 5, 264.
Braccesi, A., Zitelli, V., Bonoli, F., and Formiggini, L. 1980, Astron. Astrophys., 85, 80.
Cavaliere, A., Danese, L., DeZotti, G., and Franceschini, A. 1980 preprint.
Formiggini, L., Zitelli, V., Bonoli, F., and Braccesi, A. 1980, Astron. Astrophys. Suppl., 39, 129.
Green, R.F. and Schmidt, M. 1978, Ap.J.(Lett.), 220, L1.
Green, R.F. and Schmidt, M. 1981, private communication.
Kron, R.G. 1980, in Two Dimensional Photometry (ESO Workshop) P.O. Lindblad and H. van der Laan, eds., Geneva, p.349.
Kron, R.G. and Chiu, L.T.G. 1981, preprint.
Kron, R.G. 1981, private communication.
Marshall, H.L., et al. 1981, in preparation.
Setti, G. and Woltjer, L. 1979, Astr. and Astrophys., 76, L1.
Tananbaum, H., Avni, Y., Branduardi, G., Elvis, M., Fabbiano, G., Feigelson, E., Giacconi, R., Henry, J.P., Pye, J.P., Soltan, A., and Zamorani, G. 1979, Ap.J.(Lett.), 234, L9.
Veron, P. and Veron M.P. 1981, Astron. Astrophys., submitted.
Zamorani, G., Henry, J.P., Maccacaro, T., Tananbaum, H., Soltan, A., Avni, Y., Liebert, J., Stocke, J., Strittmatter, P.A., Weymann, R.J., Smith, M.G., and Condon, J.J. 1981, Ap.J., 245, 357.
Zamorani, G. 1981, private communication.

DISCUSSION

BARNOTHY: You mentioned that some of the UV excess objects found in the search area were not included among the quasars because of their nonstellar size. This is what one would expect if quasars were produced by gravitational lenses (J. M. Barnothy A. J. 70, 666, 1965). The effect should be particularly pronounced for high redshift objects, because in these cases a large intensification is needed to bring the apparent brightness of the object, the Seyfert galaxy nucleus, into the range of visibility. Intensification, on the other hand, is proportional to the image area, hence, very large intensification has a great propensity to produce large sizes (M. F. Barnothy & J. M. Barnothy BAAS 4, 339, 1972). I think that one should drop from the attributes of quasars the one requiring a quasar to be a "quasistellar" object, to avoid that, adhering to this characteristic, some objects which are quasars would not be included into the list of quasar candidates and hence not investigated further.

TANANBAUM: I agree that the "stellar" attribute is a requirement which may improperly exclude some quasar candidates from further study, although my concern is based primarily on the potential for excluding relatively low redshift quasar candidates associated with resolvable galaxies. In the case of the UV excess objects originally found to be extended by Bonoli et al., I have stated that further studies have been carried out demonstrating that these objects are in fact quasars with "stellar" images.

WILLS: (a) Bev Wills and I obtained some grism plates of this area earlier this year at KPNO; it will be interesting to compare these two techniques for finding QSOs. (b) Why do you classify BF 105 (with m = 19.75 and z = 0.32) as a QSO, rather than as a compact blue galaxy?

TANANBAUM: (a) I agree. (b) For BF 105 $L_{opt}^{2500Å} \approx 1.7 \times 10^{29}$ erg s^{-1} Hz^{-1}, which is above the minimum luminosity we require for a quasar.

THE MILLIARCSECOND STRUCTURE OF RADIO GALAXIES AND QUASARS

A.C.S. Readhead and T.J. Pearson
Owens Valley Radio Observatory
California Institute of Technology

Hybrid maps of the nuclei of radio galaxies and quasars show a variety of morphologies. Among compact sources, two structures are common: an asymmetric, "core-jet" morphology (eg, 3C 273), and an "equal double" morphology with two separated, similar components (eg, CTD 93). The nuclei of extended, double radio galaxies generally have a core-jet morphology with the jet directed toward one of the outer components.

1. INTRODUCTION

It is now generally accepted that one can make reliable maps with milliarcsecond resolution in VLBI by means of various hybrid mapping algorithms using closure phases and closure amplitudes [4,27,28]. Our experience with hybrid mapping and model fitting has shown that in cases where there is only a small sample of amplitudes and closure phases, or no closure phases at all, there may be significant errors in the derived structure. For this reason, we shall discuss only those objects which have been properly mapped according to the following criteria. The observations must (1) be long, continuous tracks, for reliable calibration and good (u,v)-coverage; (2) be made at four or more telescopes, for good (u,v)-coverage; and (3) include closure phases.

At the time of the IAU Symposium on Objects of High Redshift, two years ago, only a dozen objects had been mapped by VLBI [22]. The number is now about 50 (Table I). All of these objects vary, and some have now been mapped at six epochs and six frequencies, so the total number of hybrid maps is about 200. In this review, we shall discuss the morphology of these very compact objects. Table I lists the name of each object, its optical type (Quasar, Galaxy, or Unidentified), and a brief description of the milliarcsecond structure, with references to the original observations. We have been making a VLBI survey of the structure of northern sources stronger than 1.3 Jy at 5 GHz [15,16], and a number of the results we shall present are from this work. Unfortunately, as the survey is not yet complete, we do not have a

TABLE I : SOURCES MAPPED BY VLBI

0055+300	NGC315	G	0.0167	Core-jet	[8]
0133+476	OC457	Q		Slightly resolved	[15]
0212+735		Q		Core-jet?	[16]
0316+162	CTA21	U		Slightly resolved	[35]
0316+413	3C84	G	0.0177	Complex	[14,15,34]
0333+320	NRAO140	Q	1.258	Core-jet/superlum?	[10]
0355+508	NRAO150	U		Core-jet	[12]
0415+379	3C111	G	0.0485	Core-jet	[8]
0428+205		G	0.219	Slightly resolved	[20]
0429+415	3C119	Q	0.408	Core-jet?	[17]
0430+052	3C120	G	0.032	Core-jet/superlum	[2,26]
0538+498	3C147	Q	0.545	Core-jet	[33,36]
0710+439	OI417	G		Triple	[16]
0711+356	OI318	Q	1.620	Double,Core-jet?	[16]
0804+499	OJ508	Q		Slightly resolved	[16]
0814+425	OJ425	Q		Slightly resolved	[16]
0836+710		Q		Double,Core-jet?	[6,16]
0850+581	4C58.17	Q	1.322	Slightly resolved	[16]
0859+470	4C47.29	Q	1.462	Slightly resolved	[15]
0906+430	3C216	Q	0.670	Slightly resolved	[16]
0923+392	4C39.25	Q	0.698	Equal double	[15]
0945+408	4C40.24	Q	1.252	Slightly resolved	[16]
1003+351	3C236	G	0.099	Core-jet	[32]
1226+023	3C273	Q	0.158	Core-jet/superlum	[2,18,26]
1228+126	3C274	G	0.004	Core-jet	[30]
1323+321	DA344	U		Equal double	[13]
1328+307	3C286	Q	0.846	Core-jet	[17,33,35]
1518+047		U		Equal double	[20]
1607+268	CTD93	G?		Equal double	[19]
1624+416	4C41.32	U		Slightly resolved	[16]
1633+382	4C38.41	Q	1.814	Double,Core-jet?	[16]
1637+574	OS562	Q	0.745	Slightly resolved	[16]
1637+826	NGC6251	G	0.023	Core-jet	[3,23]
1641+399	3C345	Q	0.594	Core-jet/superlum	[2,26]
1642+690	4C69.21	Q		Slightly resolved	[16]
1652+398	4C39.49	G	0.0337	Slightly resolved	[16]
1807+698	3C371	G	0.050?	Core-jet	[15]
1823+568	4C56.27	Q?		Slightly resolved	[16]
1828+487	3C380	Q	0.691	Core-jet?	[15]
1845+797	3C390.3	G	0.0561	Core-jet	[8]
1901+319	3C395	G	0.635	Double,Core-jet?	[6,19]
1928+738	4C73.18	Q		Core-jet	[16]
1954+513	OV591	Q	1.230	Slightly resolved	[16]
1957+405	Cyg A	G	0.0565	Core-jet	[8]
2021+614		Q		Double,Core-jet?	[16]
2050+364		U		Equal double	[20]
2200+420	BL Lac	Q	0.069	Core-jet/superlum?	[11,15]
2251+158	3C454.3	Q	0.860	Core-jet	[5,17,35]
2351+456	4C45.51	G		Slightly resolved	[16]

well-defined, complete sample of objects, and selection effects must bias the data to some degree. Nevertheless, there are already some clear trends of great astrophysical interest.

It is well to remember that, as these objects are varying, it is in principle possible for their morphology to change quite dramatically on a time-scale of months. The rapid structural changes of the objects enable us to study the dynamical evolution. These variations have been described in detail at this Symposium by Marshall Cohen, Art Wolfe, and others, and we shall not consider them further here.

2. COMPACT RADIO SOURCES

One of the most interesting facts to emerge from the maps is that a significant fraction (perhaps as many as 50%) of the objects are one-sided jets, with a flat-spectrum core at one end of a steep-spectrum jet. Good examples are 3C 371 (Figure 1), 3C 273 and 3C 345 (Cohen and Unwin, this volume). It is perhaps misleading to use the term "jet" in this context, as the VLBI maps rarely have sufficient dynamic range to detect low-brightness structure, but the term is justified in cases like 3C 345 and NGC 6251 where the milli-arcsecond jet appears to be continuous with the larger-scale jet detected by conventional interferometry.

Observations of these one-sided jets over a range of frequencies show that the cores observed at lower frequencies can themselves be resolved into one-sided jets at higher frequencies, again with an optically thick core at one end (eg, 3C 273 [24]). Thus it appears that we are seeing continuous jets in which the core at a particular frequency is simply the region where the jet becomes optically thick at that frequency. It is therefore generally assumed that the center of activity - the "central engine" - coincides approximately with the core. The asymmetric structure itself may well be the result of relativistic beaming. This has been discussed in a number of papers [1,7,24,31], and in several contributions to this Symposium.

In some cases, objects appear at first sight to be equal doubles. However, observations at other frequencies often show that the two components have different spectra, so that the source is really asymmetric. For example, at 1.7 GHz 3C 380 looks like an equal double, but at higher frequencies it looks more like a one-sided jet [15]. Two further examples of objects that look like equal doubles at 5 GHz, but which may well turn out to be one-sided jets, because their high-frequency spectra are flat, are 0836+710 [6,16] and 2021+614 [16].

It is now clear, however, that there is a class of objects which have two equal components with very similar spectra. Phillips and Mutel [21] have observed a number of objects with spectra that peak near 1 GHz (see Figure 2) and have found that five out of six such objects are equal doubles at 1.7 GHz. Two of these (CTD 93 and

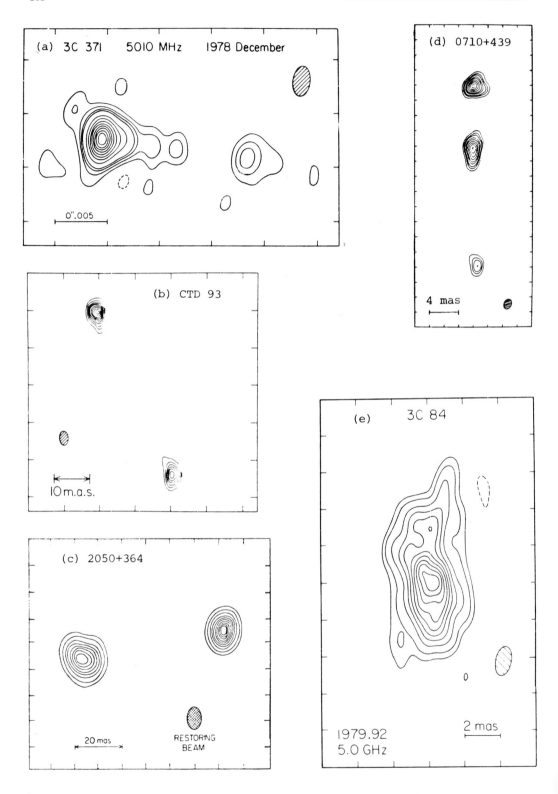

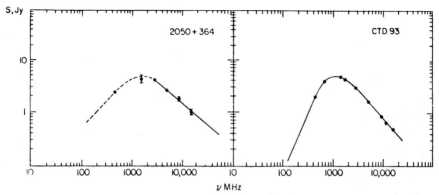

Figure 2. Radio spectra of CTD 93 [19] and 2050+364 [20].

2050+364) are shown in Figure 1. 3C 395 is rather similar to 3C 380 in that the two components have very different spectra; but the spectra of the other four objects are consistent with two very similar homogeneous synchrotron components (unlike the core-jet sources), suggesting that the components are probably nearly equal in flux density over a wide frequency range. In all four cases, at least one of the components is elongated along the source axis. In these equal double sources, there is no obvious candidate for the center of activity. Phillips and Mutel have suggested that the components straddle an invisible nucleus. They point out that the high-frequency spectra of these objects are similar to those in "classical double" sources, and they suggest that these compact doubles represent an early phase in the evolution of this class of object. A difficulty with this interpretation is that the proportion of objects showing this morphology is higher than expected: it could be as high as 15%, whereas the typical double source is expected to spend less than 1% of its lifetime at separations < 1 kpc.

A problem with present-day VLBI observations is the poor dynamic range. This is illustrated by 0710+439 (Figure 1). Here again the source is dominated by two components of almost equal brightness, one of which is extended along the source axis; but in this case there is a third component with one-tenth the flux density of the other two. If this third component had been a factor of two weaker we should not have detected it.

Some objects are much more complex. A unique example is NGC 1275 (3C 84), described at this Symposium by Jon Romney. A 5-GHz map made by Unwin et al. [34] is shown in Figure 1. It consists of a number of compact regions separated by less than a beam-width and embedded in more extended structure. The structure is difficult to interpret, especially at low frequencies where the resolution is poor and where

Figure 1 (opposite). VLBI maps of (a) 3C 371, 5 GHz [15]; (b) CTD 93, 1.7 GHz [19]; (c) 2050+364, 1.7 GHz [20]; (d) 0710+439, 5 GHz [16]; (e) 3C 84 (NGC 1275), 5 GHz [34].

the extended emission is dominant. It seems that observations at high frequencies, where the extended structure is insignificant, are needed to make sense of the structure of NGC 1275 and to determine whether there is a single center of activity or multiple centers.

There is a general tendency for the milliarcsecond structure in the compact sources to be aligned with the low-brightness outer structure when outer structure is present. The alignment is rarely perfect, however, and the objects show considerable curvature on small scales. This has been attributed to projection effects [24]. There are some cases of large misalignment: recent 23-GHz maps of 3C 345 show that the jet curves through 100 deg in the central few milliarcseconds to join on to the arcsecond-scale jet; and in three objects (3C 147 [25], 3C 395 [6], and 3C 454.3 [5]) the steep-spectrum component or jet is on the opposite side of the core from the arcsecond-scale structure.

Table II gives the numbers of objects in each morphological class, subdivided by optical identification.

3. CENTRAL COMPONENTS OF EXTENDED SYMMETRIC RADIO SOURCES

The extended symmetric double radio sources found in low-frequency surveys, like the 3C survey, are presumably selected without bias as to orientation, unlike the compact sources for which this may not be true. Most of the central components are too weak to be mapped with present VLBI networks, but a few are strong enough. So far seven central components have been mapped (those of 3C 111, 3C 236, 3C 274, 3C 390.3, NGC 315, NGC 6251, and Cyg A); they are all identified with galaxies. The VLBI maps show that the parsec-scale nuclear radio source is asymmetric in all seven cases, in spite of the high degree of overall symmetry of the outer components. In 3C 390.3, a milliarcsecond component has been detected on the side opposite the jet [9]; but the poor dynamic range of the VLBI maps is such that the presence of a "counter-jet" of one-tenth the brightness of the detected jet cannot be ruled out in the other sources. In all cases, the nuclear jet is aligned within a few degrees of the axis of the source, defined by a

TABLE II : STATISTICS OF OBJECTS MAPPED BY VLBI

	Q	G	U	Total	Fraction
Slightly resolved	10	4	2	16	33%
Core-jet	11	9	1	21	43%
Equal double	1	1	3	5	10%
Double or core-jet?	4	1	0	5	10%
Complex	0	2	0	2	4%
TOTAL	26	17	6	49	

large-scale jet or outer hot-spot. This is in marked contrast to the compact sources.

The asymmetric structure observed in these nuclei is probably not due to switching of the beam from one side to the other on a short time-scale (< 1000 years), because in all three cases where there is a large-scale jet (3C 274, NGC 315, and NGC 6251), the nuclear jet is pointing in the same direction; and the detection of weak counter-jets in NGC 315 and NGC 6251 rules out the possibility that the larger jets are switching on time-scales > 1000 years. A plausible explanation is that the material in both small-scale and large-scale jets is moving relativistically, since this allows us to reconcile the asymmetry seen in 70% of radio nuclei with the superluminal motion detected in several of them.

In a complete sample of double radio galaxies, selected without bias as to orientation, we would not expect to detect highly superluminal motion in many objects. The sources which have been mapped so far, however, are exceptional in that they have unusually strong central components. Velocities of the order of the speed of light are in principle easy to detect with VLBI observations spanning a few years. Detection of superluminal expansion in these objects would provide strong evidence that the strength of the central components is due to relativistic beaming. The detection of superluminal expansion in the quasar 3C 179, reported by Richard Porcas at this Symposium, supports this picture, as 3C 179 is a classical double source with an exceptionally strong central component.

We thank our colleagues who kindly informed us of their results before publication, especially R.B. Phillips, R.L. Mutel, and S.C. Unwin, who provided several of the figures. VLBI at OVRO is supported by the NSF (AST 79-13249).

REFERENCES

1. Blandford, R.D., and Konigl, A.: 1979, Astrophys.J., 232, pp. 34-48.
2. Cohen, M.H., Pearson, T.J., Readhead, A.C.S., Seielstad, G.A., Simon, R.S., and Walker, R.C.: 1979, Astrophys.J., 231, pp. 293-298.
3. Cohen, M.H., and Readhead, A.C.S.: 1979, Astrophys.J., 233, pp. L101-L104.
4. Cornwell, T.J., and Wilkinson, P.N.: 1981, Monthly Notices Roy. Astron. Soc., 196, pp. 1067-1086.
5. Cotton, W.D., Geldzahler, B.J., and Shapiro, I.I.: 1982, this volume, pp. 301-303.
6. Johnston, K.J., Spencer, J.H., Witzel, A., and Fomalont, E.B.: 1981, in press.
7. Kellermann, K.I., and Pauliny-Toth, I.I.K.: 1981, Ann.Rev. Astron.Astrophys., 19, in press.
8. Linfield, R.P.: 1981, Astrophys.J., 244, pp. 436-446.
9. Linfield, R.P.: 1981, in press.
10. Marscher, A.P., and Broderick, J.J.: 1981, Astrophys.J., 247, pp. L49-L52.

11. Mutel, R.L. and Philips, R.B.: 1982, this volume, p. 385.
12. Mutel, R.L. and Phillips, R.B.: 1980, Astrophys.J., 241, pp. L73-L76.
13. Mutel, R.L., Phillips, R.B., and Skuppin, R.: 1981, in press.
14. Pauliny-Toth, I.I.K., Preuss, E., Witzel, A., Graham, D., Kellermann, K.I., and Ronnang, B.O.: 1981, Astron.J., 86, pp. 371-385.
15. Pearson, T.J., and Readhead, A.C.S.: 1981, Astrophys.J., 248, pp. 61-81.
16. Pearson, T.J., and Readhead, A.C.S.: in preparation.
17. Pearson, T.J., Readhead, A.C.S., and Wilkinson, P.N.: 1980, Astrophys.J., 236, pp. 714-723.
18. Pearson, T.J., Unwin, S.C., Cohen, M.H., Linfield, R.P., Readhead, A.C.S., Seielstad, G.A., Simon, R.S., and Walker, R.C.: 1981, Nature, 290, pp. 365-368.
19. Phillips, R.B., and Mutel, R.L.: 1980, Astrophys.J., 236, pp. 89-98.
20. Phillips, R.B., and Mutel, R.L.: 1981, Astrophys.J., 244, pp. 19-26.
21. Phillips, R.B., and Mutel, R.L.: 1981, in press.
22. Readhead, A.C.S.: 1980, IAU Symp. 92, pp. 165-176.
23. Readhead, A.C.S., Cohen, M.H., and Blandford, R.D.: 1978, Nature, 272, pp. 131-134.
24. Readhead, A.C.S., Cohen, M.H., Pearson, T.J., and Wilkinson, P.N.: 1978, Nature, 276, pp. 768-771.
25. Readhead, A.C.S., Napier, P.J., and Bignell, R.C.: 1980, Astrophys.J., 237, pp. L55-L60.
26. Readhead, A.C.S., Pearson, T.J., Cohen, M.H., Ewing, M.S., and Moffet, A.T.: 1979, Astrophys.J., 231, pp. 299-306.
27. Readhead, A.C.S., Walker, R.C., Pearson, T.J., and Cohen, M.H.: 1980, Nature, 285, pp. 137-140.
28. Readhead, A.C.S., and Wilkinson, P.N.: 1978, Astrophys.J., 223, pp. 25-36.
29. Readhead, A.C.S., and Wilkinson, P.N.: 1980, Astrophys.J., 235, pp. 11-17.
30. Reid, M. et al.: 1982, this volume, p. 293.
31. Scheuer, P.A.G., and Readhead, A.C.S.: 1979, Nature, 277, pp. 182-185.
32. Schilizzi, R.T., Miley, G.K., Janssen, F.L.J., Wilkinson, P.N., Cornwell, T.J., and Fomalont, E.B.: 1981, in B. Battrick and J. Mort (eds.), "Optical Jets in Radio Galaxies", ESA SP-162, pp. 97-105.
33. Simon, R.S., Readhead, A.C.S., Moffet, A.T., Wilkinson, P.N., and Anderson, B.: 1980: Astrophys.J., 236, pp. 707-713.
34. Unwin, S.C., Mutel, R.L., Phillips, R.B., and Linfield, R.P.: 1981, Astrophys.J., in press.
35. Wilkinson, P.N., Readhead, A.C.S., Anderson, B., and Purcell, G.H.: 1979, Astrophys.J., 232, pp. 365-381.
36. Wilkinson, P.N., Readhead, A.C.S., Purcell, G.H., and Anderson, B.: 1977, Nature, 269, pp. 764-768.

DISCUSSION

EKERS: There are other classes of extragalactic radio sources which are asymmetric but which are not jets; e.g., the tail sources, and most of Perley's sample of sources dominated by a compact component. I do not think you should call any asymmetric VLBI source a jet until you have maps which demonstrate that it is jet-like.

READHEAD: There are some sources, such as NGC 6251 and 3C 273, where there is no doubt that there is a connection between the milliarcsecond "jet" and the larger scale jet. In all these sources there is a flat-spectrum core, and the "jet" component has a steeper spectrum. We have used the term "core-jet" for other sources with this morphology, though in some cases, like 3C 380, it is not clear that the description is correct, and we should perhaps simply describe these sources as asymmetric. They are marked "core-jet?" in Table I.

SCHILIZZI: The nucleus of 3C 236 is not well aligned with the outer lobes. There is a variation in position angle of about 20 degrees between the milliarcsecond and the kiloparsec scales.

READHEAD: This is a small misalignment compared with the core-dominated sources, most of which have a variation greater than 20 degrees.

LAING: The dust lane in Cyg A, which is perpendicular to the VLB jet, is seen close to edge on, suggesting that the jet is approximately in the plane of the sky. Is this consistent with the jet's asymmetry being due to Doppler beaming?

READHEAD: Owing to the poor dynamic range of VLBI maps, it is not possible to rule out a counter-jet of 1/5 of the strength of the jet. This asymmetry can be explained by relativistic beaming with $\gamma = 7$ at an inclination of 70°.

KONIGL: The two components of the compact double sources could be identified with the inner core and a detached knot in a relativistic jet, like those seen in superluminal sources. If the jet is only moderately relativistic, it could be seen at a sufficiently large angle to the axis for the core and the knot to be separately resolved and have relatively simple synchrotron spectra. This interpretation is consistent with the high measured fluxes and with the elongation of one of the components along the axis. Superluminal motions might be seen in this case. Alternatively, the emission could be from shocks behind dense clouds accelerated by the jet. The flow behind the shocks is directed sideways, so the emission is not beamed strictly along the axis of the jet, and the clouds could be seen at a large angle to the jet axis. Superluminal motions would not be expected in this case.

SHAPIRO: The existence of compact symmetric doubles is not entirely inconsistent with the presence of relativistic jets. If the jets are not purely radial outflows confined to a narrow cone, but instead represent a relativistic expansion, the surface of which is the radio source and moves outward from the center of activity in all directions, then the surface elements which move perpendicular to the collimation axis will relativistically beam their radiation at the observer, who will see only these surface elements since the rest is beamed away from the observer. An example is the relativistic blast wave model (Shapiro, P. R.: 1979, Astrophys. J., 233, 831-850).

HIGH RESOLUTION OBSERVATIONS OF THE QUASAR 3C147

E. Preuss, W. Alef, I. Pauliny-Toth
Max-Planck Institut fur Radioastronomie

K. I. Kellermann
National Radio Astronomy Observatory

We report here preliminary results of observations of the quasar 3C147 which were made at 6 cm with a resolution of about 1 milliarcsecond using a VLB interferometer system with four antennas in the USA and one in Europe. Our observations are shown in Figure 1 along with previously published maps on larger size scales. VLA observations made at 2 cm wavelength (Fig. 1a) show an extended feature lying about 0.5 arcseconds (3.5 kpc) to the northeast of a bright core (Readhead et al. 1980), while VLBI observations made at 18, 50, and 91 cm (Readhead and Wilkinson 1980, Wilkinson et al. 1977, and Simon et al. 1980) show a jet-like feature extending about 0.2 arcsec (1 kpc) in the opposite direction (Fig. 1b). The 18 cm VLB observations also indicated the presence of a smaller elongated feature extending only 3 milliarcseconds (20 pc) again toward the northeast (Fig. 1c). We have observed 3C147 in March 1978 and again in April 1981 with a resolution of 0.7 and 1.5 milliarcsec respectively. The 1981 data (Fig. 1d) clearly show the double structure of the core as well as a lower surface brightness feature which can also be seen in the 18 cm map. Our 1978 data (Fig. 1e) has better resolution and shows considerable structure in the milliarcsec component, but due to the absence of phase information in these data, the details are not reliable and the orientation is uncertain by 180°. The orientation is, however, specified by reference to the 1981 data.

Thus, on a scale of 5 milliarcsec, there is a jet-like feature which points in the same direction as a larger feature 100 times further distant. But there is also an extension pointing 0.3 arcseconds in the opposite direction! This "triply" asymmetric structure observed in 3C147 is not easy to interpret in terms of simple symmetric beaming models where the apparent asymmetric appearance is caused by the differential Doppler beaming of the approaching and receding components. This model requires that the approaching side appear stronger on all angular scales where the beaming is effective, so it may be concluded that at least to some extent there is an intrinsic asymmetry in 3C147. The observed asymmetry may reflect different external conditions affecting the radio emission from the relativistic beams or the alternation with time in the direction of a one-sided jet.

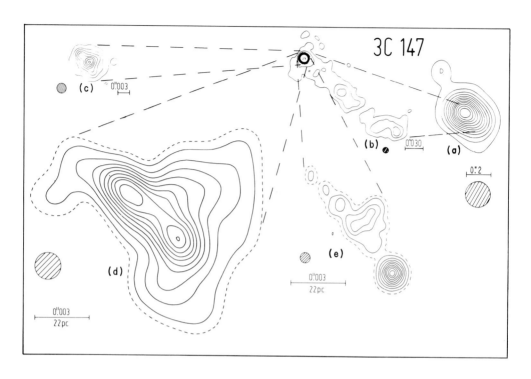

Figure 1. Structure of 3C147. HPBW is shown as shaded circles. a) Observations made with the VLA at 15 GHz. Contour resolution values are 4, 12, 20, ..., 92% of the peak value (2.1×10^5 K) (Readhead et al. 1979). b) Observations made at 1.67 GHz with 5 element VLBI networks. Contour interval 2, 6, 10, 14, ..., 114×10^6 K (Readhead and Wilkinson 1980). c) High resolution VLBI observations at 1.67 GHz. Contours 4, 12, 20, ..., 60×10^9 K (Readhead and Wilkinson). d) 5 GHz observations. Contour levels 10, 20, 30, ..., 100×10^8 K. e) 5 GHz high resolution observations, based on amplitude data only. Contour levels 25, 50, 75, ..., 250×10^8 K.

We thank B. Alef, R. W. Porcas, J. D. Romney, H. Blaschke, and W. Stursberg for help with the observations and analysis. VLBI research at Haystack, NRAO, Fort Davis, and Owens Valley Observatories is supported by the National Science Foundation.

References

Preuss, E., et al.: 1980, Ap. J. 240, L7.
Readhead, A. C. S. and Wilkinson, P. N.: 1980, Ap. J. 235, 11.
Readhead, A. C. S., et al.: 1980, Ap. J. 237, L55.
Simon, R. S. et al.: 1980, Ap. J. 236, 707.
Wilkinson, P. N., et al.: 1977, Nature 269, 764.

STRUCTURAL EVOLUTION IN THE NUCLEUS OF NGC1275

J.D. Romney, W. Alef, I.I.K. Pauliny-Toth, and E. Preuss
Max-Planck-Institut für Radioastronomie, Bonn, FRG

K.I. Kellermann
National Radio Astronomy Observatory, Green Bank, USA

The extremely powerful compact radio nucleus of NGC1275 is perhaps the most complex structure seen at milliarcsecond scales. Early attempts to determine its structure (Schilizzi et al. 1975) were unsuccessful, although it was evident that the structure consisted of several bright regions whose relative positions remained fixed while their intensities varied. Later observations and analysis (Pauliny-Toth et al. 1976; Preuss et al. 1979) confirmed this inference and revealed a slow expansion in a direction transverse to the primary axis of alignment.

We report here recent observations which manifest a new structural development. These measurements, performed at 2.8cm wavelength with VLBI arrays of seven stations (epoch 1979.1) and five stations (1981.1) in North America and Europe, yielded hybrid maps which we present in Fig.1 together with the models derived from earlier observations. The decline, since 1976, of the formerly dominant central feature is apparent, as is the steady growth of the northern component. The northeastwards extension of this latter component in the new maps has not been observed previously, and deviates from the alignment in position angle $\sim -9°$ which otherwise prevails over a wide range of angular scales in this source.

Of particular interest is the rapid increase in the separation between the central and southern components since 1976. Considering only the two most recent epochs where the separation is most accurately and unambiguously determined, we measure an angular motion of $0.34 \pm .04$ milliarcsec y^{-1}, which corresponds to an (apparent) linear velocity of $0.58c$ ($H_0 = 50$ km s^{-1} Mpc^{-1}). This velocity invites comparison with the "superluminal" sources (Cohen and Unwin 1982) which exhibit apparent velocities an order of magnitude greater, and are generally more distant (although the nearest, 3C120, is less than twice as distant as NGC1275). The overall morphology of the structures seen at epochs 1979.1 and 1981.1 does exhibit the "core-jet" configuration characteristic of the superluminal sources. We identify the active and extremely compact northern emission region as the core; this feature also has an inverted spectrum (Unwin, private communication). On the basis of this interpretation, we extrapolate the motion of the southern component back to

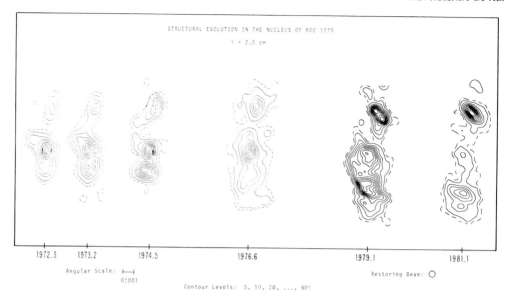

Fig. 1. Contour maps of the nucleus of NGC1275 at six epochs, aligned horizontally proportional to the epoch of observation.

coincidence with the northern, at epoch 1961.0. Studies of flux-density variations in this source (Kellermann and Pauliny-Toth 1968) show the short-centimeter-wavelength emission already rising steeply at that time; in the following 15 years the flux density increased six-fold.

The transverse expansion of the relatively more diffuse structure (Preuss et al. 1979) is seen continuing in the recent observations. We quantify this, as in earlier work, by the equivalent Gaussian width seen by a fan beam aligned with the primary source elongation. This measure increases steadily at 0.08 milliarcsec y^{-1}, corresponding to a linear velocity of $\sim$40,000 km s^{-1}; the extrapolated initial epoch of this expansion is 1961.6.

REFERENCES

Cohen, M.H., and Unwin, S.C.: 1982, this volume, p. 345.
Kellermann, K.I., and Pauliny-Toth, I.I.K.: 1968, Ann. Rev. Astron. Astrophys. 6, pp. 417-448.
Pauliny-Toth, I.I.K., Preuss, E., Witzel, A., Kellermann, K.I., Shaffer, D.B., Purcell, G.H., Grove, G.W., Jones, D.L., Cohen, M.H., Moffet, A.T., Romney, J.D., Schilizzi, R.T., and Rinehart, R.: 1976, Nature 259, pp. 17-20.
Preuss, E., Kellermann, K.I., Pauliny-Toth, I.I.K., Witzel, A., and Shaffer, D.B.:1979, Astron. Astrophys. 79, pp. 268-273.
Schilizzi, R.T., Cohen, M.H., Romney, J.D., Shaffer, D.B., Kellermann, K.I., Swenson, G.W.Jr., Yen, J.L., and Rinehart, R.: 1975, Astrophys. J. 201, pp. 263-274.

VLBI OBSERVATIONS OF M87

M.J. Reid and J.H.M.M. Schmitt
Harvard-Smithsonian Center for Astrophysics
F.N. Owen
National Radio Astronomy Observatory
R.S. Booth and P.N. Wilkinson
Nuffield Radio Astronomy Laboratories
D.B. Shaffer
NASA/GSFC and Phoenix Corp.
K.J. Johnston
U.S. Naval Research Laboratory
P.E. Hardee
University of Alabama

On 1980 February 20 we conducted an 8-station intercontental VLBI experiment in order to study the nucleus and jet of M87 at 1666.6 MHz in right circular polarization. Our array was sensitive to structures from 0.001 to 0.1 arcsec. We made a hybrid map of the nucleus of M87, and also searched for compact structures within the knots of the jet. The map (Figure 1) shows that the nucleus of M87 contains a one-sided jet. This morphology is similar to that observed in many compact extragalactic sources. The position angle of the nuclear jet is 290.5(±1) degrees, which precisely matches that of the 20 arcsec jet. No bending of the jet through an angle greater than about 2 degrees is observed. The nucleus also contains a large component (>0.1 arcsec) which is elongated along the same position angle as the jet and has a flux density of roughly 1 Jy. This component is fully resolved by the vast majority of our (u,v) points, and we could not map it with standard techniques.

Assuming that the absence of a detectable counter-jet is due to the effects of relativistic beaming, one can place limits on the flow velocity of the jet, β, in units of the speed of light and the angle, θ, the jet makes to our line of sight. The ratio, R, of the observed intensity of the jet and counter-jet is given by $(1+\beta\cos\theta)/(1-\beta\cos\theta)$. Adopting R=25, appropriate for components 15 mas down the jet, requires both β and $\cos\theta$ to exceed 0.67. Other possible explanations for the absence of an observed counter-jet are that jets are intrinsically one-sided or that the observed emission from the counter-jet is delayed (or advanced) by light travel times which exceed the lifetime of the jet emission. One crucial observation for the relativistic beaming model would be the detection proper motion within the nuclear jet. If observed components move with the jet flow, then a proper motion of 10 mas/yr would be expected and easily detected with a second epoch map.

In addition to the nucleus of M87, we mapped portions of the larger jet encompasing knots D and A which are 3.2 and 12.6 arcsec, respectively, from the nucleus. Neither knot was detected. We place upper limits of 25 mJy for any component less than 4 mas in size and of 0.2 Jy for a component less than 0.1 arcsec in size. Combining our results with recent VLA results suggests that the size of knot D is 0.2 ± 0.1 arcsec. Some theories suggests the existence of compact structures in the knots due to effects of shocks or limb brightening, whereas our observations indicate that the knot emission is fairly smooth.

Finally, we wish to point out the possibility of small wiggles in the nuclear jet. Data from our longer baselines are modeled slightly better by a jet that has a peak-to-peak oscillation of several mas perpendicular to its extent and a wavelength on the order of 10 to 20 mas than by a straight jet. Although a structure of this sort exists in the map (Figure 1), it is masked by the 8 mas restoring beam. Further observations are needed to confirm the reality of such wiggles in M87. It is interesting to note that there are published VLBI maps of extragalactic sources which also suggest wiggling jet structures (eg. 3C286 and 3C380).

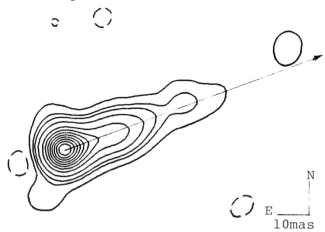

Fig. 1. Nucleus of M87 at 18 cm with an 8 mas beam. Line is $290^\circ.5$ E of N. Contours are -1, +1, 3, 5, 10, 20, 30, ... , 90% of the peak temperature of 9.9×10^9 K.

COMPACT RADIO SOURCES: THEIR USE AND SIZE

K. J. Johnston
E. O. Hulburt Center for Space Research, Naval Research
Laboratory, Washington, D. C. 20375

Compact radio sources are both interesting from an astrophysical point of view and useful for geophysics. They propose some very interesting astrophysical questions as to their energy generation and transport. Their small size allows the measurement of their positions to unprecedented accuracy (< 0$\overset{''}{.}$01) extending astrometric measurements of parallax and proper motions to further distances. These celestial "beacons" may be used to establish an almost inertial reference frame against which positions on the earth and the earth's motions such as Universal Time (UT1) and polar motion may be measured to centimeter scale accuracy. Finally these sources when used in conjunction with radio interferometry can allow precise time synchronization over global distances at subnanosecond accuracy.

For the sake of reference, let us define a compact radio source as one which is unresolved at resolutions less than or equal to 3 milliarcseconds (mas) at a radio frequency of 5 GHz. In order to be useful to the geophysicist, these sources should also display a flux density of > 0.5 Jy, allowing measurements to be made with an interferometer utilizing small diameter (< 3 meters) antennas.

In the late sixties, the development of Very Long Baseline Interferometry (VLBI) showed that radio sources exist having components of size scale a mas. The construction of the Very Large Array (VLA) allowed the determination of the spatial structure of a large number of radio sources on the scale of one arc second. In a survey of all 444 extragalactic sources of flux density > 1 Jy between -45° and 70° declination (Ulvestad, Johnston, Perley, and Fomalont 1981) 197 sources have more than 90% of their flux density in components < 1". The typical arc second scale structure displayed by high dynamic range observations is 1) a compact unresolved source with one-sided structure extending several kiloparsecs from the dominant compact source, e.g. 3C345, 2) two-sided asymmetrical structure, again with a dominant compact component, e.g. 4C55.17, or 3) complex structure surrounding a dominant compact source, e.g. 1150+497 (Perley, Fomalont, and Johnston 1981; Browne et al. 1981). There may be a correlation between source intensity and arc second structure, with the most intense sources displaying simple one-sided structure or no structure at all.

Extension of studies of detailed source structure to the mas level can only be accomplished by multistation VLBI observations of the strongest (> 1 Jy) sources using the limited bandwidth of the Mark II VLBI recording system. Systematic measurements of a large number of sources have only begun. Measurements of a large number of sources with reasonable dynamic range (Pearson and Readhead 1981; Eckart et al. 1982) demonstrate that on the mas level, sources may be classified again into the same categories as the arc second structure with the exception that there are many more one-sided asymmetric sources.

Measurements at 5 GHz with a synthesized beam of $\sim$ 1 mas using VLBI networks consisting of five and six U. S. stations and the 100 meter telescope at Effelsberg, FRG, of the 12 S5 sources ($\delta > 70°$) with S > 1 Jy have shown these sources to be very compact (Eckart et al. 1982). All the sources have at least 50% of the flux density in compact cores while the five BL Lacertae objects have 70% of their flux density in a compact core of less than 3 mas. The source 0454+844 may be the most compact source known with 95% of its flux density emanating from a core of size < 1 mas. The sizes for many of the compact cores are $\leq 0''.3$ as measured from the visibilities on the longest baselines and have brightness temperatures of $\sim 10^{11}$K. With baselines of size larger than the earth's diameter, probably many compact sources will exceed brightness temperatures of 10^{12}K which is the limit put on the size of inhomogeneous electron synchrotron sources by Compton scattering.

The only explanation for this predicted result is that the source structure and sizes seen on the mas scale are due to relativistic doppler beaming along the line of sight. This is certainly true for the strong S5 sources. An example of a source which displays relativistic jets of particles is SS433 (Hjellming and Johnston 1982). If this galactic source is scaled up from 1 $M_\odot$ to about 10^8 $M_\odot$ and the jet velocity increased to 0.9c, this source may be used to model extragalactic sources.

More maps of higher dynamic range are needed of the mas scale structure in radio sources in order to study in detail the process giving rise to these intense, compact sources. We await the full use of the VLBI technique through improved facilities.

REFERENCES

Browne, I.W.A., Orr, M.J.L., Davis, R.J., Foley, A., Muxlow, T.W.B., and
    Thomasson, P.: 1981, submitted to M.N.R.A.S.
Eckart, A., Hill, P., Pauliny-Toth, I.I.K., Witzel, A., Johnston, K.J.,
    and Spencer, J.H.: 1982, submitted to Astron. & Astrophys.
Hjellming, R.M. and Johnston, K.J.: 1982, IAU Symposium #97 "Extragalactic
    Radio Sources", D. Reidel Publishing Company, Dordrecht:Holland, p.197.
Pearson, T. J. and Readhead, A.C.S.: 1981, Ap. J. 248, pp. 61-81.
Perley, R.A., Fomalont, E.B., and Johnston, K.J.: 1982, submitted to Ap.J.

SPECTRAL SHAPES OF COMPACT EXTRAGALACTIC RADIO SOURCES

Steven R. Spangler
National Radio Astronomy Observatory
Socorro, NM 87801

ABSTRACT
    Possible mechanisms for producing the observed broad radio spectra of compact extragalactic radio sources are discussed. The explanations considered are: (a) superposition of the spectra of sub-components, (b) inhomogeneous synchrotron sources, and (c) synchrotron radiation from "non-standard" energetic electron spectra. These three models have been compared with results of spectral and VLBI observations. These comparisons indicate: (1) if the "superposition" hypothesis is correct, then the subcomponents themselves must be inhomogeneous synchrotron sources, (2) there is apparently no general inhomogeneous synchrotron source which characterizes compact extragalactic sources, and (3) many compact sources have spectra which resemble synchrotron radiation from a relativistic Maxwellian electron spectrum, but the inferred electron energies and magnetic fields have a wide range of values. VLBI observations of a selected sample of sources generally favor a single component model, but cannot distinguish between models (b) and (c).

    The opaque synchrotron source model of compact extragalactic radio sources has been quite successful in accounting for observed properties of these objects, such as the relation between angular size and frequency of flux density maximum. However, for more than a decade it has been realized that this model, in its simplest form, cannot account for a very important observed property of compact sources. A homogeneous synchrotron source with a power law energetic electron spectrum has a power law radio spectrum at frequencies below that of flux density maximum, with a spectral index of 5/2. Almost no sources are observed to have an optically-thick spectral index this large. Out of a sample of 136 sources studied by Owen, Spangler and Cotton (1980), only one (0552+398) had a low frequency spectral index consistent with this value. A typical observed value was 0.3 - 0.5. This fact has been recognized for over a decade. At least three suggestions have been made to account for this observation. (1) Subcomponent Superposition According to this viewpoint, the integrated spectra are blends of subcomponent spectra. Each subcomponent is assumed to have the spectrum of a homogeneous synchrotron source, with a different frequency of maximum. The result would be a broad integrated spectrum which contains no information on the physics of the subcomponents. (2) <u>Inhomogeneous Synchrotron Sources</u> It seems unlikely that a source would have a uniform energetic electron density and magnetic field strength throughout the source (the definition of a homogeneous synchrotron source). It may be shown (Marscher 1977) that a source possessing radial gradients in these qualities will have a power law spectrum at frequencies below maximum, with a spectral index which is in general less than 5/2 and is dependent on the functional form of the inhomogeneities. The appealing feature of this model is that measurement of a global, observationally-accessible property of a source (the optically-thick spectral index α_T) can provide information on the functional form of the source inhomogeneity, and thus the physics of the source. (3) <u>Non-Standard Energetic Electron Spectra</u> A sharply-peaked radio spectrum with α_T = 5/2 is a characteristic of a homo-geneous synchrotron source with a power law energetic electron spectrum. If the electron spectrum has a narrow spread in energy about a mean value, as is the case in a relativistic Maxwellian (Jones and Hardee, 1979), then the resultant radio spectrum can be quite different.

In an attempt to distinguish between these three possibilities, a program of multifrequency radiometric and VLBI observations was undertaken and is still in progress. The radiometric observations consisted of nearly simultaneous flux density measurements at 11 frequencies between 0.318 and 90 GHz, thus obtaining broadband "snapshot" spectra of 136 compact sources. The hope was that with such a large sample of high quality source spectra, one could investigate such questions as whether there was a preferred type of radio source inhomogeneity (manifested by preferred values of α_T), etc.

The principal feature of the subcomponent superposition model to be tested was the necessity for a fixed frequency spacing of the subcomponents. In a computer study, Cook and Spangler (1980) found that if the subcomponents had the spectrum of a homogeneous synchrotron source, then a highly regular and obviously artificial spacing of the subcomponent turnover frequencies was required to produce the observed spectra. This was not so much the case if the subcomponents possessed inhomogeneous source spectra. Thus, a hybrid of the superposition hypothesis and the inhomogeneous synchrotron source model would appear to be capable of matching the observed properties of compact extragalactic radio sources.

The principal goal of the inhomogeneous source analysis was to see if a "preferred" value for the optically thick spectral index could be found. This, in turn, would point to a predominant form of source structure. No such preferred value for the optically thick spectral index was found (Spangler 1980), giving no support to the idea that there is a general type of compact source structure. It is, of course, possible that a substantial number of compact sources are inhomogeneous synchrotron sources, but it would appear that different sources would then have to possess different types of inhomogeneity.

A comparison of observed spectra with the spectrum of synchrotron radiation from a relativistic Maxwellian electron spectrum revealed many cases in which very close agreement was found (Spangler 1980, Fig. 2). This is encouraging for the alternate electron spectrum model, but there are nonetheless difficulties. The inferred characteristic electron energies span a huge range. Furthermore, the relativistic Maxwellian synchrotron spectrum declines exponentially for sufficiently high frequencies, in apparent contradiction with observations. It seems probable that this difficulty could be overcome by a power law electron spectrum with a low energy cutoff.

The results of the analysis using only the radio spectral data were, therefore, somewhat ambiguous. All three of the hypotheses could, with certain qualifications, explain the statistical properties of the spectral sample.

It was therefore decided to make VLBI observations of a selected subsample of these sources. VLBI observations, used in concert with broadband spectral measurements, can, in principal, distinguish between the various spectral models. Ten sources were observed whose spectra indicated that they consisted of a single component. Observations were made with a four-element, continental baseline interferometer at 6 and 18 cm. Dual frequency observations were made to search for the frequency-dependent angular size expected for an inhomogeneous synchrotron source. Of the ten sources, one (2134+00) was found to definitely possess a double structure. Another object (0202+31) showed an indication of beating between subcomponents. For these two objects, it seems most probable that the observed spectrum is a blend of the subcomponent spectra. For the remaining objects, the VLBI observations were consistent with the sources consisting of a single unresolved or slightly resolved component. There therefore remains the possibility that the integrated spectra of these sources contain useful information on the physics of these objects. Three frequency observations, utilizing more stations and transcontinental baselines, are currently in progress.

REFERENCES

Cook, D.B. and Spangler, S.R. 1980, Ap.J., 240, 751.
Jones, T.W. and Hardee, P.E. 1979, Ap.J., 228, 268.
Marscher, A.P. 1977, Ap.J., 216, 244.
Owen, F.N., Spangler, S.R., and Cotton, W.D. 1980, Ap.J., 85, 351.
Spangler, S.R. 1980, Ap.Lett., 20, 123.

DISCUSSION

O'DEA: When you fit the relativistic Maxwellian spectra, what additional component is needed to explain the discrepancy at the lower frequencies?

SPANGLER: I believe the departure of the observed spectrum from the model spectrum at low frequencies is due to extended, steep spectrum components associated with the compact sources. Support for this opinion may be found in Owen, Spangler, and Cotton (A. J., 85, 351, 1980), in which a correlation was found between "decimetric excesses" and the existence of structure resolvable by the VLA.

POLARIZATION OF THE COMPACT RADIO STRUCTURE OF 3C 454.3

W. D. Cotton
National Radio Astronomy Observatory

B. J. Geldzahler and I. I. Shapiro
Massachusetts Institute of Technology

1. INTRODUCTION

The polarization structure of synchrotron sources provides information about the magnetic field structure and thermal particle distributions near the source. Such information is critical to a full understanding of compact sources in the nuclei of galaxies and quasars. VLBI maps of these objects are frequently interpreted as showing jets even though the maps are dominated by a few bright regions. If these sources do indeed contain jets then the jet should have a relatively well-ordered component of the magnetic field. The presence of this ordered magnetic field, as well as the presence of thermal plasma in and around the jet, should be revealed by the polarized radiation. We present below the results of the first successful VLBI measurements of the polarized emission from an extragalactic compact radio source.

2. OBSERVATIONS

Insufficient sensitivity and inadequate equipment have plagued previous attempts at measuring polarization structure by means of VLBI. To avoid some of these problems, we made observations at 13 cm with the 64 m antennas of the DSN at Goldstone and Madrid, which could record both right and left circular polarization, together with antennas at Onsala, Haystack, Green Bank and Ft. Davis which could record only right circular polarization.

3. RESULTS

The milliarcsecond structure of 3C 454.3 at an observing wavelength of 13 cm is dominated by two bright regions aligned at a position angle (PA) of about 115°. On the basis of VLBI observations at other frequencies, we conclude that the more compact (SE) region has a very optically thick spectrum and is probably the core. There is also a hint of an underlying jet which, for discussion, we shall assume to be present.

Since the source was not resolved in the ~N-S direction, we present the observations in Figure 1 as projected onto the line at PA = 115°. The total and polarized intensities are shown normalized to the peak brightness of 2.8 Jy/beam. The position angle of the E-vector is shown relative to PA = 115°.

The smoothness of the polarized intensity indicates the presence of a well ordered component of the magnetic field, especially since the fractional polarization reaches 20%. However, because both bright regions are weakly polarized we infer that either the magnetic field there is quite disordered or thermal plasma there is depolarizing the radiation.

The change of almost 90° in the PA of the E-vector might be due to either differential Faraday rotation, reordering of the magnetic field, or a change in the optical depth from thick to thin. Although none of these possibilities can be ruled out with certainty, the third seems the most consistent with all data including spectral information obtained from the literature. If a variable optical depth is primarily responsible for the observed rotation of the E-vector, then the magnetic field is oriented primarily along the jet. Further, if differential Faraday rotation is not important, then it is unlikely that there is much thermal plasma around the jet.

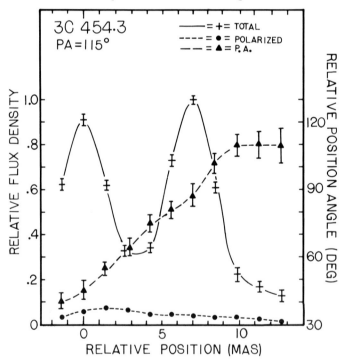

Figure 1. Polarization properties of 3C 454.3 at 13 cm

DISCUSSION

SHAFFER: What is the rotation measure of 3C 454.3?

COTTON: The integrated rotation measure is about -30 radians/m^2, which is similar to nearby sources. If the apparent rotation of the polarization E-vector is due to Faraday rotation, then internal differences 2.5 times as large are required.

LAING: In 3C 286, the degree of polarization is high and must come primarily from a region which is optically thin. The rotation measure is small, and we may reasonably infer the direction of the B-field from integrated polarization measurements. The projected B-field is <u>perpendicular</u> to the VLB structure.

COTTON: Due to various calibration uncertainties, we cannot completely exclude the possibility that the magnetic field is perpendicular to the jet, but our data are much more consistent with the magnetic field being oriented along the jet.

JONES: Rudnick and I have observed the integrated polarization of 3C 454.3 on several occasions at both centimeter and millimeter wavelengths. In the 1977-1980 interval, we saw a systematic change in the polarization angle between $\chi \sim 0°$ at long wavelengths to $\chi \sim 60°$ at 9 mm and 3 mm. Therefore, your rotation seems likely to be a structural effect.

COTTON: At 13 cm the effects you suggest should be relatively unimportant. The integrated spectrum of the jet at wavelengths near 13 cm are also consistent with the optical depth interpretation of the polarization angle rotation.

A MILLIMETRE/SUBMILLIMETRE STUDY[+] OF OPTICALLY SELECTED QUASARS

W.A. Sherwood, G.V. Schultz, E. Kreysa, H.-P. Gemünd
Max Planck Institute for Radio Astronomy

During the past decade many kinds of optical surveys have discovered hundreds of quasars most of which (~90 %) are radio quiet (<10 mJy at 5 GHz). We have observed two samples of quasars brighter than $17\overset{m}{.}6$ found by their emission lines and by their ultraviolet excess. We have also selected quasars with redshifts known to be greater than 3.00. A brief description of the observing technique is given by Sherwood et al. (1981b). We have compared our millimetre photometry of flat radio spectrum quasars with that of Ennis and Werner and find excellent agreement. For four of the 8 sources in common the data have been published: Kreysa et al. (1980) and Jones et al. (1981). In addition, analysis of our "noise" shows it to be white, gaussian distributed about zero. The three samples are summarized as follows.

1. Osmer and Smith (1980) published a list of 125 confirmed quasars discovered on objective prism plates. For our study the sole selection criterion was that the magnitude be brighter than $17\overset{m}{.}6$. There are 17 optically discovered quasars in our sample and all have now been detected. Q0420-388 was detected in September 1979 and confirmed in July 1980 and 1981 (Sherwood et al. 1981b). Twelve other quasars were detected in July 1980 and confirmed in 1981. Only 6 of the 17 are known to be radio sources (Smith and Wright, 1980; Condon et al., 1981; Sherwood et al. 1981a). If we compare the excess in the number of positive pairs (see Sherwood et al. 1981b) over the natural uncertainty in the total number of measurements as an indication of the "quality" of the detections then this sample has been detected in the mean at the 15-sigma level corresponding to a flux density of 3 Jy.

2. Green (1976), and Green and Schmidt (1978) have published a list of confirmed quasars having ultraviolet excess found in the Palomar-Green survey. All are brighter than $17\overset{m}{.}6$. From ESO, La Silla, we measured the 5 with $\delta < +10°$ plus IIIZw2 and PG0026+129. Although 4 are radio sources only two were detected for certain at 1 mm: IIIZw2 and PKS1302-102. The quality of the sample on average is only

---
[+] Based on observations made at the European Southern Observatory (ESO), La Silla, Chile

3-sigma corresponding to an average flux density of 0.7 Jy.
3. It can be seen by comparing samples 1.) and 2.) that apparent magnitude is not the key to detecting optically discovered quasars. The third sample tests the idea that the luminosity is important: with complete disregard for apparent magnitude we tried to observe (at least twice) as many quasars with Z > 3.00 as we could from ESO. The sources were taken from Osmer and Smith (1980): 6 including 4 radio sources from sample 1; Hoag and Smith (1977): 2; and from Parkes - Jauncey et al. (1978) and Wright et al. (1978). The total is 10 - all detected; 6 are radio sources. Seven were observed at 2 epochs and of these only 1 was detected only at 1 epoch. Examples of these objects may be seen in Sherwood (1981) and Sherwood et al. (1981a). This high luminosity sample is detected, on the average, at ~8.5 sigma corresponding to an average flux density of 1.9 Jy.

Conclusions
1) Among optically selected ($m \leq 17^m.5$) QSOs (most of which are radio quiet) there are certainly some quasars with strong millimetre wave emission: S(300 GHz) ~ 1-10 Jy.
2) They have inverted spectra with respect to the radio spectrum.
3) If synchrotron theory is applicable then they are compact sources 10^{-5} to 10^{-6} arc sec. in diameter with an absorbed spectrum below 300 GHz.
4) The probability of detecting an optically selected quasar at 1 mm increases with luminosity.
5) There is evidence for variability.
6) Where X-ray data is available (but not simultaneously observed) the millimetre to X-ray luminosity ratio is ≥ 1.

Acknowledgements
This work has been supported by the DFG/SFB 131, Radioastronomy. We thank Jeff Puschell and Larry Rudnick for sharing their 3.3 mm data of PKS1302-102 with us.

References
Condon, J.J. et al.: 1981, Astrophys.J. 244, pp. 5-11
Green, R.F.: 1976, Pub. A.S.P. 88, pp. 665-668
Green, R.F. and Schmidt, M.: 1978, Astrophys.J. 220, pp. L1-L4
Hoag, A.A. and Smith, M.G.: 1977, Astrophys.J. 217, pp. 362-381
Jauncey, D.L. et al.: 1978, Astrophys.J. 223, pp. L1-L3
Jones, T.W. et al.: 1981, Astrophys.J. 243, pp. 97-107
Kreysa, E. et al.: 1980, Astrophys.J. 240, pp. L17-L19
Osmer, P.S. and Smith, M.G.: 1980, Astrophys.J.Suppl. 42, pp. 333-349
Sherwood, W.A. et al.: 1981, Mitt.Astron.Ges. 52, pp. 138-139
Sherwood, W.A. et al.: 1981, Nature 291, pp. 301-303
Sherwood, W.A.: 1981, ESO Messenger, No. 24, pp. 15-17
Smith, M.G. and Wright, A.E.: 1980, Mon.Not.R.astr.Soc. 191, pp. 871-886
Wright, A.E. et al.: 1978, Astrophys.J. 226, pp. L61-L64

DETECTION OF A BROAD HI ABSORPTION FEATURE AT 5300 km SEC^{-1} ASSOCIATED WITH NGC 1275 (3C84)

P. C. Crane
National Radio Astronomy Observatory

J. M. van der Hulst
Department of Astronomy, University of Minnesota

A. D. Haschick
Haystack Observatory

Observations of NGC 1275 at $\sim$ 1396 MHz with the NRAO line interferometer in 1974 and 1976 suggest the presence of a very broad, shallow HI absorption feature centered at $\sim$ 5300 km sec^{-1}. These observations were repeated in 1981 June with the Very Large Array using a greater bandwidth to determine a satisfactory baseline.

The line-interferometer observations were calibrated baseline by baseline for instrumental gain and phase variations and for the bandpass shape, using 3C84 and 3C147. The spectra for all baselines and epochs have been averaged and Hanning-smoothed; a bandwidth of 5 MHz and 48 complex frequency channels were used; after Hanning smoothing the effective resolution is 44.8 km sec^{-1}. The VLA observations, which used baselines > 500 m to resolve the extended structure present, were calibrated on an antenna basis using 3C48 to remove instrumental gain and phase variations. All spectra for NGC 1275 and 3C48 were averaged and Hanning-smoothed; the bandpass calibration was done by dividing the NGC 1275 spectrum by the normalized 3C48 spectrum. For 12.5 MHz bandwidth, and 32 complex frequency channels, the effective resolution is 167.9 km sec^{-1} after Hanning smoothing.

Despite the difference in resolution and the lack of a reliable baseline for the interferometer observations, the two spectra (shown in Figure 1) agree very well. The parameters determined from the VLA observations are given in Table 1, and show that this is the broadest and weakest HI absorption feature yet reported. The velocity and width are very similar to those of the narrow nuclear emission lines reported by Heckman et al. (1981), which they suggest come from the extended low-velocity system of filaments mapped by Rubin et al. (1975). They interpret their observed blueward asymmetries in the line profiles and the net blueshifts with respect to the system velocity of NGC 1275 in terms of radial outflow and attenuation by dust. Cowie et al. (1980) identify the same system of filaments as gas accreting onto NGC 1275 from cooling gas from the Perseus cluster. Our observation of a redshift of $\sim$ 100 km sec^{-1} supports the latter model.

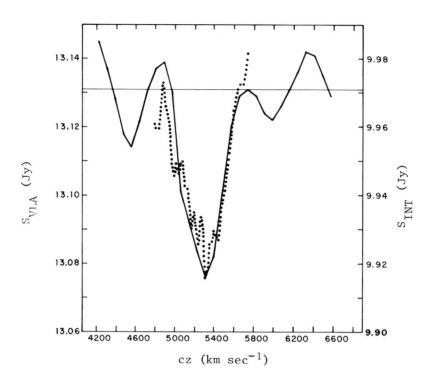

Figure 1. Comparison of interferometer (···) and VLA (——) spectra of NGC 1275.

Table 1

Central velocity	(Heliocentric)	5320 km sec^{-1}
Velocity width	(FWHM)	450 km sec^{-1}
Optical depth		0.0041
Column density	(N_H/T_S)	4×10^{18} H cm^{-2} K^{-1}

References

Cowie, L. L., Fabian, A. C., and Nulsen, P. E. J.: 1980, M.N.R.A.S. 191, 399.
Heckman, T. M., Miley, G. K., van Breugel, W. J. M., and Butcher, H. R.: 1981, Ap. J. 247, 403.
Rubin, V. C., Ford, W. K., Jr., Peterson, C. J., and Lynds, C. R.: 1978, Ap. J. Suppl. 37, 235.

A SEARCH FOR HI IN ELLIPTICAL GALAXIES WITH NUCLEAR RADIO SOURCES

L.L. Dressel
National Research Council and Goddard Space Flight Center

T.M. Bania and R.W. O'Connell
University of Virginia

Many searches have been made in the last few years for 21 cm emission from neutral hydrogen in elliptical galaxies. Emission has been detected in several galaxies, which have 10^8 to 10^9 $M_\odot$ of HI (for H_0 = 100 km s^{-1}Mpc^{-1}). Upper limits between 10^6 and 10^9 $M_\odot$ have been set for the HI mass in about 40 other galaxies. Why most E galaxies have so little gas, and why some few have detectable gas, remains a matter of great interest. Two of the galaxies with large HI mass, NGC 1052 and 4278, are known to have powerful nuclear continuum radio sources ($P_{2380} \sim 10^{22}$ WHz^{-1}). Since both of these attributes are fairly rare among elliptical galaxies, their coexistence in these galaxies is not likely to have occurred by chance. We have therefore observed twelve other elliptical galaxies with nuclear radio power $P_{2380} > 10^{22}$ WHz^{-1} at Arecibo Observatory, to determine whether a large mass of HI is a necessary auxillary to nuclear continuum emission.

In the Arecibo observing program, we detected one emission line and possibly one absorption line. An emission line ~ 330 km s^{-1} wide was convincingly detected in UGC 09114. The implied HI mass is 7×10^8 $M_\odot$, which is one of the highest HI masses detected in an elliptical galaxy. We are fairly confident that 09114 is not a misclassified early spiral galaxy: 1) Photographs reveal no strong central concentration (K. Kingham, private communication). 2) The optical spectrum is that of a normal E galaxy, except for a fairly strong [OII]λ 3727 emission line; the spectrum is thus typical of E galaxies with compact radio sources (O'Connell and Dressel 1978). 3) Finally, the radio spectrum of 09114 is remarkably inverted, with a spectral index of about +1.3 from 1400 to 5000 MHz. Flat and inverted spectra are typical of nuclear sources in E and S0 galaxies, but are rare among nuclear sources in spiral galaxies. We have possibly detected HI in absorption in UGC 06671, but a curved baseline makes this detection uncertain. No emission or absorption was detected in the remaining galaxies in the program. Upper limits between 1.5×10^8 and 7×10^8 $M_\odot$ of HI were determined for UGC 01308, 02112, 03063, 04859, 07378, 08779, 11718, 12269, and 12727.

To examine the relationship between HI content and nuclear radio

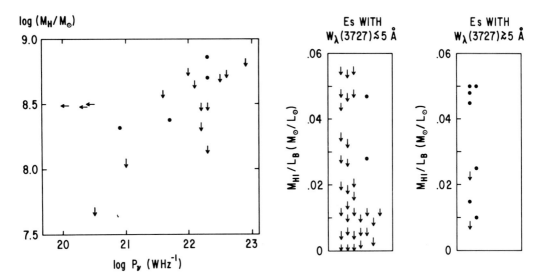

Fig. 1: HI mass as a function of nuclear radio continuum power

Fig. 2: HI mass to B luminosity ratios for Es with $W_\lambda(3727) < 5\text{Å}$, $> 5\text{Å}$

emission in E galaxies, we have combined our data with data for other galaxies (NGC 1052, 3904, 4105, 4278, 4552, 4636, 4649, 5322, 5846) which have detections or good upper limits for both HI content and nuclear radio power. (See Sanders 1980 for references.) HI mass versus nuclear 2380 MHz power is plotted for all of the galaxies in Figure 1. No correlation is evident in this figure. However, one cannot quite rule out the possibility that E galaxies with powerful nuclear sources are hydrogen-rich relative to most E galaxies: most galaxies with $P > 10^{22}$ WHz^{-1} have still not been observed with enough sensitivity to detect the amount of HI found in NGC 4278 ($\sim 2 \times 10^8$ $M_\odot$).

Our statistical investigations have revealed one parameter that is well correlated with HI content in E galaxies. This parameter is [OII] $\lambda 3727$ line strength, which is indicative of the amount of ionized gas in a galaxy. Our sample of galaxies is drawn from Sanders' (1980) compilation of E galaxies searched for HI and from the Arecibo program. For most of these galaxies, the data of Humason et al. (1956) or of O'Connell and Dressel (1978 and in preparation) can be used to determine whether the equivalent width of the 3727 line is $\gtrless$ 5Å. The strong correlation between large HI mass and prominent 3727 emission is evident in Figure 2.

REFERENCES

Humason, M.L., Mayall, N.U., and Sandage, A.R.: 1956, Astron. J. 61, 97.
O'Connell, R.W. and Dressel, L.L.: 1978, Nature 276, 374.
Sanders, R.H.: 1980, Astrophys. J. 242, 931.

CHANGES IN THE HI ABSORPTION LINE SPECTRUM OF AO 0235+164

Michael M. Davis
Arecibo Observatory

Arthur M. Wolfe
University of Pittsburgh

The BL Lac object AO 0235+164 is an extreme example of a QSO containing a compact, non-thermal energy source which is violently variable. In 1975 it brightened simultaneously at optical and radio frequencies, with a hundred-fold increase in optical brightness on a time scale of a few weeks (Rieke et al. 1976; Ledden, Aller and Dent 1976). During the outburst 0235 was among the most luminous objects in the universe. It exhibits longer term variations at meter wavelengths (Condon and Dennison 1978), and has experienced repeated optical and short wavelength radio outbursts since 1975 (Pica, Smith and Pollock 1979). Very long baseline interferometry shows it to be the most compact known extragalactic radio source.

The optical spectrum has two well-defined redshift systems, at $z = 0.852$ and $z = 0.524$ (Rieke et al. 1976; Burbidge et al. 1976). The lower optical redshift has also been detected in 21 cm absorption (Roberts et al. 1976). A faint nebulosity 2" SW of the BL Lac exhibits emission lines at the $z = 0.524$ redshift (Smith, Burbidge and Junkkarinen 1978). A plausible case can be made that this absorber is a spiral galaxy with an HI disk, and has a cosmological redshift $z = 0.524$. However, the nebulosity is unusual in having emission line luminosities at least ten times greater than the most luminous spiral galaxies, so that the hypothesis of chance alignment of the nebulosity with the BL Lac object must be treated with some caution. The question arises, whether AO 0235+164 is cosmologically distant from the nebulosity, at or beyond a distance corresponding to $z = 0.852$, or whether the two objects are physically co-located, with the redshift difference attributable perhaps to an ejection velocity.

If the two objects lie near each other, the 21 cm continuum flux will compete with collisions in determining the hyperfine level populations of the hydrogen ground state. In this case we might expect to detect a decrease in the HI optical depth following an increase in the continuum flux density. On the basis of this expectation, we instituted a monitoring program to search for variation in the HI profile, using the Arecibo telescope at 932 MHz.

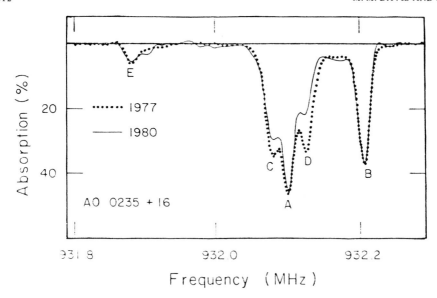

Figure 1. Sample of the observed variation in the HI profile.

The results of the monitoring program were somewhat unexpected. In spite of a nearly constant continuum flux density, we have seen highly significant variations in the absorption profile. Since early 1980, the profile has returned to its earlier shape, but monthly monitoring has revealed smaller but significant variations in the line depths, on a time scale of a few months. An upper limit of 0.7 km/s can be set on the variation in component velocities, but there are systematic trends in the residuals of order 0.2 km/s which are probably real rather than instrumental. In addition, a recent VLBI observation showed what may be a real change in the fringe phase of component B (Frank Briggs, private communication). In the following paper we assess the significance of these results.

References

Burbidge, E.M., Caldwell, R.D., Smith, H.E., Liebert, L., and Spinrad, H. 1976, Astrophys. J. 205, L117.
Condon, J.J., and Dennison, B. 1978, Astrophys. J. 224, 835.
Ledden, J.E., Aller, H.D., and Dent, W.A. 1976, Nature, 260, 752.
Rieke, G.H., Grasdalen, G.L., Kinman, T.D., Hintzen, P., Wills, B.J., and Wills, D. 1976, Nature, 260, 754.
Pica, A.J., Smith, A.G., and Pollock, J.T. 1979, Bull. Amer. Astron. Soc., 11,457.
Roberts, M.S., Brown, R.L., Brundage, W.D., Rots, A.H., Haynes, M.P., and Wolfe, A.M. 1976, Astron. J., 81,293.
Smith, H.E., Burbidge, E.M., and Junkkarinen, V.T. 1977, Astrophys. J., 218,611.

THEORETICAL MODELS TO EXPLAIN THE VARIABLE 21 cm ABSORPTION SPECTRUM IN AO 0235+164

A. M. Wolfe
Department of Physics, University of Pittsburgh,
Pittsburgh, PA 15260

The variability of the highly redshifted 21 cm absorption spectrum in AO 0235+164 presents us with two essential facts. First, rapid changes occur in the line depths of foreground clouds with redshift $z \simeq 0.524$. Second, the background radio source is among the most violently variable objects known. Thus it is reasonable to suppose that changes in the lines stem from activity in the background BL Lac object. The question is how?

We (F. H. Briggs, M. M. Davis, and I) have considered two alternatives. The <u>intrinsic</u> scenario assumes that activity in 0235 somehow alters the 21 cm opacity of absorbing HI within its neighborhood. Since the redshift of 0235 is $z > 0.85$, we are confronted with the task of explaining how objects with redshifts differing by $\Delta z \simeq 0.3$ can by physically associated. In the <u>extrinsic</u> scenario we assume that activity in 0235 results in time-dependent changes of the brightness centroid of the 932 MHz radio source. When viewed through a non-uniform distribution of absorbing clouds the shifting light path brings about changes in line depth, without changes in opacity. In this case we assume that 0235 is cosmologically distant from the foreground HI.

Before comparing the data with the predictions of either model it is useful to reiterate some model-independent facts. First, the absence of significant variations of the 932 MHz continuum, VLB studies at 2.7 GHz (Cotton 1981), and synchrotron self-Compton considerations suggest that the background source contains a non-variable component with diameter $\theta_s \simeq 2.5$ to 3.0 mas. Second, the shift in brightness centroid discussed in the previous talk is plausibly explained by relativistic bulk motion of a compact "knot" component. If the knot moves with the same type of rectilinear motion detected in known superluminal sources (Cohen and Unwin 1982), it must propagate through an angular displacement $\delta\theta > 2$ mas in the 3.25y between previous VLB observations. Third, the appearance of the phase-shift spectrum implies that <u>each</u> absorption line forms in a multi-cloud configuration in which the diameter of each cloud $\theta_{CLOUD} < 0.3 \theta_s$. As a result the source illuminates 4 or more multi-cloud regions having diameters of ∼ 25 pc.

In the <u>extrinsic</u> model, variations in line depth result from the projected motion of the knot across cloud boundaries. To estimate variations expected in observable quantities we have constructed a Monte Carlo model in which cloud parameters such as location, individual velocity, optical depth, and internal velocity dispersion are selected with a random-number generator. Single-dish and VLB spectra are calculated for a sequence of parameters until agreement with the initial (i.e., before 1978 August) spectra is obtained. We then displace the knot through its own diameter in a series of radial excursions to the boundary of the non-variable source. Each displacement results in variations in line depth and in the velocity centroids of the line profiles. We find that models adjusted to simulate the large changes in the line depth of feature D (see previous paper) also predict that the velocity centroid of this feature varies by more than 0.75 kms^{-1}, the 3-σ upper limit, in 2 out of 10 diameter displacements. In the 4.25y monitoring period there would have been ≃ 10 such displacements, and so this model predicts at least <u>two shifts in velocity centroid that are larger than observed</u>. For this reason, we are currently investigating modifications in which the knot undergoes non-linear motions behind a <u>single cloud</u> of each configuration. In this case changes in line depth would be due to gradients in optical depth.

In the <u>intrinsic</u> scenario, we place 0235 within ∼15 kpc of the absorbing HI. We have not considered in any detail how this physical association came about, although ejection of 0235 from the nucleus of the galaxy containing the HI is a possibility. In any case the 21 cm continuum radiation will be so intense at the HI that radiative excitations will compete with collisions in determining the hyperfine level populations of the hydrogen 1s state. Variations of the continuum will then alter the level populations with a consequent variation in 21 cm optical depth. We have investigated these effects by solving the coupled, time-dependent transfer and population rate equations for a wide range of separation distances d, gas densities n, and fractional increases in continuum strength x. In order to reproduce the variations observed in the line depths of the 4 deepest features without introducing outbursts stronger than x≈0.8 we had to restrict d between 2.5 and 5.0 kpc, and n between 100 and 400 cm^{-3}. The resulting "light curves" for the line depths explain the synchronous behavior observed in these features between 1980 August and October quite naturally. Moreover, the absence of detectable fluctuations in velocity centroid is a natural consequence of this model. But, the lack of a continuum increase during those epochs when the line depth of feature D had decreased substantially is difficult to understand. With the increased sampling frequency of the current monitoring program we will be able to rule out this model if future decreases in line depth are not preceded by continuum outbursts.

In summary, comparison of the model predictions with the data leads to the following conclusions: 1) The lack of significant fluctuations in the velocity centroids of the line profiles indicates that the "knot" source has not propagated across the expected angular displacement, $\delta\theta > 2$mas. Rather, the motion of the knot is limited behind a single cloud of each of 4 configurations. The question for the extrinsic model is whether motions in which $\delta\theta \ll 2$mas can result in the changes observed in line depth and in VLB phase-shift.
2) An important distinction between the extrinsic and intrinsic models is that in the extrinsic case, changes in VLB phase-shift will probably be accompanied by changes in line depth, while in the intrinsic case no such correlations are expected.

References

Cohen, M H. and S. C. Unwin, 1982, this volume, p. 345.
Cotton, W., 1981, private communication.

VARIABLE RADIO SOURCES

R.Fanti and L.Padrielli
Istituto di Fisica and Istituto di Radioastronomia,Bologna

M.Salvati
Istituto di Astrofisica Spaziale,Frascati

1. INTRODUCTION

Flux variations are a common feature of flat spectrum compact extragalactic radio sources. Detailed analysis and quantitative comparisons with theoretical models (e.g. van der Laan,1966) are difficult due to the complex characteristics of the flux variations,which generally appear to consist of different outbursts blended together in time. Nevertheless, the general consensus is that the basic process has been correctly identified and consists in an expansion of a synchrotron radiating plasma cloud of relativistic electrons and magnetic field partially opaque to its own radiation. The main differences between data and predictions of the theory are that the variations propagate too fast and with too large amplitude toward lower frequencies. This behaviour however may be indicative of continuous energy supply and consequent accelerated expansion.

At the moment the most troublesome problems are the flux variations at low frequencies (L.F.V.).After several years of doubts concerning their reality,in the past five years evidence has been accumulated on their reality (see e.g.: Cotton 1976;Mc Adam 1978,1979,1980;Condon et al.1979; Fanti et al.1979,1981;Fisher and Erickson 1981). So far there are more than 100 sources known or suspected to be low frequency variables.However,no definite explanation has been presented yet. First it seems that generally there is no correlation between the classical high frequency variability and that at low frequency. On the contrary,the two sets of events seem discontinuous,justifying the use of the terms "high frequency variability" ($\lambda<20$ cm.) and "low frequency variability" ($\lambda>20$ cm.).

A second crucial question is the source size. Since it is usually supposed that the time scale for variations cannot be less than the light travel time across the source,an upper limit on the linear dimensions of the variable region can be estimated. When combined with the distances calculated in the usual way from the red-shift,this generally indicates, for sources varying at low frequencies,angular dimensions which are so small that the inverse Compton limit for an incoherently radiating synchrotron source is exceeded.

2. STATISTICS ON OCCURRENCE OF L.F.V.

The occurrence of L.F.V. in complete samples of radio sources has been examined by several authors, notably Cotton (1976), Condon et al. (1979), Mc Adam (1980), Fanti et al. (1981). The phenomenon appears very common in samples of flat spectrum sources (30% to 50% showing fractional flux changes $\Delta S/\bar{S} > 5\%-10\%$). It is important to realize that here by flat spectrum source we mean a source with spectral index $\alpha < .5$ around the frequency at which variability is searched for. This definition includes sources with a really flat spectrum over a broad frequency range and sources whose spectrum, flat around the search frequency, steepens at $\nu > 1$ GHz, reaching the typical slope of a transparent synchrotron radiating radio source. There is no difference in the occurrence of L.F.V. between these two classes of sources. The best, at the moment, estimates of occurrence of L.F.V. among normal straight spectrum sources are those of Cotton (1976) and Mc Adam (1980). They indicate that fractional changes with $\Delta S/\bar{S} > 20\%$ are found with a probability of 2%-5%, or a factor > 4 less frequent than for flat spectrum sources.

In general quasars account for the largest fraction of the reported L.F.V. Four BL Lac objects (AO 0235+164, 0735+178, OJ 287 and BL Lac itself) also show this behaviour. The situation is less clear for radio galaxies. At present there are $\sim$ 30 low frequency variables identified with galaxies. One of them is 3C 120. Also 3C 84 is a possible one. Most of the remaining ones are generally optically weak (m > 18) and little is known of their optical properties.

Finally we note that at least three sources (3C 120, 3C 279, 3C 345) out of the four reported to exhibit superluminal motions (Cohen et al. 1977) also show low frequency variability.

3. CHARACTERISTICS OF THE L.F.V.

Generally the pattern of the variations is similar to that seen at centimeter wavelengths. In a minority of cases the light curve can be described as a superimposition of one or more non overlapping outbursts, lasting several months, over a relatively stable flux level. In the majority of the sources, however, the variations are more complex, although they can still be described in terms of partly or totally overlapped outbursts. In these cases it is very difficult to establish the level of an eventual underlying stable component.

Typical time intervals from relative maxima to minima range from several months to a few years, although in some cases flux changes are seen to occur within 1 - 3 months. There seems to be a relative scarcity of flux variations lasting more than 2 - 3 years (Mc Adam 1978). The frequency of the outbursts, corrected to the source proper frame, is about 0.8 years^{-1}, similar to that found at short wavelengths.

Besides the case of PKS 0736+01 (discussed by Mc Adam, 1978), there are no other cases of flux variations such to suggest negative dips instead of positive outbursts.

A number of sources show long periods of stable flux level, lasting a few or several years. This is the case for 3C 454.3 during 1973-1974 and CTA 102 from 1975 to the present epoch. Note that while in its quiescent period CTA 102 was at its minimum flux density level for the last 15 years, 3C 454.3 in its quasi stable flux level was definitely higher, by 2 - 3 Jy, than the well defined deepest minima observed in 1970 and 77.

4. FLUX DENSITY VARIABILITY AT DIFFERENT FREQUENCIES.

Although sources varying at centimeter wavelengths appear to have higher probability of varying also in the decimeter domain, this however does not mean that variability is due to the same event occurring at all frequencies with frequency dependent amplitude. The only case of this type known at present and conforming to the standard model of the opaque synchrotron emitting and expandind cloud is BL Lac. This source had three periods of strong activity from 1970 to 1980, during which a reasonable good flux monitoring was obtained in the range from 10 GHz to .4 GHz at several frequencies. In these three periods outbursts were first seen at very high frequency and subsequently drifted to lower frequencies, with somewhat reduced amplitude. In several other sources, however, the low frequency outbursts seem clearly unconnected with the higher frequency ones, as is often indicated by the quiescent level of the light curve at an intermediate frequency (e.g.:2.7 GHz) and by the lack of any preceding comparable amplitude event at higher frequencies. Furthermore, the evidence gained from studies of severall sources (see e.g.: Cotton and Spangler 1978; Fisher and Erickson 1981) shows that the low frequency variability extends over a broad band, confined to $\nu<2$ GHz and occurs roughly simultaneously at all frequencies, with amplitude decreasing, although in a not well known way, at increasing frequency.

In some sources a close coincidence in time has been noted of high frequency and low frequency maxima (Spangler and Cotton, 1981), which however were disconnected throughout the frequency range. While it is difficult at present to evaluate the statistical significance of these coincidences, it is certainly important to pursue such a kind of search on a larger sample of sources over a longer period of time.

5. THE TIME SCALES OF VARIABILITY AND THE INVERSE COMPTON PROBLRM

The generally adopted definition for the time scale variability is:

$$\tau_{var} = (1+z)^{-1}(d \ln S(\nu)/dt)^{-1} \sim (1+z)^{-1}\Delta S_{max} \Delta t/\Delta S$$

where $\Delta S/\Delta t$ is the rate of change of the flux and ΔS_{max} the maximum flux of the varying component. There is considerable uncertainty on ΔS_{max}, depending on whether the emission of the varying source is seen superimposed on an underlying stable component or not. In the first case ΔS_{max} can be obtained by subtracting the flux level of the stable component to the maximum observed flux S_{max}. The flux level of the stable component can be estimated tentatively from the deepest minima of the light curve, and $\Delta S_{max} \sim \Delta S$, the observed flux change. In the second case ΔS_{max} is taken equal to S_{max}. With the second assumption we obtain longer time scales, the ratio between the two estimates being $S_{max}/\Delta S$. In the following we adopt the second hypothesis, which minimizes the problems associated with fast variability, although we should keep in mind that the most realistic situation is certainly intermediate between the two cases. If more than one distinct burst is present in one source we consider a variability time scale for each of them. The distribution of τ_{var} for sources taken from the Bologna (Fanti et al. 1979,1981) and Molonglo programs (Mc Adam 1979), has a mean value around 2 - 3 years, although variations as fast as 1 - 3 months are seen (see, e.g., Fanti et al. 1979; Mc Adam 1979; Spangler and Cotton 1981). The observed time scale distribution is similar to the corresponding one at centimeter wavelengths.

If the variability is intrinsec to the source, on the basis of the "causality argument", we expect that the linear radius (Jones and Tobin 1977) r should be :

$$r/\delta < c \tau_{var} = r_{var}$$

(δ is a numerical factor of the order of few units, which describes the flux changing law : $S \propto t^\delta$; $\delta = 3$ for the standard expansion model). We take $\delta = 1$, since we have no clear information on the process producing the variability. The "variability diameters" computed in this way are not taken as representative of the whole source but only of the variable component seen at that particular epoch ($\pm \tau_{var}$). The angular diameter of the varying component ($q_o = 1$) is given by :

$$\theta < \theta_{var} = 2 r_{var}/distance = 2 r_{var}(1+z)^2 H(cz)^{-1}$$

Angular sizes computed in this way, using an Hubble constant $H = 100$ Km $s^{-1}Mpc^{-1}$ are mostly in the range of 0.1 to 1 m.a.s. These angular sizes can be used to compute the source brightness temperature T_b. Typical values are $\sim 10^{14} - 10^{15}$ °K, occasionally up to 10^{16} °K. These values exceed by 2 to 4 orders of magnitude the 10^{12} °K limit for incoherent synchrotron radiation.

6. DIRECT MEASURES OF ANGULAR SIZES OF LOW FREQUENCY VARIABLES

VLBI observations at relatively low frequencies (Readhead et al.1977 at 50 cm.; Romney et al.1981, at 18 cm.) all show that low frequency variables are generally resolved, often with jet-like morphology with overall sizes of few m.a.s. to several tens of m.a.s. The components are often partially resolved as well with sizes of 1 - 2 m.a.s. They show that a substantial fraction of the flux originates in regions of brightness temperature $T_b < 10^{12}$ °K, although we cannot exclude the existence of much smaller subcomponents with sizes comparable with that implied by the time scale variability.

A much higher resolution ($10^{-4} - 10^{-3}$ m.a.s.) is achievable by means of interstellar scintillation studies. A number of low frequency variables have been studied by means of this technique (e.g. Dennison and Condon 1981),always with negative results. For several of these sources the upper limits to the interstellar scintillation index have been considered an evidence against the existence of high brightness radio components. However a choice of conservative values of H and q_o, as those we have taken, or of the interstellar scintillation angular scale, or invoking intergalactic angular size broadening, leaves some margin for the existence of high brightness components. Furthermore,the lack of interstellar scintillation does not necessarily apply to the rapidly variable fraction of the flux (actually the one which needs to be small),since the scintillation measures generally do not coincide with major outbursts. Even with our conservative assumptions on τ_{var}, the sources for which the interstellar scintillation measurements coincide, within τ_{var}, with an outburst, are half a dozen at most. For each of these the choice of conservative parameters allows for the presence of high brightness components. If we were to have taken the other less conservative assumption for τ_{var}, only for DA 406 would there have been a coincidence between an outburst and the observation of interstellar scintillation (august 1977). In this case the less conservative assumption on τ_{var} would imply $\theta_{var} \sim 7 \cdot 10^{-5}$ arcsec against the lower limit, from absence of interstellar scintillation, of $6 \cdot 10^{-5}$ arcsec, and the presence of a high brightness component would be just marginally possible.

7. X RAY OBSERVATIONS OF LOW FREQUENCY VARIABLES

If the low frequency variables are as small as implied by their τ_{var} and if they radiate at radio frequencies by incoherent synchrotron process, the large brightness temperatures implied would require an enormous inverse Compton flux during a radio outburst. Evidence that this is not the case is obtained from the X ray literature data (e.g. HEAO-A observations in the band 3 to 17 kev of several radio sources during active phases, reported by Marscher et al.1979). No X ray flux was detected. The deduced brightness temperatures are generally $< 5 \cdot 10^{11}$ °K.

8. DISCUSSION

There are essentially two ways to solve the inverse Compton problem which arises in the case of L.F.V. One way is to look for alternative radiation mechanisms allowing higher brightness temperatures. The other is to invoke explanations which, assuming the ordinary synchrotron process, are nevertheless able to reconcile big sizes and fast variations. To this second class belong models based on relativistic expansion or models involving phase effects. The explanations should also account for other characteristics of the variability, namely the spectral behaviour, the statistics on the occurrence of L.F.V.,the source structure,etc..

A third possibility would be to consider the variations as due not to changes in the emission region,but rather to some kind of modulation of a constant source by variable scattering (scintillation) or absorption in an intervening medium. However considerations based on time scales and correlation bandwidth indicate that ionospheric,interplanetary,interstellar and intergalactic scintillation are unlike processes. The suggestion by Shaperovskaya (1978) that L.F.V. are due to scintillation in the weak focusing regime,occurring in nearby galactic structures (loops,spurs,ridges,..) is no longer supported by the sky distribution of the known or suspected variables. Absorption models can work only if phase effects are present and therefore will be considered in the class of phase models.

The inverse Compton problem is solved if small pitch angle synchrotron radiation or coherent processes are assumed to be responsible for the variation (e.g.: Cocke and Pacholczyk 1975; Cocke et al.1978). Of course only the varying flux needs to be explained by these processes, while the more stable underlying component could still be due to the ordinary synchrotron process. For instance,the sources with a normal high frequency spectrum flattening at low frequencies,where variability is seen,have the typical characteristics of a synchrotron radiating source. The variations would be due to a distinct component radiating by a different process at low frequencies only and responsible for the variation. In this hypothesis one would not expect a one-to-one correspondence between high and low frquency events,with the former evolving continuously into the latter;but a correlation between levels of activity in the two radio domains would be possible if the ultimate engine is the same. Also there would be no difficulties in understanding simultaneous high and low frequency variations. The evidence against this class of models lies essentially in the absence of interstellar scintillation,which however still leaves some marginal possibilities for them. Furthermore it would be difficult to understand the correlation between occurrence of L.F.V. and the shape of the radio spectrum near the frequency of variability.

Apparent violation of the causality argument occurs if the emitting regions have bulk motions at speed close to the speed of light. In this case the intrinsecally long history of the source is compressed into a small parcel of the observer's time. This class of models was proposed a long time ago (Rees 1967) just to explain the fast variations known at that time and somewhat later has been used to explain apparent superluminal motions found by VLBI observations. Many variants have been proposed (see,e.g.,Blanford and Mc Kee,1977;Salvati,1979;Blanford and Konigl 1979,and references therein) and we examine only a few difficulties. Relativistic motions could be of two types : i) relativistic isotropic expansion of one component; ii) relativistic collimated expansion (or beamed ejection). In both cases apparent brightness temperatures deduced from time scales are $\sim \Gamma^3$ larger than the values measured in the source frame (Γ is the Lorentz factor of the expansion). Typical Lorentz factors needed are around 10 (18 for 3C 454.3 in the 1978 outburst;27 for PKS 1524-13 in the 1977 outburst;21 for PKS 1504-16 in 1975). In the first case (isotropic expansion) one would naturally explain the large occurrence of L.F.V. in samples of flat spectrum radio sources since no preferred orientation between source and observer is necessary. Difficulties are essentially due to the energetics implied, which is generally > 10^{58} ergs and in some cases exceeds 10^{60} ergs. The energetics can be reduced by a factor Γ^2 if the expansion is beamed (case ii). Such a model would also fit the generally accepted schemes for superluminal motions (Blanford and Konigl 1979). However in this case one requires a preferred orientation between the source and the observer's line of sight, at variance with the large fraction of sources showing variability in complete samples, and the same explanations should be invoked as for superluminal motions (Scheuer and Readhead 1979). In this class of models, which maintain the synchrotron process as that responsible for radiation, the spectral characteristics of the varying component and the simultaneity at the various frequencies would indicate that the emitting regions are transparents and the variations would be due, for instance, to an increase in the number of radiating particles, followed by rapid expansion. The typical linear sizes would be of the order of 10 pc (case ii) or 100 pc (case i) and might be also situated at large distances from the core of the source.

Phase effects have been discussed in connection with flux variability in Fanti and Salvati (1980). The merit of these models resides in reducing the energetics involved in the phenomena, compared with the relativistic models. They may also be coupled with relativistic expansion models, but require more moderate values of Γ. However, if associated with jet-like source morphology they still require the same preferred orientation as for beamed relativistic motion.

The phase models class includes also absorption models where a suitable screen orthogonal to the line of sight changes opacity when impinged on by a signal coming from a distant center. The most discussed model of this type is that of Mc Adam (see Marscher 1979 and Condon et al. 1979) where the screen is a neutral hydrogen layer which is ionized by the signal and becomes opaque. This model would naturally produce superluminal "negative" outbursts, whereas the general situation is more complex. Furthermore this model has difficulties in explaining the simultaneity of high frequency and low frequency events. An absorption model capable of producing positive outbursts instead of negative ones can be envisaged if the screen is usually partially opaque and the triggering signal makes it more transparent, followed by a reincrease of opacity. In case of simultaneity of high frequency and of low frequency events, the high frequency one would be closely connected with the triggering signal. A model of this type, where the screen is a partially ionized hydrogen layer, has been suggested by Spangler and Cotton (1981). This version of the absorption model is strongly suggested by the finding that the L.F.V. occur at a frequency where the radio spectrum flattens, a clear indication of the presence of an absorption process. Since the low frequency flattening is generally believed to be due to synchrotron self absorption, we are led to think that the varying absorber should be looked for more in the relativistic particles - magnetic field system

REFERENCES

Blanford,R.D.,and Mc Kee,C.F.,M.N.R.A.S.,180,343
Blanford,R.D.,and Konigl,A.,1979,Astrophys.J.,232,34
Cocke,W.J.,and Pacholczyk,A.G.,1975,Astrophys.J.,195,279
Cocke,W,J,et al.,1978,Astrophys.J.,226,26
Cohen,M.H.,et al.,1977,Nature,268,405
Condon,J.J.,et al.,1979,Astron.J.,84,1
Cotton,W.D.,1976,Astrophys.J.,204,L63
Cotton,W.D. and Spangler,S.R.,1978,Astrophys.J.,228,L63
Dennison,B.,and Condon,J.J.,1981,Astrophys.J.,246,91
Fanti,R.,et al.,1979,Astron.Astrophys.Suppl.,36,359
Fanti,R.,et al.,1981,Astron.Astrophys.Suppl.,45,61
Fanti,R.,and Salvati,M.,1980,Proc.5theurop.meeting astron.,Liège
Fisher,J.R.,and Erickson,W.C.,1981,Astrophys.J.,242,884
Jones,T.W.,and Tobin W.,1977,Astrophys.J.,215,474
van der Laan,H.,1966,Nature,211,1131
Marscher,A.P.,1979,Astrophys.J.,228,27
Marscher,A.P.,et al.,1979,Astrophys.J.,233,498
Mc Adam,W.B.,1978,Proc.Astron.Soc.Austral.,3,283
Mc Adam,W.B.,1979,Proc.Astron.Soc.Austral.,in press
Mc Adam,W.B.,1980,Proc.Astron.Soc.Austral.,in press
Readhead,A.C.S. et al.,1977,Astrophys.J.215,L13
Scheuer,P.A.G.,and Readhead,A.C.S.,1979,Nature,227,182
Rees,M.J.,1967,M.N.R.A.S.,135,345
Romney,J.,et al.,1981,in preparation
Salvati,M,1979,Astrophys.J.,233,11
Shaperovskaya,N.Ya.,1978,Soviet Astron.,22,544
Spangler,S.R.,and Cotton,W.D.,1981,Astron.J.,86,730

DISCUSSION

O'DELL: I have two relevant remarks based upon recent work by J. J. Broderick, J. J. Condon, B. K. Dennison, J. E. Ledden and myself:
(1) Using our three-epoch observations at 318 MHz, we also find that the typical duration of events appears to be < 5 years in low-frequency variables. (2) Regarding the association of low-frequency variability with variability at other frequencies, we have found a statistically significant ($> 99\%$ confidence) correlation of low-frequency variability with optically violent variability as phenomena (not event-to-event correlation). In particular, of 24 quasars common to our complete sample and the sample monitored, by the Florida group, for optical variability, we find the following results:

	Not Opt. Var.	Opt. Var.	Opt. Violent Var.
Low-freq. var.	0	2	6
Possible low-freq. var.	1	0	1
Not low-freq. var.	7	7	2

INTERSTELLAR SCINTILLATIONS AS A TOOL FOR INVESTIGATIONS OF HYPERFINE STRUCTURE IN EXTRAGALACTIC RADIO SOURCES

L. M. Ozernoy and V. I. Shishov
Lebedev Physical Institute, Academy of Sciences of USSR, Moscow

Attempts to use interstellar scintillations (ISS) to study the fine angular structure of quasars and galactic nuclei have not yet met with any success. The main reason for this lies in the comparatively large angular size of the nuclear structure which has been investigated. If dependencies on wavelength for the scintillation index, flux and angular size of a source are taken into account (Ozernoy and Shishov, 1980), it appears that centimeter wave band rather than decimeter wave range used in previous work is most appropriate for observing ISS.

ISS and Angular Structure. A radio source can be regarded as a point one if its intrinsic angular size, ϕ_o, is less than $\phi_p \equiv a/L \approx 10^{-5}$ sin b. Here $a \approx 3 \times 10^{10}$ cm is the characteristic size of fine scale irregularities in the electron density of interstellar medium, and $L \approx 150 (\sin b)^{-1}$ pc is the thickness of the scattering layer for a source with galactic latitude b.

VLBI Observations of Scintillations. Observations of a radio source by VLBI allow the isolation of a compact component. Figure 1 shows the theoretical wavelength dependence of the scintillation index m both in the case when the angular size of a source, ϕ, does not depend on λ (dashed line) and in the case when ϕ is due to scattering by irregularities intrinsic to the source (heavy line), i.e. $\phi = \phi_s^{(o)}(\lambda cm/3)^2$. Measurements of ϕ's for a number of quasars at meter wavelengths indicate that $\phi_s^{(o)} \lesssim 10^{-5}$ arc sec. If the scattering is not negligible, an optimum wavelength band for observing ISS is 3-30 cm. The contribution of the compact component, as compared with the extended one, is presumably the greatest at the short-centimeter side where m is expected to be 1-10%. The time-scale of the scintillations is expected to be $T = a/v \sim 3$ hours, v being the earth's orbital velocity.

VLA Observations of Scintillations. A system with angular resolution worse than about 0".01 cannot separate the scintillating from the non-scintillating halo (the latter is usually responsible for the main part of the radio flux from a quasar). The resulting scintillation index $m(\lambda)$ to be expected is shown in Fig. 2 both in an idealized case

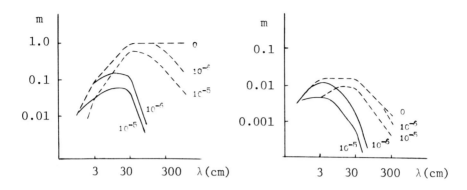

Figure 1. (left). Scintillation index as a function of wavelength when a source consists of the core alone.
Figure 2. (right). The same as in Fig. 1 when a source consists of the core and halo.

of the kernel whose angular size is independent of λ (dashed line) and in a more realistic case when the kernel size is determined by scattering (heavy line). In the latter case the ratio of the core flux, I_c, to the total flux, I, increases toward shorter wavelengths (Andrew et al. 1978) and can be fitted by the relation $I_c/I = (\lambda cm/3)^{-1}$. The optimum wave range to observe ISS is, as before, a band around $\lambda = 3$ cm; the characteristic scintillation time remains the same ($T \sim 3$ hours). But the scintillation index is decreased to roughly 1%. Nevertheless, the VLA system offers a unique opportunity to measure the ISS by the differential method. Simultaneous measurements of the flux both from the kernel and from a non-scintillating component (e.g. a halo) allow us, in principle, to avoid the multiplicative inferences.

Implications. ISS can be used successfully for study of hyperfine ($3 \lesssim \phi \lesssim 100$ microarc sec) structure in radio sources and, therefore, would nicely complement the VLBI at cm wavelengths. Investigations of the substructure of compact radio components in QSS and galactic nuclei seem to be very effective in a combination ISS + VLBI. A combination ISS + VLA would allow to observe simultaneously several compact components and measure the scintillations of the hyperfine structure relative to the nonscintillating one of a larger size. Search for ISS from Sgr A West is of particular interest.

References

Andrew, B. H., MacLeod, J. M, Harvey, G. A., and Medd, W. J.: 1978, Astron. J. 83, 863.
Ozernoy, L. M., and Shishov, V. I.: 1980, Pis'ma Astron. Zh. 5. 171 (Sov. Astron. Lett. 5, 92).

RADIO FLUX FLICKER OF EXTRAGALACTIC SOURCES

D. S. Heeschen
National Radio Astronomy Observatory, Charlottesville, VA

Compact sources (compactness evidenced by flat/complex spectra) display a "flicker" in their intrinsic centimeter wavelength radiation, with an amplitude of about 2% and a characteristic time scale of a few days.

Two hundred thirty sources were observed daily for 10-25 days, on three occasions at 9 cm and on one occasion at 6 cm wavelength. The 9 cm observations were made with the 92 m telescope at Green Bank; the 6 cm observations with the 43 m telescope. Sources were selected principally from the 5 GHz surveys (Pauliny-Toth et al. 1972a,b, 1978). They were selected only on the basis of flux density and position. Most are stronger than 1 Jy at 6 cm, and lie between 35° and 60° declination; all are stronger than 0.4 Jy. Observations consisted of transits nominally through the midpoint of a dual-feed configuration. The ratio of response in the two feeds is then a measure of pointing error in declination, which was in turn used to correct the observed antenna temperature. For each source and each observing period, a mean temperature T and its standard deviation σ_T were calculated from the daily observations. Finally, a fluctuation index $\mu = \sigma_T/T$ was obtained for each source in each observing period.

Only nighttime observations will be considered in the analysis. It was found that daytime observations have significantly larger scatter, probably due primarily to thermal effects in the telescope structure. The sources were separated into two classes: (s) = steep, well defined radio spectra; (f) = flat, complex, or poorly defined spectra.

The distributions of fluctuation indices, $\psi(\mu_s)$ and $\psi(\mu_f)$, are shown in Table 1 and Figure 1. In all observing periods, at both wavelengths, the mean value of μ for (f) sources is significantly greater than that for (s) sources. The distribution $\psi(\mu_s)$ can be taken as representing the system performance; flux density of a strong source can be measured with an accuracy of about 1-1/2% in one transit with the 92 m telescope. The distribution $\psi(\mu_f)$ then indicates that flat spectra sources have an additional scatter in their measured fluxes. The indices, μ, are not correlated with pointing errors, elevation, flux density, solar elongation, or galactic latitude. For these and other reasons, instrumental or atmospheric effects can be ruled out as the cause of the difference between $\psi(\mu_s)$ and $\psi(\mu_f)$, as can interplanetary and interstellar scintillation. Thus the difference is intrinsic to the sources themselves.

Table 1
Mean Values of Fluctuation Indices, μ

period	wavelength	μ (No.) (s)	μ (No.) (f)
May-June 79	9 cm	0.016 (52)	0.027 (49)
May-June 80	9	0.014 (52)	0.024 (51)
Nov-Dec 80	9	0.016 (27)	0.024 (36)
June 80	6	0.014 (18)	0.024 (17)

An autocorrelation analysis was done on the data of the May-June 1979 observing period. The flux density fluctuations of (s) sources had a characteristic time of about one day, consistent with the sampling rate. The characteristic time scale for fluctuations of the (f) sources, however, was 3-4 days.

I conclude that compact extragalactic sources exhibit a "flicker" in their intrinsic centimeter wavelength radiation, with an average amplitude of about 2% and a characteristic time scale of 3-4 days.

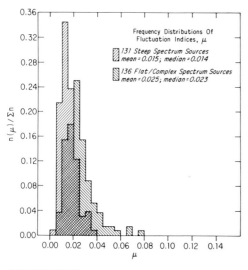

Fig. 1. Frequency distributions of fluctuation indices, μ. Nighttime data from the three observing periods at 9 cm wavelength are combined. The two distributions are significantly different. The 131 values of μ for steep spectrum sources have a mean of 0.015 and a median of 0.014. The 136 values of μ for flat/complex spectrum sources have a mean of 0.025 and a median of 0.023

REFERENCES

Pauliny-Toth et al. 1972a. A.J. 77, 265.
Pauliny-Toth and Kellermann. 2972b. A.J. 77, 797.
Pauliny-Toth et al. 1978. A. J. 83, 451.

BROADBAND STUDIES OF COMPACT SOURCES

T.W. Jones and L. Rudnick
Department of Astronomy
School of Physics and Astronomy
University of Minnesota

ABSTRACT

Progress is reported on a program to study the polarization, spectral and variability properties of compact sources at centimeter and millimeter wavelengths. Source characteristics divide according to the broadband shapes of their spectra.

We report here progress on an extensive program to study the physical properties of compact radio sources through their polarization, spectral and variation characteristics. The program includes simultaneous polarimetry at centimeter and millimeter wavelengths for a sample of 20 active sources. When possible, optical and infrared photometry and polarimetry have been obtained, as well. In addition, we have obtained VLA observations (at 2, 6 and 20cm) of an unbiased sample of 40 flat spectrum ($\alpha > -.5$, $S \sim \nu^\alpha$) sources in the S4 survey and a sample of 20 strong nuclear components of classical doubles. This observational work is supported by theoretical calculations of the appearance and evolution of compact sources. An analysis of the spectral and variability characteristics of the active sources has been reported elsewhere (Jones et al. 1981). A preliminary study of the S4 sources also has been reported (Rudnick and Jones 1982).

The study has shown a strong relationship between the polarization and variability of sources and their spectral shape. If we divide the S4 sources into groups with straight or power spectra (╲, e.g., 0954+55), simple convex or "humped" spectra (⌒, e.g., 0923+39), and complex spectra (∿, e.g., 1641+39), we find that only the complex sources commonly exhibit large amplitude variability on timescales of a few years. The simple convex and straight spectrum sources appear relatively constant, although they are mostly unresolved at 6cm on the VLA ($\leq 0''.2$) and, therefore, of kiloparsec or smaller dimensions. The straight spectrum sources have a median polarization ~3% at 2cm, but the value decreases to ~0.3% at 20cm indicating considerable Faraday depolarization. This can be explained if these sources have internal electron densities ~10^{-2}cm^{-3} as if, for example, they are confined in the inner regions of their parent galaxies. In contrast, both sets of simple convex and complex spectrum sources show little, if any, wavelength dependence to their <u>median</u> polarizations (~1% and ~2% respectively). Individually the simple convex spectrum polarizations also show little wavelength dependence. VLBI observations will be especially helpful in determining if these sources are indeed a physically distinct class of object. Preliminary analysis of polarization data on double radio source nuclei shows them to be unpolarized at the 1% level. We have not yet investigated whether this is associated with their spectral properties, or requires different physical conditions than the other compact sources.

Individually, the complex spectrum sources show large polarization changes from one wavelength to another. Our observations of active sources at 9mm and 3mm using the NRAO 11m telescope indicate a continuation of this pattern to very high frequencies. On the other hand, the plane of polarization seems to vary only slowly with wavelength in these sources. We interpret these data as evidence for a composite source structure with distinct regions visible at different wavelengths. Although the magnetic field is apparently tangled by different degrees at different locations, there must be some large scale ordering to the field as well. Data are now being analyzed which should allow us to explore magnetic fields in the longer wavelength emitting regions by separating out effects of galactic Faraday rotation.

This work is supported by the NSF through grant AST 79-00304. Radio observations were performed at facilities of the National Radio Astronomy Observatory and in collaboration with H. Aller at the University of Michigan Radio Observatory.

TABLE 1

Source Characteristics Summary

Spectrum or class	10 year Variability	6cm Polarization	Wavelength Dependence of Polarization
╲	≤10%	~2%	Faraday depolarized
⌒	≤10%	~1%	little
∼	≥50%	~2%	individually complex
nuclear comp.	small?	<1%	

REFERENCES

Jones, T.W., Rudnick, L., Owen, F.N., Puschell, J.J., Ennis, D.J. and Werner, M.W.: 1981, Ap.J. 243, pp. 97-107.
Rudnick, L. and Jones, T.W.: 1982, Ap.J. (submitted).

POLARIZATION VARIABILITY OF SOME COMPACT RADIO SOURCES

M.M. Komesaroff, D.K. Milne, P.T. Rayner, J.A. Roberts and D.J. Cooke
Division of Radiophysics, CSIRO, Sydney, Australia

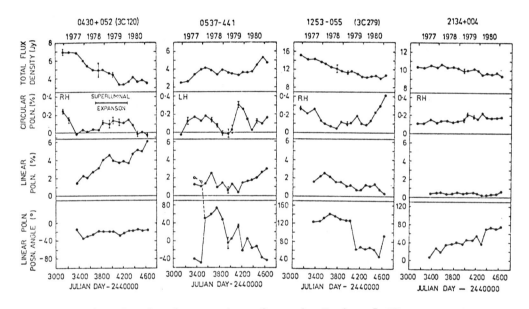

Figure 1. Sample observations from the Parkes 5 GHz programme.

Figure 1 shows observations for four sample sources from the Parkes 5 GHz polarization monitoring programme. Interesting features illustrated include

- Sudden changes of the position angle of the linear polarization by $\gtrsim 70°$ in PKS 0537-441 and 1253-055 (3C279).

- A linear increase in the position angle of the polarization of PKS 2134+004 through 70° over $3\frac{1}{2}$ years.

- Distinct bursts of circular polarization in PKS 0430+052, 0537-441 and 1253-055. In PKS 0430+052 (3C120) such a burst coincides with the possible superluminal expansion (Walker et al., 1981). In PKS 1253-055 (3C279) a burst of circular polarization is currently occurring at a time of very low linear polarization.

Sudden changes in the direction of linear polarization of up to 90° can result from changes in the relative intensity of two source components having nearly orthogonal directions of polarization. With this process the *degree* of linear polarization (relative to one component) necessarily drops to a very low level at the time of the sudden change of position angle (see Fig. 2, where for simplicity the two

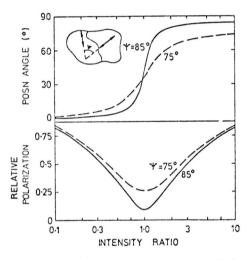

Figure 2. Two-component model.

components are assumed to have the same degree of polarization). In the case of the sudden large changes of position angle in PKS 0537-441 and 1253-055 it is by no means clear that the degree of linear polarization dropped to a low level.

The behaviour of PKS 2134+004 suggests that it may belong to the class of sources, discovered by Ledden and Aller (1979), in which a linear swing in position angle occurs simultaneously at several frequencies (ruling out Faraday effects) and with little change in the degree of linear polarization. Since in one case the position angle rotated through 360° this surely implies a physical rotation of the emitting region or at least a rotary motion of a pattern of excitation, e.g. along a helical magnetic field.

Figure 3 illustrates a class of model in which the direction of polarization is the projection on the sky of the generator of a cone as the generator rotates in azimuth about the axis of the cone. The

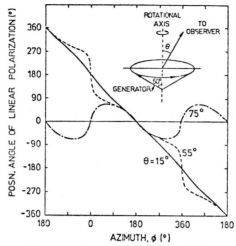

Figure 3. Rotating model.

variation of polarization position angle is shown for a cone semi-angle of 60° and for viewing angles of 15°, 55° and 75° from the rotational axis. The same curves apply to a helix of pitch angle 90°-60° = 30°. Notice that if the swing of position angle is to *exceed* ±90° then the direction of the observer must lie *inside* the cone traced out by the polarization axis. When the direction of viewing lies close to this cone there is a sudden change in the position angle near the time when the polarization axis is most nearly directed at the observer. On some theories a maximum in the degree of circular polarization would be expected at this same time.

References

Ledden, J.E., and Aller, H.D.: 1979, Astrophys. J. Lett. 229, L1.
Walker, R.C., Seielstad, G.A., Simon, R.S., Unwin, S.C., Cohen, M.H., Pearson, T.J., and Linfield, R.P.: 1981 (personal communication).

CM-WAVELENGTH FLUXES AND POLARIZATIONS OF COMPACT EXTRA-GALACTIC X-RAY SOURCES

Margo F. Aller, Hugh D. Aller and Philip E. Hodge
The University of Michigan, Radio Astronomy Observatory

Abstract: Cm-wavelength observations of 15 BL Lac objects are presented. The degree of radio-wavelength variability is compared with the strength of the emission at optical and X-ray wavelengths.

We present observations of the total flux density and linear polarization at 4.8, 8.0 and 14.5 GHz for 15 BL Lac and related type objects obtained with the University of Michigan 26-meter telescope since fall 1979. Almost all of the sources in our sample are well-established X-ray emitters and both active and quiescent objects are represented. The average fluxes and polarizations are summarized in Table I.

Table I
Average Radio Flux and Polarization Data

Source	4.8 GHz S(Jy)	P(%)	χ	8.0 GHz S(Jy)	P(%)	χ	14.5 GHz S(Jy)	P(%)	χ	α
3C 66A	1.68	1.9	87	0.98	2.4	78	0.68			-0.81
0521-365	7.3	2.5	72	6.15	2.6	70	4.2	1.7	66	-0.89
OJ 287	2.43	5.4	86	3.31	6.1	81	4.10	6.1	89	-1.16
Mrk 421	0.71	2.0	128	0.66			0.71	8.0	100	-0.74
Mrk 180				0.34	12.	158	0.40			-0.65
1400+162	0.43	4.4	6	0.41	<4		0.33			-0.91
1413+135	1.07			1.98			2.87	0.9	67	-1.14
1418+546	1.54	1.2	171	2.08	1.6	163	2.19	1.5	92	
4C 14.60	1.35	5.2	153	1.59	7.0	137	1.68	4.7	133	-1.01
Mrk 501	1.21	1.7	147	1.26	2.8	152	1.16	2.5	104	-0.74
I Zw 187	0.21	3.0	54	0.24			0.21			-0.72
3C 371	1.86	1.5	38	2.06	3.0	0	2.08	3.1	139	-0.89
2155-304	0.39	4.9	158	0.36			0.34			-0.57
BL Lac	6.8	1.5	23	6.2	0.3	23	7.4	1.0	45	<-0.92
OY 091	0.4			0.46			0.47			-0.95

The flux densities were generally determined to an accuracy of 0.02 Jy or better; in cases where one decimal digit is indicated, the values were determined to an accuracy of only 0.1 Jy. The degrees of polarization given satisfy the criterion $P/\sigma_p \geq 3$, with the exception of the value for I Zw 187 which is a 2σ measurement.

Large and well-defined radio outbursts in total flux density were observed in OJ 287, the unusual red BL Lac object 1413+135, 1418+546 and BL Lac; 0521-365, 4C 14.60 and 3C 371 showed moderate variability. The well-studied X-ray sources Mrk 421 and Mrk 501 exhibited little variability.

While at optical wavelengths the degree of polarization is by definition high, the degrees of polarization shown here are generally less than 8% and are typically 2-3%. These values are comparable to those we have observed in active quasars but are significantly lower than those observed at optical wavelengths. The radio position angles for 4C 14.60 and 0521-365 are related to the published optical position angles (parallel and perpendicular, respectively). Any possible relation for the other sources is less clear.

Spectral indices formed between 14.5 GHz and approximately 2 keV using published X-ray data are shown in the last column of Table I. The mean value of these indices is -0.86. There is a wide spread in the values of this parameter which may in part be due to the fact that the radio and X-ray data were not taken at the same epoch. The sources which are most highly variable at 14.5 GHz (OJ 287, 1413+135) have spectral indices steeper than -1.1 while in contrast well-studied sources which have shown little or no variability at 14.5 GHz (Mrk 421, Mrk 501) have spectral indices of approximately -0.7.

We conclude that a relatively high degree of polarization at radio wavelengths, while sometimes present, is not a characteristic property of all BL Lac objects. Although some of these objects are extremely active, others are quiescent for periods of at least several years. Based on our small sample of objects there is an apparent anticorrelation between cm-wavelength variability and the strength of the X-ray emission.

We thank the staff of the University of Michigan Radio Astronomy Observatory for their assistance in obtaining the data shown here. This work was supported by the National Science Foundation under grant AST 80-21250.

ROTATING STRUCTURES IN EXTRAGALACTIC VARIABLE RADIO SOURCES

Hugh D. Aller, Philip E. Hodge, and Margo F. Aller
The University of Michigan Radio Astronomy Observatory

Abstract: Four sources have now been found by the Michigan variability program which exhibit large amplitude rotations in polarization position angle with time. The most straightforward explanation for the phenomenon is a physical rotation in the radio emitting region.

Since the discovery of an apparently linear rotation with time in the polarization position angle of 0235+164 during the 1975 outburst (Ledden and Aller 1979), our observations at 4.8, 8.0, and 14.5 GHz have found three other sources, 0607-157, 0727-115 and BL Lac, which have exhibited large-amplitude polarization rotations with time (Aller, Aller and Hodge 1981; Aller, Hodge, and Aller 1981); also, a second rotation event appears to have occurred in 0235+164 starting in late 1980. The parameters for these rotations are summarized below. Evidence for rotation in four additional objects has been presented by Altschuler (1980).

TABLE I

Source	Epoch	Rotation Rate	Observed Range	Comments
0235+164	1975	$-3°.3$/Day	180°	Linear rotation at 8.0 and 14.5 GHz
	1980-81	$-2°.1$/Day	$\simeq 180°$	Rotation at 4.8, 8.0 and 14.5 GHz; gap in data when the source was near the sun makes range uncertain
0607-157	1977-78	$-0°.6$/Day	$\simeq 180°$	Linear rotation; observations only made at 8.0 GHz; range uncertain
0727-115	1977-80	$+0°.3$/Day	390°	Rotation at 4.8, 8.0 and 14.5 GHz; initial constant rotation rate followed by jumps of 55°, 70°, 82° and 68°. A modulation of the polarized flux density is apparent
BL Lac	1980	$-12°$/Day	440°	Absence of rotation at 4.8 GHz: evidence for opacity effects

In the three sources for which we have multi-frequency data the observations cannot be explained by any frequency dependent mechanism (such as Faraday rotation). The observed rates of rotation have ranged over a factor of 40; and only 0235+164 and 0607-157 have exhibited an apparently constant rate of rotation. In 0727-115 the changes occurred in a series of steps. The large size of the change in position angle (more than 360 degrees in 0727-115 and in BL Lac) appears to rule out the acceleration-aberration process suggested by Blandford and Königl (1979) to explain the apparent rotations; the most straightforward explanation appears to be a true physical rotation in the emitting region such as the accretion-disk scenario discussed by Pineault (1980, 1981) or a helical motion in a jet structure (Hodge, Aller, and Aller 1979).

An important characteristic of the large amplitude rotation event in BL Lacertae is that it was not observed at the lowest frequency, 4.8 GHz. This fact together with the time of appearance of the rotation during the evolution of an outburst have led us to suggest that the rotation arose deep in the emitting region and that self-absorption prevented us from observing the phenomenon at 4.8 GHz (Aller, Hodge, and Aller 1981). Except for the periods when 0235+164 and BL Lac exhibited rotations, their polarization position angles were almost constant over several outbursts in total flux density. During these stable periods the polarization position angle may be determined by a geometrical effect, such as emission from an extended jet structure, while during the rotation phenomenon the emitting region deeper in the core may dominate the polarization of the source.

We thank the staff of the University of Michigan Radio Astronomy Observatory, T. Seling, G. Latimer, and K. Holmes, who were directly responsible for obtaining much of the data presented here. This work was supported by the National Science Foundation under grant AST-8021250.

REFERENCES

Aller, H.D., Aller, M.F. and Hodge, P.E. 1981, Astron. J., 86, 325.
Aller, H.D., Hodge, P.E. and Aller, M.F. 1981, Astrophys. J. (Letters), in press.
Altschuler, D.R. 1980, Astron. J., 85, 1559.
Blandford, R.D., and Königl, A. 1979, Astrophys. J., 232, 34.
Hodge, P.E., Aller, H.D., and Aller, M.F. 1979, Bull. Am. Astron. Soc., 11, 709.
Ledden, J.E. and Aller, H.D. 1979, Astrophys. J. (Letters), 229, L1.
Pineault, S. 1980, Astrophys. J., 241, 528.
_____ 1981, Astrophys. J., in press.

DEPOLARIZATION OF EXTRAGALACTIC RADIO SOURCES

M. Inoue
Department of Astronomy, University of Tokyo, Tokyo, Japan
and
H. Tabara
Faculty of Education, Utsunomiya University, Mine 350,
Utsunomiya, Japan

In the last decade correlations of the depolarization parameter λ_d with several parameters, i.e., the radio luminosity L, redshift z, and source size D have been investigated by several authors, where λ_d is the wavelength at which the percentage polarization drops to its half maximum. Kronberg et al. (1972) first pointed out that λ_d decreases with increasing z, while Morris and Tabara (1973) showed that λ_d increases with L and suggested that the depolarization is due to the internal Faraday rotation within radio sources. Conway et al.(1974), on the other hand, suggested that the λ_d-(1+z) relation is primary. Recently, Cohen (1979) showed that the λ_d-z relation may be the remnant of the physically meaningful relation λ_d-L, though the former is statistically real.

In fact, the diagrams of λ_d-L, λ_d-z, and λ_d-(1+z) shown in Figures 1, 2, and 3 all indicate some correlations. This is mainly due to the (implicit) variable z. (λ_d is corrected by a factor $(1+z)^{-1}$ from the observed frame to the emmited frame.) In this paper, we have analysed correlations between these parameters which are all the function of z.

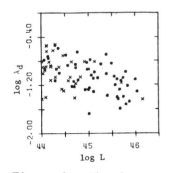
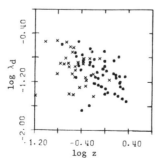
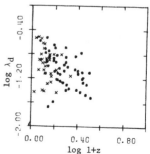

Figure 1. The depolarization λ_d vs. the radio luminosity L for radio galaxies (x) and quasars (•).

Figure 2. The same as Figure 1 but for λ_d vs. redshift z.

Figure 3. The same as Figure 1 but for λ_d vs. (1+z).

The reality of the correlation is tested as follows: λ_d's in the set of parameters (λ_d, L, z) for sample sources are randomly rearranged. From each set of the rearranged parameters, the slope of the straight line derived by linear regression, the standard deviation (SD), and the correlation coefficient were obtained and then compared with those for the original sample sources.

The sample sources are (1) spectral types of S and C_ with the spectral index $\alpha > 0.5$ ($S \propto \nu^{-\alpha}$), (2) regular depolarization type (see Conway et al. 1974), and (3) the radio luminosity $L > 10^{44}$ erg/s, assuming $q_0 = 0.5$ and $H = 50$ km/s/Mpc. Condition (3) is based on the fact that low luminous sources show different polarization properties from higher luminous sources (Morris and Tabara 1973).

Figures 4 and 5 show the slope vs. SD plots for 100 sets of rearranged parameters together with those of the original one. In figure 4, SD of the original sample sources is significantly smaller than those of rearranged sets. However, SD's of the original set for log λ_d -log z (Figure 5) and log λ_d-log (1+z) relations are not so significant. In case of the distribution of the correlation coefficients, the situation is almost the same, and hence we conclude that the log λ_d-log L relation is physically meaningful, and indicates $L \propto \lambda_d^{-4.6}$. This is consistent with the ram pressure model of radio sources.

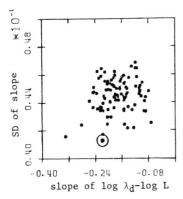
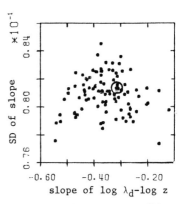

Figure 4. The slopes of linear regression for the rearranged sets of log λ_d-log L and those standard deviation. The original set is shown by the dotted circle.

Figure 5. The same as Figure 4 but for log λ_d-log z.

REFERENCES

Cohen, N. L. 1979, Astron. Astrophys., **71**, 362.
Conway, R. G., Haves, P., Kronberg, P. P., Stannard, D., Vallee, J. P., and Wardle, J. F. C. 1974, Mon. Not. R. Astr. Soc., **168**, 137.
Kronberg, P. P., Conway, R. G., and Gilbert, J. A. 1972, Mon. Not. R. Astr. Soc., **156**, 275.
Morris, D., and Tabara, H. 1973, Publ. Astron. Soc. Japan, **25**, 295.

THE OPTICAL POLARIZATION OF QSOs

Richard L. Moore
California Institute of Technology, Pasadena, CA, U.S.A.

Recent optical polarization studies indicate that there are two distinct types of QSOs - "normal" QSOs with $P \lesssim 1\%$, and highly polarized QSOs (HPQs) with $P > 3\%$. The HPQs are very similar to BL Lac objects, yet still share some properties of normal QSO's (e.g. strong emission lines). Our results generally support the relativistic beaming model for QSOs; however, certain key predictions of this model are not observed.

I. POLARIMETRIC PROPERTIES OF QSOs

An extensive survey of the optical linear polarization of QSOs has recently been completed by the author, in collaboration with Drs. H.S. Stockman and J.R.P. Angel. The results of this survey and the implications of the results are briefly summarized in this paper.

A primary conclusion from our survey is that the great majority of QSOs have very low (but significant) intrinsic optical polarization (Stockman, Moore, and Angel 1982). The distribution of polarization is dominated by QSOs with $P < 1.5\%$ ($\bar{P} \sim 0.6\%$). Among these low polarization ("normal") QSOs, there are no significant polarimetric differences between radio-loud and radio-quiet QSOs.

A small fraction of QSOs exhibit distinctly higher polarization ($P \sim 4-20\%$). Essentially no QSOs have intermediate polarizations ($2\% < P < 4\%$). The discontinuity in the distribution of polarization suggests that there are two basic types of QSOs - the normal QSOs and the highly polarized QSOs (HPQs).

The variability and wavelength dependence of polarization provide further evidence for a distinction between HPQs and normal QSOs. HPQs exhibit strong variability on time scales of days and only slight wavelength dependence (Moore and Stockman 1981). In contrast, the polarization of normal QSOs is only weakly variable on time scales of years and exhibits significant wavelength dependence (Stockman, Moore, and Angel 1982).

II. CORRELATIONS BETWEEN POLARIZATION AND OTHER QSO PROPERTIES

The objectives of our survey are not only to define the polarimetric properties of QSOs, but also to systematically examine the correlations between high polarization and other properties. Previous analyses of the properties of the HPQs have been limited by the fact that only four examples were known. There is now a large sample of HPQs (27 known) available for examining correlations. Details of this discussion will be presented by Moore and Stockman (1982).

Our results confirm previously suspected correlations (e.g. Visvanathan 1973) between high polarization, strong rapid photometric variability, and steep optical continua. We would note that the optical/infrared continua are not only steeper, but also better approximate a straight power law than do the continua of normal QSOs. A possible correlation is that the HPQs may have a higher ratio of X-ray to optical luminosity than normal QSOs (both radio-loud and -quiet).

An important aspect of the HPQs is their radio properties (Moore and Stockman 1981). With one exception, all of the HPQs are radio-loud QSOs. Only one of 50 ($2 \pm 2\%$) optically-selected QSOs in our survey has $P > 2\%$, this compares with 26 HPQs among 181 radio-selected QSOs surveyed ($14.4 \pm 2.8\%$). The difference is statistically significant. Also, the one radio-quiet HPQ (PHL 5200) is an atypical HPQ in nearly every respect (for example, it is not variable); the origin of its high polarization ($P \sim 4\%$) is probably different from that of other HPQs. The occurence of (variable) high polarization is the one polarimetric distinction we have found between radio-loud and radio-quiet QSOs.

The radio-loud HPQs are nearly all compact, flat spectrum variable radio sources (see Moore and Stockman 1981). The HPQs also frequently exhibit extreme radio properties such as superluminal motion and low-frequency variability. Two of the four known superluminal sources, 3C 279 and 345, are HPQs. Also, of seven definite HPQs which have been monitored at low radio frequencies, five have exhibited variability.

The general picture of the HPQs as a class is, thus far, that they are very similar to BL Lac objects. However, analysis of our survey results shows that the HPQs also share some properties with normal QSOs. There are no apparent correlations between high polarization and redshift, optical luminosity, and emission line equivalent width. These null correlations have direct implications for the theoretical model discussed below.

III. THEORETICAL IMPLICATIONS

There are several models which address the relationship between normal QSOs and HPQs. For brevity, we consider here the implications of our results for only the relativistic beaming model (e.g. Blandford and Rees 1978, Scheuer and Readhead 1979).

In general, the beaming model provides a straightforward explanation of many of our results. In this model, normal QSOs and HPQs have the same fundamental structure; however, the observer's orientation with respect to the anisotropic jet emission determines whether a QSO is observed as a normal QSO or an HPQ. The emission characteristics of the HPQs define the properties of the jet emission, while the characteristics of normal QSOs represent the properties of the isotropic emission. This model naturally accounts for the fact that essentially all HPQs are compact radio sources. Relativistic beaming eases theoretical difficulties imposed by the rapid variability and high luminosity of HPQs (both at radio and optical frequencies); beaming can also account for superluminal expansion.

However, there are two important predictions of this theory which are not supported by our results. In the HPQs, we are viewing both the jet emission (variable, highly-polarized, steep continuum) and the isotropic component (emission lines, low polarization, hard continuum). This implies that the HPQs should be systematically more luminous and should exhibit weaker (lower equivalent width) emission lines. Our correlation analyses do not confirm these predictions. Because the HPQs and normal QSOs have similar luminosities and emission line strengths, one must conclude (in this model) that the anisotropic component contributes a small fraction of the continuum. However, other characteristics of the HPQs (e.g. the smoothness of the energy distribution, the lack of polarimetric wavelength dependence, and the amplitude of variability) argue that the anisotropic component dominates the optical/infrared continuum.

While the relativistic beaming model is very attractive for explaining our results, it is not clear how to resolve the apparent contradiction described above. Perhaps the isotropic continuum (but not the emission lines) is suppressed in the HPQs. A program of spectropolarimetric monitoring of the HPQs during bright and faint phases would provide an excellent test of whether there are two components present.

REFERENCES

Blandford, R.D., and Rees, M.J. 1978, Pittsburgh Conference on BL Lac Objects, ed. A.M. Wolfe, p. 328 (Pittsburgh: University of Pittsburgh).
Moore, R.L. and Stockman, H.S., 1981, Ap. J., 243, 60.
Moore, R.L. and Stockman, H.S., 1982, in preparation.
Scheuer, P.A. G., and Readhead, A.C.S., 1979, Nature, 277, 182.
Stockman, H.S., Moore, R.L., and Angel, J.R.P., 1982, in preparation.
Visvanathan, N. 1973, Ap. J., 179, 1.

DISCUSSION

REES: How does 3C 273 (a superluminal radio source) fit in to your classification scheme?

MOORE: 3C 273 is superluminal, but is a low polarization QSO. This could still fit into the beaming model if, for example, the opening angle of the optical beam were smaller than the radio beam. There are other possible explanations, but not a clear answer.

WILSON: PHL 5200 in which you find high polarization, has broad optical absorption lines, presumably intrinsic to the QSO. Is there any correlation between the polarization and the presence of such absorption lines?

MOORE: We have surveyed a sample of other absorption-trough QSOs like PHL 5200 and all of them have low polarization.

MARSCHER: I have two comments. First, the flux which is responsible for ionization and hence for the luminosity of emission lines, comes from the far ultraviolet. Optical continuum observations generally tell you little about the presence or absence of an isotropic, flat-spectrum, non-varible component dominant in the far UV, so that we should not discard relativistically beamed models on the basis of strong emission lines. Also, 3C 446 looks like a BL Lac object when it's bright, and a quasar when weak. Hence, inclusion of the (unknown) fraction of BL Lac's which are in fact quasars with weak broad lines swamped by the continuum, might give HPQs weaker mean emission lines than other QSOs.

MOORE: I would not discard relativistically beamed models on the basis of strong emission lines in HPQs; there is a good deal of other evidence which supports this type of model. However, the prediction that the presence of the anisotropic optical continuum should systematically decrease the emission line equivalent widths (and increase the optical luminosities) of HPQs is based on the assumption that the isotropic emission (e.g., the optical/far-UV flux ratio) is similar among HPQs and normal QSOs. The results do require refinement of this simplifying assumption in the beaming models.

It is true that distinguishing HPQs and BL Lac objects is not straightforward without good spectroscopic observations over a range of brightness. I would like to see more monitoring of HPQs and BL Lac objects, particularly at faint phases. It is my opinion that the selection effect you mention, while present, does not strongly influence the results.

SUPERLUMINAL RADIO SOURCES

M.H. Cohen and S.C. Unwin
Owens Valley Radio Observatory
California Institute of Technology

ABSTRACT

Several compact radio sources studied with VLBI show apparent transverse velocities much greater than the speed of light. This review is a summary of VLBI observations of these "superluminal" sources as of August 1981. The physical models proposed to explain this phenomenon are also discussed.

INTRODUCTION

The only direct observations of superluminal motion (1) are made using VLBI, and six examples are now known. These are shown in order of redshift z in Table I. All except 3C 279 also are the subject of individual contributions in this volume (2-6). We also carry along 4C 39.25 as an example of a non-superluminal source which nevertheless has many features in common with the others. The fourth column of Table I shows the transverse scale in light-years per milli-arcsec (mas) assuming the normal cosmological interpretation of redshift, and with $H_o = 100\ \underline{h}$ km/sec/Mpc and $q_o = 0.05$. Column 5 gives a recent average value of flux density at centimeter wavelengths (all these sources are variable) and the last column is the corresponding luminosity at an emitted wavelength of 3 cm, assuming a flat spectrum.

Note that the distribution of sources with flux density (Table I) is inverted; i.e. there are many more strong sources than weak sources. There evidently are very many more such sources waiting to be found, with $S_{cm} > 0.5$ Jy. However, it is difficult to discover superluminal sources, mainly for lack of observing time needed to map candidates at several epochs. The best candidates seem to be BL Lac (7) and 3C 446 (8). Although the brightness distributions of these two change rapidly, a model with variable but stationary components is not excluded. It may be that the sampling interval has been too long to see systematic motions, if they exist.

MORPHOLOGY AND KINEMATICS

The morphology and kinematics of the superluminal sources are summarized in Table II. Columns 2 and 3 indicate the projected size of large-scale radio emission; the size of the VLBI structure (column 4) is only approximate, since it is frequency- and time-dependent. Δ PA (column 5)

TABLE I
Superluminal sources

Name	IAU	z	scale (ly/mas)	S_{cm} (Jy)	L_{cm} (10^{32} erg/s/Hz)
3C 120	0430+052	0.033	1.5 h^{-1}	4	0.5
3C 273	1226+023	0.158	6.0	35	95
3C 279	1253-055	0.538	13.5	10	315
3C 345	1641+399	0.595	14.1	12	465
3C 179	0723+679	0.846	16.3	0.5	40
NRAO 140	0333+321	1.258	18.3	2	370
4C 39.25	0923+392	0.698	15.2	8	430

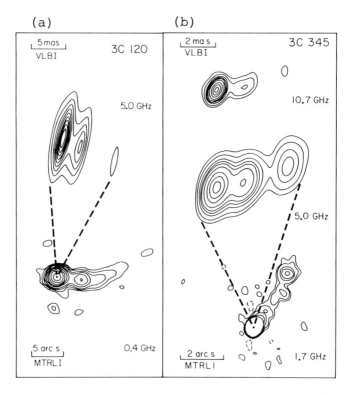

Figure 1. Large- and small-scale radio structure in (a) 3C 120 and (b) 3C 345. Dashed lines indicate the location of the VLBI structure within the compact core on each low-resolution map. The MTRLI maps are from Ref. 27; VLBI maps from Refs 16 and 26.

gives the total curvature at 2.8 cm, from the compact (VLBI) core to the outermost radio structure. Column 6 gives the measured proper motion μ in milli-arcsec (mas) per year. We emphasize that μ and z (Table I) are the only observed quantities. They show a strong inverse correlation, such as might be expected if z were a simple measure of distance and the intrinsic values of μ had only moderate dispersion. The other quantities are derived, and depend on h as indicated in Table II. Column 7 gives the apparent transverse velocity, calculated as v/c = μ s(1+z), where s is the scale in column 4 of Table I. References (column 10) are mainly since 1977; for earlier ones see the reviews by Cohen et al. (9), Kellermann and Shaffer (10), and Preuss (11).

Table II
Morphology and kinematics

Name	Outer structure jet (kpc)	Outer structure double (kpc)	Inner VLBI (pc)	Δ PA (deg)	μ (mas/y)	v/c	γ_{min}	$\theta(\gamma_{min})$ (deg)	References
3C 120	$5\ h^{-1}$		$3\ h^{-1}$	$(45)^a$	1.35	$2.1\ h^{-1}$	$3\ h^{-1}$	20 h	2,16,23,27
3C 273	40		$>15^b$	$(35)^a$	0.76	5.3	5	11	3,13,20,23,27
3C 279	(10)	$65\ h^{-1}$	$(12)^c$	$(15)^{ac}$	(0.5)	(10)	10	6	13,14,25,27
3C 345	15		20	107	0.36	8.2	8	7	4,26,27,29,30
3C 179	8	75	$(6)^c$	$(6)^c$	0.14	4.2	4	14	5,13,47
NRAO 140	$(40)^d$		$>10^b$	20	0.13	5.4	5	10	6,13,48
4C 39.25		20^e	9	20	<0.02	<0.5			25,49-51

Notes: (a) VLBI accuracy limited by poor N-S resolution on low-declination sources. (b) Poorly-defined low-level extension also exists. (c) Double model. (d) See text. (e) Complex, but nearly linear.

Three of the superluminal sources (3C 120, 3C273 and 3C345) have several features in common: (i) each has a large-scale one-sided radio jet which is misaligned with the VLBI structure (12); (ii) a bright "core" at one end of the compact structure dominates these sources at most epochs; (iii) strong gradients in spectral index are found along both the compact and extended jet features; (iv) at every epoch for which good hybrid maps of these sources exist, there is no emission on the opposite side of the core from the jet, to a level of $\simeq$ 5% (2-4). Figure 1 shows the large- and small-scale structures in 3C 120 and 3C 345, and it illustrates the above points.

We believe that the other superluminal sources have a similar morphology, although the details have not been seen in all cases. VLA maps of 3C 279 are more complicated (13). In addition to a core-jet structure in PA -157°, approximately aligned with the VLBI structure (14), there is an extended component at 50 h^{-1} kpc, in PA -34°; it is unclear how these structures are related. The two newly-discovered sources 3C 179 and NRAO 140 as yet have no hybrid maps; the results are discussed in Refs (5) and (6) respectively. The outer structure of

NRAO 140 is ambiguous. A weak (1%) unresolved component which shows on the 20-cm VLA map may be a hot spot, but the jet, if it exists, is below the dynamic range of 500:1 (13).

3C 120 shows a misalignment of 45° between the VLBI structure and the MTRLI map (Fig. 1a), part of which is accounted for by a bend in the compact structure. Figure 2 shows most of the published history on the size of 3C 120. 3C 120 has such large internal proper motions, and the components evolve so rapidly, that it has not been possible to follow individual events except for two limited periods of time when the structure was dominated by two strong components. Between 1972.5 and 1974.4 the data were well-fitted by an expanding-double model, with proper motion μ = 1.51 ± 0.13 mas/yr in PA -115° (15). During 1979 the structure was again dominated by a close double, with μ = 1.35 ± 0.3 mas/yr in PA -115°, although the hybrid maps showed additional low-level structure (2,16); these two expansion phases were thus very similar. Between 1975 and 1978 there were insufficient observations to follow the structure changes properly. Some of the scatter in Figure 2 is probably due to fitting a double-model to structure which was actually more complicated.

3C 120 is classified as a Seyfert or N-galaxy (17). The emission-line axis, separating positive and negative velocities and thus possibly a rotation axis, is at -108° ± 15° (17), or very close to the inner VLBI axis. Optical images also show a "bent jet" (18,19) with curvature in the same general direction as the radio jet (Fig. 1a).

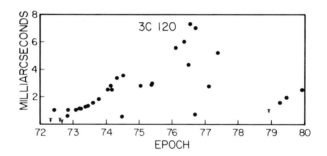

Figure 2. Angular size of the 3C 120 core against time. Points prior to 1978 are from Ref. 15, and represent the component separation of a double-model, found by least-squares fitting to amplitude data. Later points are measured from hybrid maps (16).

3C 273 has a core-jet VLBI structure with knots in the jet which move out from the dominant core; for the outermost knot, μ = 0.76 mas/yr (3,20). Earlier results based on fitting the visibility data from 1972 to 1977 with a double-model gave μ = 0.41 mas/yr (15,21-23). The difference probably arises from the complexity of the source, since recent hybrid maps show that it is usually at least triple. With limited amplitude data it is possible in this way to deduce expansion, but to underestimate the proper motion. There is no need to invoke acceleration (20).

3C 279 was the first suspected superluminal source (24) but it still has no modern hybrid maps. The most convincing data are in Ref. 14, visibility amplitudes from 1972-73 which, when modeled with a double, give $\mu \simeq 0.5$ mas/yr. 3C 279 varies rapidly and its visibility function has looked very different on different occasions (14,15,25). The very rapid motions reported in Ref. 25 are illusory if the separations do not represent the same pair of components at all epochs.

3C 345 The morphology and kinematics are best known for 3C 345 since good hybrid maps have been made for it since 1977 (23,26), and it is at high declination where the (u,v) coverage is good, and the restoring beam is nearly circular. Figure 1b shows 3C 345 on two angular scales. The large-scale jet has a projected length 15 h^{-1} kpc, and ends in a hot spot of intensity 2% of the peak (27). The VLBI maps of the core, taken near the same epoch (26), show an inner "core-jet" structure which is common on this angular scale (28) and repeats the outer morphology but on a scale 1000 times smaller. There are 3 inner components, with overall projected separation 20 h^{-1} pc. Comparison of 10.7 and 5.0 GHz maps shows that the eastern component ("core") has a spectral index $\alpha \simeq 1.0$, and variable flux density. This core appears to be the end of the system closest to the active center, because two prominent knots in the jet are separating from the core with the same proper motion (4). These components evolve rapidly, with half-lives $\simeq 3$ years at 5.0 GHz, and spectra which steepen to $\alpha \simeq -1$ at the detection limit of the maps. As with 3C 273, there is no evidence for changes in the separation rate of a component. 3C 345 shows strong curvature in the jet close to the core; Figure 1b shows that the curvature continues along the outer jet. At 10.7 GHz the jet starts at PA $-85°$, and at 22 GHz the starting PA is $\simeq -130°$ (29,30). The hot spot at the end of the jet is at PA $-31°$ and the tangent there is at PA $-23°$; thus Δ PA $\simeq 107°$.

In contrast with the early data (9,31,32) which interpreted the source as a double, the recent data provide two separations, since it as essentially a triple now. It is not clear how these can be reconciled, but as with 3C 120, it is likely that the source was poorly represented by a double for at least part of the time.

OTHER SOURCES

A few sources have been studied enough to say that they do not show rapid expansions in the manner of the superluminal sources. 4C 39.25 is shown in Tables I and II. Apart from the proper motion, it has most of the characteristics of the superluminals except for the lack of an outer jet. Even in this regard, however, it may not be unique, for the evidence for the outer jet in NRAO 140 is only circumstantial. 4C 39.25 has variations in total flux density, but they are slower and less pronounced than for the superluminals. 3C 84 (NGC 1275) is much more complex than the structures in Table II. It contains several components which vary in strength and which appear to separate slowly. Romney (33) reports evidence for expansion at $v \lesssim c$. The high-redshift (z = 1.94) object 2134+004 has $\mu \lesssim 0.1$ mas/yr, which places a weak limit on the transverse velocity of $v/c \lesssim 5.8$ (21). NRAO 150 is a double at 18-cm wavelength, and between 1974 and 1979 the strength of one component increased, but the spacing did not (34). At 3.8 cm NRAO 150 consists of a very close double, presumably located in the 18-cm variable component, which was stable from 1972 to 1975. This source has no redshift but is very interesting and deserves further study.

A number of rapid variables have been measured several times with VLBI (11,35). BL Lac, 3C 446, OJ 287, 3C 454.3, and others have variable brightness distribution and superluminal motions are suspected. However, in all these cases there are insufficient data to see if systematic expansions occur. The rapid variables present difficulties because they must be sampled frequently. The slower variables are being vigorously pursued, however, and it seems likely that the classes of confirmed superluminals and non-superluminals will increase rapidly.

PHYSICAL MODELS

The three well-studied sources 3C 120, 3C 273, and 3C 345 are very similar and any physical model must explain their essential characteristics: (i) core-jet structure, with several moving components, (ii) spectral gradient, (iii) evolution and decay of outer component, (iv) outer one-sided jet, (v) spatial curvature, mainly near core, (vi) superluminal motion, (vii) spacing and proper motion independent of wavelength (2-6 cm). The best data (4,20) show straight-line motion with no acceleration, but these data runs are only a few years long. The moving components can be at different position angles, corresponding to the general curvature of the source, but no component has yet been tracked around a bend. The polarization in the components of these sources is unknown; but high-resolution polarization measurements are starting, and in 3C 454.3 the magnetic field is along the jet, at 13-cm wavelength (36). Earlier studies (9) suggested that the extrapolated time of separation of a component from the origin was clse to the onset of a major outburst in total flux density. Recent work, however, has not borne this out. There even is a suggestion for 3C 345 that the outer component did not come from the origin, but from a region $\simeq 10$ pc away (32).

Many suggestions for the origin of the superluminal effect have been made; these have been summarized by Blandford, McKee, and Rees (1), and by Marscher and Scott (37). Enhancement of proper motion and intensity by a gravitational lens (38) seems unlikely because 3C 120 is close and the chance of an undetected foreground object is small; also, the lens has to be arranged so that the primary image that we see is at least 20 times brighter than any secondary images. Schemes which depend on variations in opacity seem suspect because they most naturally lead to separations which vary with frequency, contrary to observations. The magnetic dipole model (39) has difficulty explaining the curvature; also, the separation velocity of 3C 120 is too low unless $h \lesssim 0.5$, and the core-jet morphology does not match the expected double very well.

We feel that at present the relativistic jet model (40) is the best because (i) it is an economical unifying idea, with one jet responsible for the superluminal effects, the outer jet, and the outer hot spots, (ii) the one-sided core-jet morphology fits into it naturally, with the core being deep in the jet and moving components being generated by shock waves or instabilities, or perhaps being swept-up material (40,41), (iii) the common occurence of the superluminal effect (9) could be due to selection effects and the strong Doppler brightening of emitting material moving at a small angle to our line of sight (40,42,43). (iv) the strong curvature close to the core could be explained as a small intrinsic curvature amplified by the small angle to the line-of-sight (12,42), and (v) the near equality of the core and the jet components is reasonable if both are Doppler-enhanced but the core is the quasi-stationary region where the moving material becomes optically thin (40).

Table II contains values of the minimum Lorentz factor γ_{min} for the given value of v/c. (γ does not scale exactly with h so the values are approximate.) The angle θ to the line of sight corresponding to γ_{min} is also given. The solid angles are small, but this is not disturbing; in this model a radio-selected sample will have most of its objects with $\theta < 2\gamma^{-1}$, even though the objects have random orientations (42).

Over a long-term average the jet must be two-sided, to account for the outer double structure in 3C 179 and 3C 279, but there is no way at present to tell whether the retreating side is invisible because of the Doppler effect, or whether the jet is episodic and alternates sides. Unless the intrinsic curvature is rather large, the outer doubles are seen nearly end-on, and their true separation is up to 500 kpc. This is large, but not exceptionally so, for double sources.

An important statistical problem involving the large Doppler boost in flux density ($\sim \gamma^2$) does remain for the relativistic jet hypothesis. The most powerful way of stating it is for 3C 273: the optical continuum is unboosted (because the line/continuum ratio is normal) and so there should be 25 quasars as bright as 3C 273, in some z-range around 0.158. At radio wavelengths the many sources at large θ are much weaker, and indeed should be the radio-quiet quasars (42), but recent observations seem not to bear out the predictions (44). Also, 4C 39.25 becomes by far the most powerful source in Table I if it is at large θ and all the others are at small θ; even if its radiation is isotropic, 4C 39.25 is still much stronger than the others.

These problems are alleviated if the implications of the Doppler boost are changed. Rees (45) has proposed a "spray" model with a wide ejection cone. A strong gradient of intensity with cone-angle within the spray can alter the relation between the boost and the kinematic effects. The statistical objections to the relativistic jet model are ambiguous because the numbers are small and the samples are incomplete; if they persist, then the spray model and others will have to be carefully investigated.

We thank many colleagues for giving us material in advance of publication, and are grateful to T.J. Pearson for comments on the manuscript. MHC thanks the John Simon Guggenheim Memorial Foundation and the Institute of Astronomy, Cambridge, for their generosity and hospitality.

REFERENCES

1. Blandford, R.D., McKee, C.F., and Rees, M.J.: 1977, Nature, 267, pp. 211-216.
2. Walker, R.C.: 1982, IAU Symp. No. 97.
3. Pearson, T.J. et al.: 1982, IAU Symp. No. 97, this volume, p. 355.
4. Unwin, S.C.: 1982, IAU Symp. No. 97, this volume, p. 357.
5. Porcas, R.W.: 1982, IAU Symp. No. 97, this volume, p. 361.
6. Marscher, A.P., and Broderick, J.J.: 1982, IAU Symp. No. 97, volume, p. 359.
7. Mutel, R.L.: 1982, IAU Symp. No. 97, this volume, p. 385.
8. Brown, R., Johnston, K.J., and Wolfe, A.M.: 1981, (private communication).
9. Cohen, M.H., et al.: 1977, Nature, 268, pp. 405-409.
10. Kellermann, K.I., and Shaffer, D.B.: 1977, "L'Evolution des Galaxies et ses Implications Cosmologiques" CNRS Colloquium No. 263, Paris (IAU Colloquium No. 37), pp. 347-363.
11. Preuss, E.: 1981, "Optical Jets in Galaxies", ESO/ESA Workshop ESA SP-162, Paris, pp. 97-105.
12. Readhead, A.C.S., Cohen, M.H., Pearson, T.J., and Wilkinson, P.N.: 1978, Nature, 276, pp. 768-771.
13. Perley, R.: 1981, (VLA maps, private communication).
14. Cotton, W.D., et al.: 1979, Astrophys. J., 229, L115-L117.
15. Seielstad, G.A., Cohen, M.H., Linfield, R.P., Moffet, A.T., Romney, J.D., Schilizzi, R.T., and Shaffer, D.B.: 1979, Astrophys. J., 229, pp. 53-72.
16. Walker, R.C., Seielstad, G.A., Simon, R.S., Unwin, S.C., Cohen, M.H., Pearson, T.J., and Linfield, R.P.: 1981, Ap.J., (submitted).
17. Baldwin, J.A., Carswell, R.F., Wampler, E.J., Smith, H.E., Burbidge, E.M., and Boksenberg, A.: 1980, Astrophys. J., 236, pp. 388-405.
18. Wlerick, G.: 1981, "Optical Jets in Galaxies", ESO/ESA Workshop ESA SP-162, Paris, pp. 29-35.
19. Arp, H.: 1981, ibid. pp. 53-61.

20. Pearson, T.J., et al.: 1981, Nature, 290, pp. 365-368.
21. Schilizzi, R.T., et al.: 1975, Astrophys. J., 201, pp. 263-274.
22. Kellermann, K.I., et al.: 1977, Astrophys. J., 211, pp. 658-668.
23. Readhead, A.C.S., Pearson, T.J., Cohen, M.H., Ewing, M.S., and Moffet, A.T.: 1979, Astrophys. J., 231, pp. 299-306.
24. Gubbay, J., Legg, A.J., Robertson, D.S., Moffet, A.T., and Seidel, B.: 1969, Nature, 222, pp. 730-733.
25. Pauliny-Toth, I.I.K., Preuss, E., Witzel, A., Graham, D., Kellermann, K.I., and Rönnäng, B.: 1981, A.J., 86, pp. 371-385.
26. Unwin, S.C., et al: 1981, (in preparation).
27. Wilkinson, P.N., et al.: 1981, (MTRLI maps, in preparation).
28. Readhead, A.C.S., and Pearson, T.J.: 1982, IAU Symp. No. 97, this volume), pp. 279-288.
29. Bååth, L.B., et al.: 1981, Astrophys. J., 243, pp. L123-L126.
30. Readhead, A.C.S., Hough, D.H., Ewing, M.S., Walker, R.C., and Romney, J.D.: 1981, (in preparation).
31. Cohen, M.H., Pearson, T.J., Readhead, A.C.S., Seielstad, G.A., Simon, R.S., and Walker, R.C.: 1979, Astrophys. J., 231, pp. 293-298.
32. Schraml, J., Pauliny-Toth, I.I.K., Witzel, A., Kellermann, K.I., Johnston, K.J., and Spencer, J.H.: 1981, (in preparation).
33. Romney, J.D. et al.: 1982, IAU Symp. No. 97, this volume, pp. 291-292.
34. Mutel, R.L., and Phillips, R.B.: 1980, Astrophys. J., pp. L73-L76.
35. Shaffer, D.B.: 1978, "Pittsburgh Conference on BL Lac Objects", Pittsburgh, pp. 68-81.
36. Cotton, W.D., Geldzahler, B.J., and Shapiro, I.I.: 1982, IAU Symp. No. 97, this volume, pp. 301-303.
37. Marscher, A.P., and Scott, J.S.: 1980, P.A.S.P., 92, pp. 127-133.
38. Barnothy, J.M.: 1982, IAU Symp. No. 97, this volume, pp. 463-464.
39. Bahcall, J.N. and Milgrom, M.: 1980, Astrophys. J., 236, pp. 24-42.
40. Blandford, R.D., and Königl, A.: 1979, Astrophys. J., 232, pp. 34-48.
41. Marscher, A.P.: 1980, Astrophys. J., 239, pp. 296-304.
42. Scheuer, P.A.G., and Readhead, A.C.S.: 1979, Nature, 277, pp. 182-185.
43. Kellermann, K.I., and Pauliny-Toth, I.I.K.: 1981, Ann. Rev. Astron. and Astrophys, (in press).
44. Condon, J.J., Condon, M.A., Jauncey, D.L., Smith, M.G., Turtle, A.J., and Wright, A.E.: 1981, Astrophys. J., 244, pp. 5-11.
45. Rees, M.J., 1981, IAU Symp. No. 94, pp. 139-164.
46. Spencer, J.H., Johnston, K.J., and Witzel, A.: 1981 (in prep.).
47. Porcas, R.W.: 1981, Nature, (submitted).
48. Marscher, A.P., and Broderick, J.J.: 1981, Astrophys. J., 247, pp. L49-L52.
49. Shaffer, D.B., et al.: 1977, Astrophys. J., 218, pp. 353-360.
50. Bååth, L.B., et al.: 1980, Astron. Astrophys., 86, pp. 364-372.
51. Pearson, T.J., and Readhead, A.C.S.: 1981, Astrophys. J., 248, pp. 61-81.

DISCUSSION

DAISHIDO: The spectral variations in quasars and Seyfert galaxies have been explained by the expanding models of Shklovsky, van der Laan, Kellermann, and others about 10 years ago. However, the observed "superluminal" clouds move rapidly, but do not seem to expand themselves. How then do you explain the spectral evolution?

COHEN: The moving components in 3C 345, for which we have the best N-S resolution, are unresolved and may have been expanding. The range of allowed sizes could give rise to substantial evolutionary effects.

TERRELL: Is there a critical number of superluminal cases beyond which the usual explanation (ejection toward the observer) will be in trouble?

COHEN: There is no critical number of sources, because the statistics depend on the distribution of radio luminosities and relativistic γ-factors. At present, the numbers of confirmed superluminals and non-superluminals is too small to exclude this explanation.

REES: Two comments: (i) The "Doppler enhancement" problem which occurs for $\gamma \gtrsim 10$ could be eased by assuming an intrinsic speed $\gtrsim \gamma^{-1}$ in the velocity vector of the ejecta. (ii) The internal pressure within individual superluminal "blobs" exceeds the likely gas pressure (estimated at a "deprojected" distance of ~ 50 pc). This supports the view that these blobs are generated by shock fronts in the jet; they expand and disperse as they move out, rather than having emerged directly from the core.

ABELL: What is the prognosis of getting absolute astrometry of these sources to 1 mas precision, so that we can pin down what is moving?

COHEN: Relative astrometry, e.g., 3C 345 vs NRAO 512 (Shapiro, et al. 1979, Astron. J., 84, pp. 1459-1469) already gives accuracy substantially better than 1 mas. In a few years it should be possible to make a series of maps of 3C 345 and tell whether or not the core is stationary with respect to NRAO 512. Absolute positions are much harder to measure to comparable accuracy.

SUPERLUMINAL EXPANSION OF 3C 273

T.J. Pearson, S.C. Unwin, M.H. Cohen, R.P. Linfield,
A.C.S. Readhead, G.A. Seielstad, R.S. Simon, and R.C. Walker
Owens Valley Radio Observatory
California Institute of Technology

Figure 1 shows hybrid maps of the core of 3C 273B at five epochs, made with arrays of 4 or 5 VLBI antennas. The maps span a period of 3.5 years. They all show a bright eastern peak and a lower-brightness extension to the west. There is a local maximum in the western extension between 6 and 8 milliarcsec from the main peak. This "blob" moves steadily further away from the main peak along a roughly straight line in PA $-116° \pm 2°$. Compare this with the position angle of the 25-arcsec optical jet, $-137°$. The maps show a slight curvature to the south with increasing separation from the main peak. Lower-resolution VLBI maps at lower frequencies show that this curvature continues at greater separations, suggesting a smooth connection between the milli-arcsecond position angle and the position angle of the optical jet. In our latest map (1981.09) the blob is no longer detectable with the limited dynamic range of the VLBI network (about 20:1).

The separation of the western blob from the eastern peak increased steadily between 1977 and 1980, with no evidence for acceleration or deceleration. The angular expansion rate is 0.76 ± 0.04 milli-arcsec/year, corresponding to a linear expansion rate of $v/c = (5.3 \pm 0.3)/h$, assuming $z = 0.158$, $H_0 = 100h$ km/(s Mpc), and $q_0 = 0.05$; 1 milliarcsec = $6.0/h$ l.y. = $1.9/h$ pc.

In the relativistic jet model the expansion is presumed to represent a physical motion almost directly towards the observer; we are looking down a jet of relativistic material expelled from the quasar. The eastern end of 3C 273B is the "core", the point at which the jet becomes optically thick. In a steady flow, this point remains fixed even though the radiating material has a high velocity. The moving (western) component is a "knot" propagating outwards along the jet. In order to account for an apparent transverse velocity v as large as 5.3c, the angle ϕ between the jet and the line of sight must be small. For h=1, ϕ must be less than 21°, and the space velocity is minimized at $\phi = 11°$. These angles will be smaller for smaller values of h. The probability that ϕ should be as small as 11° in a randomly selected quasar is 1%, or 2% if the jet is two-sided.

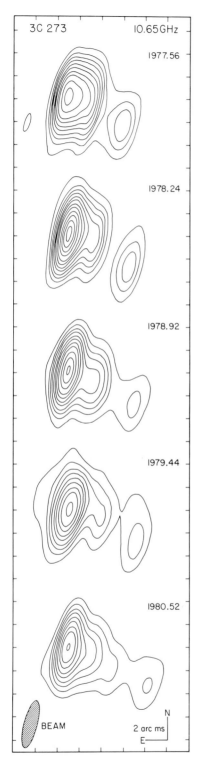

3C 273 is far from being a randomly chosen quasar: it is the closest quasar in the 3CR sample and has an intrinsic optical intensity about 4 times the intensity of the other quasars in the sample. It is surprising that it should also be exceptional in having a jet pointing almost directly at us. A possible explanation is that the optical radiation, like the radio, is not isotropic. There is no direct evidence for this, but beaming of the optical continuum cannot be ruled out. If the optical radiation is isotropic, then the small angle between the radio jet and the line of sight causes grave difficulties for the simple relativistic theories. For example, if the difference in radio intensity between optically selected and radio-selected quasars is due to the relativistic beaming of the radio radiation, because radio-selected quasars point towards us, we would expect to see 50 to 100 radio-quiet quasars of comparable optical magnitude to 3C 273, which we do not.

This work was supported by the NSF (AST 79-13249).

REFERENCE

Pearson, T.J., Unwin, S.C., Cohen, M.H., Linfield, R.P., Readhead, A.C.S., Seielstad, G.A., Simon, R.S., and Walker, R.C.: 1981, Nature, 290, pp. 365-368.

Figure 1.
Maps of 3C 273B at 10.65 GHz at five epochs, from Pearson et al. (1981), who give details of the observations. Contour levels and restoring beam are the same for all maps. (Reproduced by permission of Macmillan Journals Ltd.)

SUPERLUMINAL EXPANSION OF THE QUASAR 3C 345

S. C. Unwin
Owens Valley Radio Observatory
California Institute of Technology

VLBI hybrid maps of the compact radio structure in 3C 345 at 5.0 and 10.7 GHz show a core-jet morphology, with two components in the jet separating from the core with apparent transverse velocities $\simeq 8/h$ c (for $H_o = 100$ h km/s/Mpc and $q_o = 0.05$). These "knots" decay as they move down the jet, with lifetimes $\simeq 2 - 3$ years.

We have been monitoring the structure of 3C 345 at 5.0 and 10.7 GHz using arrays of 4 and 5 telescopes, at approximately 6-month intervals beginning in 1977. Hybrid maps (Readhead and Wilkinson 1978) were made from each observation, and excellent fits to the visibility amplitudes and closure phases were obtained. Since 1979, the increased angular resolution provided by the telescope at Effelsberg FRG has enabled individual components in the jet to be mapped at 5.0 GHz for the first time. At most epochs the source comprised a dominant compact core, with a curved one-sided "jet" extending to the west. Any compact emission on the eastern side is at most 5% of the core. The hybrid maps show very clearly that the jet is curved, with the innermost parts of the jet in P.A. $-85°$, and the outer components in P.A. $-73°$. The jet continues to bend out to $\simeq 2$ arcsec (Cohen and Unwin 1982), although most of the rotation occurs within a few milliarcsec (mas) of the core.

Figure 1(a) shows 5.0-GHz profiles along the jet of 3C 345. It illustrates the triple structure found at most epochs, and the expansion of the two jet components relative to the core. These knots decay as they move out, with half-lives $\simeq 3$ years at 5.0 GHz and $\simeq 1$ year at 10.7 GHz; the spectrum of a knot steepens with time from $\alpha \simeq -0.5$ at the point where it is resolved from the core to $\alpha \simeq -1.0$ at the sensitivity limit. Figure 1(b) shows the increasing separation of the knots from the core with time; there is no significant difference in the slopes: 0.33 ± 0.07 mas/yr beginning in 1975.2 ± 1.0 for the inner knot, 0.39 ± 0.07 mas/yr beginning in 1969.5 ± 1.9 for the outer. If the implied expansion speed of $\simeq 8/h$ c is the result of relativistic beaming (Scheuer and Readhead 1979) then the jet must lie within $6°$ of the line of sight. There were no obvious features in the flux-density

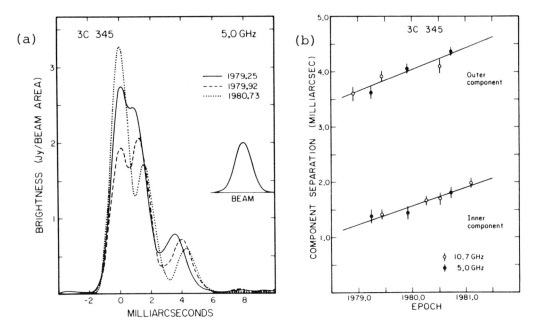

Figure 1. (a) Profiles across the 5.0-GHz hybrid maps of 3C 345 in P.A. $-78°$. Zero separation is taken as the eastern "core", to which the profiles are aligned. The restoring beam is a Gaussian with FWHM 1.2 mas; 1 Jy/beam area = 3×10^{10} K. (b) Core-knot separations as a function of time, measured directly from hybrid maps. The estimated error bars allow for finite dynamic range and resolution of the components. Straight lines represent least-squares fits to the expansion.

history of 3C 345 at the extrapolated epochs of zero separation. The spectrum of a knot probably evolves rapidly at first, delaying its contribution to the total flux. There is no evidence for acceleration during the period in which these knots have been monitored. The component separations are frequency-independent, and hence there are no strong spectral gradients across a knot.

Many people have contributed to the observations discussed here; among them are M.H. Cohen, R.P. Linfield, T.J. Pearson, A.C.S. Readhead, G.A. Seielstad, R.S. Simon, and R.C. Walker. This work at OVRO is supported by the NSF under contract AST 79-13249.

REFERENCES
Cohen, M.H., and Unwin, S.C.: 1982), IAU Symp. No. 97, this volume, p. 345
Readhead, A.C.S., and Wilkinson, P.N.: 1978, Ap. J., 223, pp. 25-36.
Scheuer, P.A.G., and Readhead, A.C.S.: 1979, Nature, 277, pp. 182-185.
Unwin, S.C., et al.: 1981, (in preparation).

SUPERLUMINAL MOTION IN NRAO 140 AND A POSSIBLE FUTURE METHOD FOR CONSTRAINING H_o AND q_o

Alan P. Marscher
Center for Astrophysics and Space Sciences, UCSD

John J. Broderick
Department of Physics, VPI & SU

NRAO 140 is a quasar (z = 1.258) which is among only 3 or 4 such objects (and the one with the highest z) which were detected at X-ray energies prior to the operation of the Einstein Observatory (Marscher et al. 1979). We obtained contemporaneous X-ray and radio VLBI observations of the source in early 1980, to determine whether Compton scattering within the radio source is the primary X-ray emission mechanism (Marscher and Broderick 1981b). Instead, we found that the radio parameters predicted more than 10^3 times more X-ray flux than was observed. Since the Compton calculation is independent of distance, and since the troublesome component was partially resolved (and hence not a high-brightness-temperature emitter), we found that relativistic motion aimed nearly directly toward the observer with Lorentz factor exceeding 4, needed to be invoked in order to bring the predicted Compton flux down to the observed level (Marscher and Broderick 1981a, b). Since relativistic motion is also the preferred explanation for the apparent superluminal expansion seen in some compact radio sources (e.g., M. Cohen, this volume; Marscher and Scott 1980; Kellermann and Pauliny-Toth 1981), we predicted that the compact components in NRAO 140 should appear to separate at a speed exceeding about 4c.

We have obtained further VLBI observations of NRAO 140 at 2.8 cm in February 1981 and June 1981, in order to test this prediction. The correlated flux densities and closure phases show clear, systematic changes compared with the April 1980 data. We find that these changes are modeled (both by hybrid mapping and by model fitting) very well by an increase in the separation of the compact components by 0.09 to 0.16 milliarcseconds (mas), which corresponds to an angular separation rate of 0.08 to 0.14 mas/yr. For cosmological distances and H_o = 50 and q_o = 0, these rates in turn correspond to velocities of separation which range from 6.7c to 12c; for H_o = 100 and q_o = 1, the range is 2.1c to 3.7c. Since our prediction that motions should exceed 4c can be taken only as an approximation owing to the possibility that the viewing angle may not be the optimum one for superluminal motion, even H_o and q_o as high as these latter values are consistent with the constraints imposed by the observed-versus-predicted X-ray flux. We also stress that the observed increase in separation of the components is less than 10%; further VLBI monitoring is necessary in order to determine the separation rate more precisely.

From these results, we can outline a prescription for the determination of upper limits to the cosmological parameters H_o and q_o. To use this method, one needs a dedicated VLB array and an orbiting X-ray observatory (such as the proposed AXAF) with sensitivity at least as good as that of the Einstein satellite. One first observes, contemporaneously with the X-ray satellite, a large number of quasars over a wide range of redshifts, with the VLB array at several frequencies. One then picks out those quasars (one hopes that a large enough group exists!) for which the predicted Compton X-ray flux far exceeds the observed value. One can calculate the minimum values of bulk Lorentz factors γ needed to bring the predicted X-ray fluxes down to the observed values. One then observes this group of objects for at least two years to determine the apparent velocity of separation of the components for each source. After one does a suitable average over angles of ejection relative to the line-of-sight, the mean ratio (min. γ from Compton arguments)/ (γ required to explain separation velocity; function of H_o, q_o), can be determined. Quasars of similar redshift would yield an upper limit to H_o by the requirement that this ratio exceed unity. Quasars at different redshifts would allow one to place an upper limit on q_o by this method.

A.P.M. was an Einstein (HEAO 2) Guest Investigator, and was supported by NASA and the NSF. J.J.B. also gratefully acknowledges funding by the NSF.

REFERENCES

Kellermann, K. I., and Pauliny-Toth, I.I.K. 1981, Ann. Rev. Astron. Astrophys., in press.
Marscher, A. P., and Broderick, J. J. 1981a, Ap. J., 247, L49.
Marscher, A. P., and Broderick, J. J. 1981b, Ap. J., 249, in press.
Marscher, A. P., et al. 1979, Ap. J., 233, 498.
Marscher, A. P., and Scott, J. S. 1980, P.A.S.P., 92, 127.

SUPERLUMINAL EXPANSION IN 3C 179

R.W. Porcas
Max-Planck-Institut für Radioastronomie, Bonn, F.R.G.

ABSTRACT

The detection of apparent faster-than-light motion (v=7.6c) in the core of 3C 179 poses some problems for the simple relativistic jet explanation.

I wish to report the detection of superluminal expansion in the core of the 18^m quasar 3C 179 (z=0.843). Unlike the other known superluminal sources, 3C 179 has pronounced outer double-lobed structure (LAS = 14" in PA $\sim$ 80°) which produces the dominant radio emission at frequencies $\leq$1 GHz. The compact, flat-spectrum core of 3C 179 is only $\sim$0.4 Jy. Transatlantic VLBI observations at 10.7 GHz, using the antennas at Effelsberg, Haystack, Green Bank and Owens Valley have revealed an inner double structure in PA = 92°, with flux ratio 2:1. The separation of these components, however, has changed between 2 observing epochs, spaced 1.2 years apart, increasing from 1.07 to 1.24 mas, corresponding to an apparent transverse relative velocity of 7.6c (with "traditional" H_0=55 km/s/Mpc, q_0=0.05). The stronger, eastern component has also increased in flux density by $\sim$15%. If one interprets this apparent superluminal velocity as due to bulk relativistic motion in a jet directed at the observer[1] then the minimum Lorentz factor, γ, of the motion is 7.6 and the required small angle to the line of sight, θ, is $1/\gamma$=7.5°.

Scheuer and Readhead[1] have emphasized the significance of outer double-lobed radio emission because such sources in flux-limited samples can be expected to have random orientations. Thus estimates of the likelihood of selecting a given source orientation are not confused by the possibility of relativistically beamed emission from a one-sided jet morphology as with previous superluminal sources.

3C 179 was selected from the 30 quasars mapped by Owen et al.[2] which have flux densities $\geq$0.7 Jy at 0.97 GHz, $m_B \leq$ 19 and LAS $\geq$ 10". Since no beaming of the optical emission is thought to occur, there is no selection effect operating to produce preferential alignment to the line of sight for sources in this sample. The a priori probability of a source having θ ≤7.5 is 1-cosine (θ) = 1/110. 3C 179 was chosen from the sample because of its strong core (which made possible the VLBI observations reported here) and its clearly defined outer double-lobed morphology. This selection within the sample favours a small value of θ because of the associated Doppler enhancement of core flux density, and we might expect (possibly) one of the 30 sources to have the requisite 7.5° alignment. We cannot also expect, however, that this source will have an unusual value of γ, and thus one finds that γ of 7.6 must be typical for sources in the sample. This is in direct conflict with the argument of Scheuer and Readhead that the typical value of γ in the cores of double sources is ≤2, based on the statistics of the ratio of core (beamed) to lobe (unbeamed) flux densities in samples of double sources. We must conclude either that (i) 3C 179 is a freak, (ii) the argument of Scheuer and Readhead is incorrect, (iii) a revaluation of H_o even more dramatic than the presently fashionable value 100 is required, or (iv) some mechanism other than collimated, relativistic bulk motion is responsible for the superluminal effect.

REFERENCES

1. Scheuer, P.A.G. and Readhead, A.C.S.: 1979, Nature 277, pp. 182-5
2. Owen, F.N., Porcas, R.W. and Neff, S.G.: 1978, Astron. J. 83, pp. 1009-20

DISCUSSION

COHEN: The history of 3C 345 and 3C 273 shows that recent maps yield a proper motion greater than that obtained in the early days when a double was forced into limited data. It seems likely that this will be true for 3C 179 also and thus your present value should be regarded as a lower limit.

RELATIVISTIC JETS AS RADIO AND X-RAY SOURCES

Arieh Königl
Astronomy Department, University of California
Berkeley, California 94720

ABSTRACT: Various predictions and implications of the relativistic-jet model for compact extragalactic sources are reviewed in the light of recent radio and X-ray observations.

According to the relativistic-beaming hypothesis (Scheuer and Readhead 1979, Blandford and Königl 1979), the radio emission in most of the bright, compact extragalactic sources originates in relativistic jets, and can be detected only when the observer is located within the beamed "emission cone" of the jet. In this picture, the sequence: radio-quiet quasars, radio-loud quasars, and optically violently variable quasars (OVVs) corresponds to similar sources which are viewed at progressively smaller angles to their jet axes. Except for the absence of strong emission lines, BL Lac objects are similar to OVVs in their spectra, high degree of variability and strong linear polarization, and may represent the strongly beamed sources (the analogs of radio-loud quasars) among giant elliptical galaxies. An immediate consequence of the beaming hypothesis is that, for every relativistic core-dominated source with Lorentz factor γ, there are $\sim \gamma^2$ extended double sources in which the central components are radio quiet. In sources which are viewed nearly along the jet axis, the extended outer lobes should appear as low-surface-brightness, steep-spectrum halos around the bright flat-spectrum cores. Recent low-frequency observations (e.g., Browne et al. 1981) have verified that most core-dominated sources are indeed surrounded by such halos, and that the source statistics are consistent with the values of $\gamma (\geqslant 6)$ that are inferred in a number of these sources from apparent superluminal motions (e.g., Cohen 1982). It is suggestive that Lorentz factors in this range can also explain the frequent occurrence of low-frequency variability in core-dominated sources (Blandford and Königl 1979).

The relativistic-jet model provides a natural interpretation of the phenomenon of apparent superluminal expansion, and particularly of the fact that the separating components have comparable flux densities. In this picture (Blandford and Königl 1979), one of the components is identified with the stationary, optically thick "core" of the jet, while the other components are associated with shock waves which either propagate in the jet or which form behind dense clouds that are accelerated by the flow. The core component is expected to have a flat radio spectrum and a relatively constant flux density, whereas the other components could have steeper spectra and would diminish in brightness as they separated from the core. These predictions appear to be borne out by recent high-resolution VLBI observations (e.g., Unwin 1982).

Radio-loud quasars are different from radio-quiet quasars also in their X-ray properties, having on the average a higher ratio of X-ray to optical luminosity (e.g., Ku, Helfand, and Lucy 1980). The differences are most noticeable for OVVs which, together with BL Lac objects, appear to form a distinct class of sources, much as they do in the radio and optical regimes. In the context of the relativistic-jet model, these observations suggest that the X-rays in these strongly polarized and highly variable sources could be associated with the beamed emission component which produces the strong radio flux. It is, in fact, possible to account for the main spectral characteristics of this class -- the flat radio spectrum and the two-component X-ray spectrum -- with a simple model of an unresolved, inhomogeneous jet, in which the local emission spectrum breaks as a result of synchrotron radiation losses (Königl 1981). This model implies that all of these sources will be found to have a hard X-ray component (attributed to synchrotron self-Compton emission) at sufficiently high frequencies. On the basis of this model one can, in principle, deduce various source parameters from the detailed shape of the spectrum. An example of a source for which a reasonable set of parameters could be inferred in this fashion is the BL Lac object PKS 2155-304 (Urry and Mushotzky 1981). Relativistic beaming in this source is suggested independently by arguments involving the inverse-Compton "catastrophe."

Quasars are now believed to account for a large fraction of the diffuse X-ray background in the few-keV range (e.g., Tananbaum 1982). The radiation is most likely associated with the unbeamed emission component. The only quasar identified so far as a high-energy ($\geq$ 100 MeV) γ-ray source is 3C273, which is also a radio-loud "superluminal" quasar. If 3C273 is a typical quasar, then it is possible to argue on statistical grounds (Königl 1981) that the high-energy γ-rays in quasars also originate in the beamed emission component, and that radio-loud quasars contribute most of the diffuse $\geq$ 100 MeV background.

REFERENCES

Blandford, R.D., and Königl, A.: 1979, Astrophys. J. 232, pp. 34-48.
Browne, I.W.A., et al.: 1981, Monthly Notices Roy. Astron. Soc. (submitted).
Cohen, M.H.: 1982, this Symposium.
Königl, A.: 1981, Astrophys. J. 243, pp. 700-709.
Ku, W.H.-M., Helfand, D.J., and Lucy, L.B.: 1980, Nature 280, pp. 323-328.
Scheuer, P.A.G., and Readhead, A.C.S.: 1979, Nature 277, pp. 182-185.
Tananbaum, H.: 1982, this Symposium.
Unwin, S.C.: 1982, this Symposium, p. 357.
Urry, C.M., and Mushotzky, R.F.: 1981, Astrophys. J. (in press).

COMPTON ROCKETS: RADIATIVE ACCELERATION OF A RELATIVISTIC FLUID

S. L. O'Dell
Physics Dept., Virginia Tech
Blacksburg, VA 24061 USA

Summary. Unless a plasma moves at relativistic bulk speeds, the Compton radiative lifetime for relativistic electrons near a luminous object is less than the transit time from the source. Acceleration by adiabatic decompression is too slow to preserve much of the electrons' energy. However, the Compton-rocket thrust and the radiatively induced pressure gradient can accelerate a relativistic fluid to relativistic bulk speeds on a time scale comparable to that for radiative loss. Consequently, severe Compton losses are not only reduced by relativistic bulk motion, but can in fact effect such motion.

Relativistic electrons of proper Lorentz factor γ_{e*}, in a fluid moving with speed $c\beta$ away from a source of luminosity L, lose energy via single Thomson scattering at an average rate

$$\frac{d \ln \langle \gamma_{e*} - 1 \rangle}{d \ln r} = -\frac{2}{3} \frac{\langle \gamma_{e*}^2 - 1 \rangle}{\langle \gamma_{e*} - 1 \rangle} \Lambda_1 \left(\frac{r_1}{r}\right) \frac{(1-\beta)}{\beta(1+\beta)} ,$$

where r_1 is a fiducial distance and the parameter

$$\Lambda_1 \equiv \frac{\sigma_T L}{2\pi mc^3 r_1} = \left(\frac{m_H}{m}\right)\left(\frac{r_S}{r_1}\right)\frac{(L/M)}{(L/M)_E} = 4.3 \, (L_{45}/r_{15}) ,$$

with r_S and $(L/M)_E$, respectively, the Schwarzschild radius and Eddington light-to-mass limit for Thomson scattering, and $L_{45} \equiv L/(10^{45} \text{ erg/s})$ and $r_{15} \equiv r/(10^{15} \text{ cm})$. Clearly, Compton radiative losses near r_1 are quite severe (cf. Hoyle, Burbidge, and Sargent 1966) unless, or until, $\gamma^3(1+\beta)^2\beta \gtrsim \Lambda_1 \langle \gamma_{e*}^2 - 1 \rangle / \langle \gamma_{e*} - 1 \rangle \rightarrow \Lambda_1 \langle \gamma_{e*}^2 \rangle / \langle \gamma_{e*} \rangle$ for $\langle \gamma_{e*}^2 \rangle \gtrsim \langle \gamma_{e*} \rangle \gg 1$.

Relativistic bulk motion thus mitigates the Compton problem (e.g., Shklovsky 1964; Woltjer 1966; Rees and Simon 1968). Adiabatic decompression is, however, too slow since it requires a distance $\sim r$ to be effective. On the other hand, if the enthalpy of the fluid is concentrated in or can be transferred sufficiently rapidly to the relativistic electrons, the dynamics are very different. First, the rapid radiative loss of internal kinetic energy induces a steep pressure gradient which accelerates the fluid in the direction of flow. Second, the Compton-rocket thrust (O'Dell 1981), resulting from the anisotropic loss of internal energy via Thomson scattering, accelerates the fluid away from the source of incident radiation (Cheng and O'Dell 1981). These two effects are usually comparable and can accelerate an ultrarelativistic fluid to relativistic bulk speeds over a radiative-loss time scale.

To see this explicitly, consider a spherically symmetric, steady-state, relativistic wind starting with speed $\beta_1 = \sqrt{1/3}$ at r_1. For simplicity, take $(\langle \gamma_{e*}^2 \rangle / \langle \gamma_{e*} \rangle^2) = (\langle \gamma_{e*}^2 \rangle_1 / \langle \gamma_{e*} \rangle_1^2)$ fixed and define $\mathcal{L}_1 \equiv \Lambda_1 (\langle \gamma_{e*}^2 \rangle_1 / \langle \gamma_{e*} \rangle_1)$ and $\Delta r \equiv (r - r_1)$. Figure 1 shows that the combined effects of the radiatively induced pressure gradient and the Compton-rocket thrust substantially reduce the Compton losses compared with those sustained if the electrons are confined to an adiabatic mesh. The initial parameter $\mathcal{L}_1 = 10^3$ is about the maximum appropriate to observed radio sources near a quiescent quasar. However, during an outburst of an OVV quasar or a blazar, or in the formation of a compact radio component or jet, larger values of $\mathcal{L}_1$ can occur. Figure 2 shows $\langle \gamma_{e*} \rangle_2$ and $(\gamma \beta)_2$ at $r_2 = 2 r_1$ for a range of $\mathcal{L}_1$.

References:
Cheng, A.Y.S., and O'Dell, S. L. : 1981, Ap. J. (Letters), in the press.
Hoyle, F., Burbidge, G. R., and Sargent, W.L.W. : 1966, Nature, 209, pp. 751-753.
O'Dell, S. L. : 1981, Ap. J. (Letters), 243, pp. L147-L149.
Rees, M. J., and Simon, M. : 1968, Ap. J. (Letters), 152, pp. L145-L148.
Shklovsky, I. S. : 1964, Soviet Astr.--AJ, 8, pp. 132-133.
Woltjer, L. : 1966, Ap. J., 146, pp. 597-599.

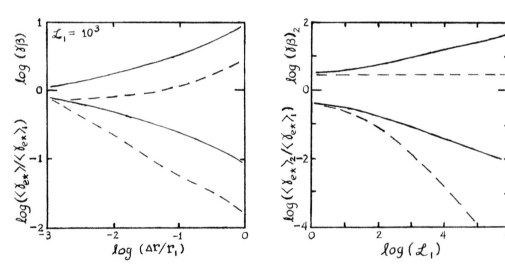

Fig. 1. Dependence of fractional residual energy $(\langle \gamma_{e*} \rangle / \langle \gamma_{e*} \rangle_1)$ and bulk unitary speed $(\gamma \beta)$ upon distance $\Delta r = (r - r_1)$ for $\mathcal{L}_1 = 10^3$. Dashed and solid lines denote adiabatic and radiative dynamics, respectively.
Fig. 2. Dependence of fractional residual energy $(\langle \gamma_{e*} \rangle_2 / \langle \gamma_{e*} \rangle_1)$ and bulk unitary speed $(\gamma \beta)_2$ at $r_2 = 2 r_1$ upon $\mathcal{L}_1$. Symbols are as in fig. 1.

VLA OBSERVATIONS OF THE PALOMAR BRIGHT QUASAR SURVEY

David B. Shaffer
Phoenix Corporation

Richard F. Green
Steward Observatory, University of Arizona

Maarten Schmidt
California Institute of Technology

Green and Schmidt (Green 1976, and in preparation) have optically surveyed some 10000 square degrees of the northern sky to search for bright quasars. Their final sample contains about 100 quasars. The B magnitudes of the sample range from 13.1 to 16.5, with most in the range 15.0-16.2. The redshifts range from 0.03 to over 2, considerably concentrated toward smaller values (median of 0.18).

We observed 94 of these quasars with the partially complete VLA in November/December 1979, and detected radio emission from 27 of them, or 29%, to a limit of 1-2 mJy. Our succes rate is considerably higher than in most previous surveys, which typically detected about 10% of the observed quasars (Smith and Wright 1980; Sramek and Weedman 1980). Our detection percentage is comparable to that in the smaller sample studied by Condon et al. (1981) and the brightest quasars in Smith and Wright's sample: about 35%. (there is some overlap between our sample and that of Condon et al.) We conclude that bright quasars are definitely more likely to be detectable radio sources.

Some 15 (or 16%) of the sample were already known to be radio sources, including five 3C sources and several 4C and Parkes sources. The number and strength of the previously known sources emphasizes one result from other quasar radio surveys: the detected sources are generally quite strong and the detection percentage goes up very slowly as the sensitivity is increased. Only 6 of our detections were weaker than 10 mJy.

The shape of our integral source count curve is essentially the same as that of Strittmatter et al.(1980): a steep rise (slope about -1.5) for the few strongest sources, with a knee around 500 mJy below which the slope is exceedingly flat, -0.2 or less. Such source counts are difficult to reconcile with the prediction of Scheuer and Readhead's (1979) beaming model for quasar radio emission, as discussed by Strittmatter et al. The extended structure of many of our sources also argues against the beaming hypothesis. We found at least 2/3 of the quasars to be discernibly resolved at 1" resolution. Many of them are classical double and triple sources, a few have a bright radio nucleus with one-sided extended emission ("class D2" quasars), and a few are only slightly resolved. All but one of the extended sources have a radio core coincident with the optical quasar. This high incidence of extended structure, which is most prominent in sources with the highest ratio of radio to optical luminosity, with prominent cores is very difficult to reconcile with beaming theories. Additionally, the source counts for the extended structures and radio cores have essentially the same shape as noted above for the total counts. This indicates similar over-all behaviour of radio luminosity for both cores and extended emission, contrary to core enhancement/reduction that depends on fortuitous alignment.

Although our sample and those of Condon et al. and Smith and Wright indicate a much higher (about three times) probability of detection of radio emission from bright quasars, we see little evidence of a strong correlation with apparent magnitude within our sample. We detect 3 of the 4 brightest quasars, but the next detection is for brightness rank #15. If we divide our sample into quarters, the detection rate in the first 3 quarters is essentially the same (32%), while the rate in the last quartile is rather lower (17%). The difference is about 2 sigma. Condon et al. and Smith and Wright find a similarly high detection rate down to B about 17.5, or about one magnitude fainter than our sample. There must, however, be a change in the ratio of radio to optical luminosity as one goes to still fainter quasars. Otherwise, there would be higher detection rates for fainter, optically selected samples since many of the bright quasars have radio luminosities such that they would be easily detectable if they were several magnitudes fainter.

Detectability and absolute optical luminosity are strongly correlated within our sample. In the most luminous quartile, the detection rate is 35%, in the second quartile the rate is almost 60% (13 of 23). The weakest half of the sample contains only 6 radio sources (out of 47 quasars, or 13%). The strong correlation between luminosity and detectability partially accounts for the enhanced probability of radio emission from bright sources.

REFERENCES

Condon, J. J., O'Dell, S. L., Puschell, J. J., and Stein, W. A. 1981, Ap. J. **246**, pp.624-646.
Green, R. F. 1976, P.A.S.P., **88**, pp.665-668.
Scheuer, P. A. G. and Readhead, A. C. S. 1979, Nature, **277**, pp.182-185.
Smith, M. G. and Wright, A. E. 1980, M.N.R.A.S., **191**, pp.871-886.
Sramek, R. A. and Weedman, D. W. 1980, Ap. J., **238**, pp.435-444.
Strittmatter, P. A., Hill, P., Pauliny-Toth, I. I. K., Steppe, H. and Witzel, A. 1980, A&A, **88**, pp.L12-L15.

OPTICAL SPECTRA OF RADIO-LOUD AND RADIO-QUIET ACTIVE GALACTIC NUCLEI

Donald E. Osterbrock
Lick Observatory, Board of Studies in Astronomy and Astrophysics
University of California, Santa Cruz

Many radio galaxies have strong emission lines in their optical spectra. The fraction with such lines is much larger than in "normal" galaxies. Radio galaxies generally also have very bright nuclei; thus those with strong emission lines are similar in both respects to Seyfert galaxies. Hence radio and Seyfert galaxies are both generally considered to be similar physical objects: active galactic nuclei. Their observational properties show they are closely related to quasars (quasi-stellar radio sources) and (radio-quiet) QSOs. A short table of the space density of these objects, culled from many sources, chiefly Schmidt (1978) and Simkim, Su and Schwarz (1980) is given below. Although all the numbers are quite uncertain, there is no doubt that the radio-loud objects are relatively rare. With less certainty, it appears that the ratio of numbers of radio galaxies to Seyfert galaxies is about the same as the ratio of numbers of quasars to QSOs.

Table 1. Approximate Space Densities Here and Now

Field Galaxies	10^{-1} Mpc^{-3}
Luminous Spirals	10^{-2}
Seyfert Galaxies	10^{-4}
Radio Galaxies	10^{-6}
QSOs	10^{-7}
Quasars	10^{-9}

Many of the strong radio sources are broad-line radio galaxies (BLRG), having H I emission lines with widths from 5000 to 30,000 km s^{-1} in their optical spectra. They are mostly extended double radio sources. They tend to have a greater fraction of their radio luminosity concentrated in a central compact component than the narrow-line radio galaxies do (Hine and Longair 1979). The optical spectra of the BLRG also show very strong featureless continua with approximately power-law forms. These featureless continua apparently extend to high frequencies in the ultraviolet and provide the source of ionizing photons responsible for much of the observed optical emission. Nearly all the BLRG are N galaxies in form.

The corresponding radio-quiet objects are Seyfert 1 galaxies. They are observed to have dense broad-line emission regions much the same as the BLRG. However the Seyfert 1 galaxies are mostly spirals or related to spirals, often barred or having companions.

There are differences in the optical broad emission-line spectra of the BLRG and Seyfert 1's. Each class contains a wide range of objects, but on the average the BLRG have much weaker Fe II/Hβ emission-line ratios and stronger Hα/Hβ ratios. The radio galaxies' line profiles often appear to be more irregular and flatter than the Seyfert 1 galaxies' line profiles. The same correlation, that the Fe II optical emission lines tend to be weak, holds up for most but not all radio-loud quasars, while the few radio-quiet QSOs observed in the optical region tend to have strong Fe II optical emission, like Seyfert 1 galaxies (Phillips 1978a and unpublished private communication; Peterson, Foltz and Byard 1981). Those quasars that do have strong optical Fe II emission are mostly compact radio sources (Miley and Miller 1979).

The Fe II emission thus must contain a clue as to the difference in structure between radio-loud and radio-quiet active galactic nuclei. Physically, Fe II is a very low stage of ionization. Analysis of the observations shows that the Fe II resonance lines in the ultra-violet must have very large optical depths, but the dimensions are reasonable in terms of our knowledge of active galactic nuclei (Phillips 1978b; Collin-Souffrin et al. 1979). The optical spectra of quasars with large redshifts show they do have strong Fe II emission in the ultra-violet, indicating that some but not as much Fe II is present as in the radio-quiet objects (Wills et al. 1980; Gaskell 1981; Grandi 1981). The relative intensities of the Fe II lines show that $T \approx 10^4 K$ in the dense region in which they are emitted. The excitation is probably collisional. The source of ionization of the partly ionized region in which the Fe II arises is still controversial, perhaps primary photo-ionization by hard photons followed by Auger and collisional ionization (Netzer 1980), perhaps mechanical heating (Collin-Souffrin et al. 1980).

What then is the significance of the relative weakness of the Fe II emission in the radio-loud BLRG? One possibility is that there is not as much gas in the dense broad-line regions of BLRG, so that there is only a small partly ionized region. Another possible interpretation would be that the gas is present, but that the extended region of low ionization does not exist. On the photoionization mechanism this would mean that there is a lack of hard ionizing photons in the spectrum of the central sources of the BLRG. There is no evidence to support this from X-ray measurements, nor from the observed optical lines emitted by the outer narrow-line region. On the mechanical input hypothesis the BLRG would differ from the Seyfert 1's in having less mechanical input of heat, presumably from dissipation of mass motions or magnetohydrodynamic waves.

The narrow-line active galactic nuclei are narrow-line radio galaxies (NLRG) and Seyfert 2 galaxies. Members of both classes generally have weaker featureless continua than BLRG and Seyfert 1's. The narrow-line objects are not as "active." Perhaps they are not as close to the critical Eddington luminosity, and gas is not ejected from their accretion disks or turbulence is not generated by some instability within them. There is no major observed difference in the optical spectra of NLRG and radio-quiet Seyfert 2 galaxies (Cohen and Osterbrock 1981), although in form the former are mostly cD, D, and E galaxies, while the latter are mostly spirals. There is a strong correlation between the strength of the narrow lines with radio emission. All BLRG have relatively strong forbidden lines, more like Seyfert 1.5 galaxies than like Seyfert 1's. Among the Seyfert galaxies that have been detected as weak radio sources, Seyfert 2 galaxies on the average have much stronger radio emission. The few so-called Seyfert 1 galaxies that have been detected as radio sources are in fact mostly Seyfert 1.5 galaxies. This correlation suggests that the escape of the radio plasma from its source, presumably at the center of the active nucleus, occurs along the same channels as the escape of the ionizing photons out into the narrow-line region. The interaction of the plasma with the ionized gas in radio galaxies must produce some effects which might be most easily observed in the form of heating effects; it will be worthwhile to look for them observationally.

I am greatly indebted to J. M. Shuder and R. D. Cohen for their help in obtaining, reducing, and interpreting the data that went into this paper, and to them and G. R. Blumenthal, W. G. Mathews, and J. S. Miller for many most useful discussions. I am also grateful to the National Science Foundation for support of this research, most recently under grant AST 79-19227.

REFERENCES

Cohen, R. D. and Osterbrock, D. E.: 1981, Ap.J. 243, 81.
Collin-Souffrin, S., Dumont, S., Heidmann, N. and Joly, M.: 1980, Astr. Ap. 83, 190.
Collin-Souffrin, S., Joly, M., Heidmann, N. and Dumont, S.: 1979, Astr. Ap. 72, 93.
Gaskell, C. M.: 1981, Ph.D. Thesis, UCSC.
Grandi, S. A.: 1981, Ap.J. submitted.
Hine, R. G. and Longair, M. S.: 1979, M.N.R.A.S. 188, 111.
Miley, G. K. and Miller, J. S.: 1979, Ap.J. (Letters) 228, L55.
Netzer, H.: 1980, Ap.J. 236, 406.
Peterson, B. M., Foltz, C. B. and Byard, P. L.: 1981, Ap.J. submitted.
Phillips, M. M.: 1978a, Ap.J. Supp. 38, 187.
Phillips, M. M.: 1978b, Ap.J. 226, 736.
Schmidt, M.: 1978, Phys. Scripta 17, 135.
Simkin, S. M., Su, H. J. and Schwarz, M. P.: 1980, Ap.J. 237, 404.
Wills, B. J., Netzer, H., Uomoto, A. K. and Wills, D.: 1980, Ap.J. 237, 319.

OPTICAL SPECTRA AND RADIO PROPERTIES OF QUASARS

Beverley J. Wills
McDonald Observatory and Department of Astronomy
The University of Texas at Austin

Relations between observed optical and radio properties can, in principle, constrain the geometry and physical conditions of the broad-line regions in quasars and active nuclei. Osterbrock and colleagues (see this symposium) and J.E. Steiner (preprint) have noted differences between Hα/Hβ, [OIII]/Hβ and optical Fe II emission for Seyfert 1 galaxies, broad-line radio galaxies (BLRG's) and quasars. Stockman et al. (1979) discovered a tendency for optical continuum polarization angles for quasars to be aligned with the direction of the outer radio lobes. Setti and Woltjer (1977) and Miley and Miller (1979) noted that the quasars with strongest Fe II are among the most compact radio sources (e.g. 3C 48, 0736+01, 1510-08), and Miley and Miller also note that the distribution of line widths is narrower for the more compact than for extended radio sources.

In Fig. 1, using high quality spectrophotometric scans obtained at McDonald Observatory, and data from the literature I show that, for quasars, the relative strength of optical Fe II emission (the broad blended feature λ4570) may be roughly inversely proportional to line widths (full width at half maximum, FWHM). Fig. 2 shows a similar relation between the relative intensity of the UV Fe II blend between 2300 and 2600 Å (the λ2500 feature) and the widths of Mg II and Hβ. Perhaps half the quasars are common to Figs. 1 and 2, but Fig. 2 extends the relation to higher redshifts. I distinguish between compact and extended radio sources and include radio quiet quasars, Seyfert 1 galaxies and BLRG's. Note that the quasars associated with extended radio sources have the broadest emission lines and the weakest Fe II, falling close to the region occupied by BLRG's which also have extended radio structure. Those quasars with strong Fe II and compact radio structure are most similar to the Seyfert 1 galaxies. The correlation for radio-compact quasars alone is not completely convincing. It may be that the various classes just occupy different regions of the diagram.

My results and those of Osterbrock and others suggest that it may be fruitful to investigate the relationships between the intensity ratios Hα/Hβ and [OIII]/Hβ as functions of line widths and radio structure for

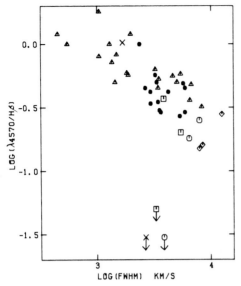 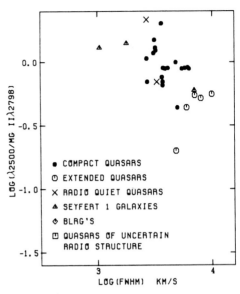

Figure 1. Relative FeII λ4570 intensities as a function of line width. Symbols are as in Figure 2.

Figure 2. Relative FeII λ2500 intensities as a function of line width, for broad-line objects.

quasars alone and for broad emission line objects as a whole.

Osterbrock (this symposium) has suggested some reasons why optical line strengths and radio structure may be related. It has been suggested at this symposium and elsewhere that compact sources may often be extended double radio sources viewed end-on, and that we could be viewing the energy machine (rotating massive body) from the pole. A picture in which the broad line emitting material tends to be confined to a thick plane (disc) perpendicular to the radio axis would explain 1) the compact radio objects being associated with smaller mass motions (narrower lines) and 2) the greater relative strengths of those emission lines having highest optical depths (e.g., the UV Fe II resonance lines). This model is not without serious objections, but is at least amenable to further, quite simple, observational tests.

I gratefully acknowledge D. Wills' help in obtaining the observations. This research is supported by NSF grant AST 79-01182.

REFERENCES

Miley, G.K., and Miller, J.S.: 1979, Ap.J. (Letters), 228, L55.
Setti, G., and Woltjer, L.: 1977, Ap.J. (Letters), 218, L33.
Stockman, H.S., Angel, J.R.P., and Miley, G.K.: 1979, Ap.J. (Letters), 227, L55.

CHARACTERISTICS OF NEBULOSITY ASSOCIATED WITH PARKES QUASARS

P. A. Wehinger
Arizona State University and Northern Arizona University
S. Wyckoff
Arizona State University
T. Gehren
Max-Planck-Institut für Astronomie, Heidelberg

A sample of 13 out of 15 low redshift ($0.1 \leq z \leq 0.6$) radio-loud quasars have been resolved on large-scale (19 arcsec mm^{-1}), sky-limited ($\mu_R \sim 26.5$ mag sec^{-2}) Kodak IIIa-F photographs obtained with the ESO 3.6-m telescope. The QSO images were analyzed by digitally subtracting the plate background and the point-spread function defined by images of nearby (≤ 1 arcmin) field stars having magnitudes comparable to the quasars ($\Delta m \lesssim 0.3$ mag). The resolved nebulosities underlying the QSO images have isophotal diameters in the range $\theta \sim 7$ to 40 arcsec, with surface brightnesses $\mu_R \sim 22\text{-}26$ mag sec^{-2}, and integrated apparent red magnitudes $\sim 16\text{-}21$.

For the 13 resolved quasars, the nebulosity was found to have an average metric diameter, $<M_{R1}> \sim -21.8 \pm 0.8$, assuming $z_{neb} = z_Q$ cosmological, $H_o = 60$ kms^{-1} Mpc^{-1}, and $q_o = +1$. The isophotal diameters of the resolved nebulosity are correlated with quasar redshifts, where $\theta \sim z^{-1}$ within the observational errors. The integrated apparent red magnitudes and the isophotal diameters are also correlated where $R \sim -5 \log \theta +$ constant within the observational errors. The correlations are indicative of roughly constant linear diameters and constant surface brightnesses for the associated QSO nebulosities. The redshift-magnitude relation for the underlying nebulosity can be interpreted as a Hubble law with the integrated absolute magnitudes of the nebulosity being ~ 2 mag fainter than first-ranked giant elliptical galaxies, with a dispersion in absolute magnitude of ~ 3 mag.

The physical and statistical properties of the nebulosity surrounding the quasars inferred from the observations lend significant support to the hypothesis that quasars are the luminous nuclei of distant galaxies. The observations further suggest that: 1) quasars are the nuclei of galaxies of average luminosity; and 2) the quasar nuclei are considerably more luminous relative to their host galaxies than other types of active galactic nuclei.

Further imagery of two high redshift quasars, analyzed in the same manner, but not part of the above sample, exhibit unresolved images at $z = 1.048$ and $z = 2.568$.

This research is supported in part by the European Southern

Observatory and the Max-Planck Gesellschaft. For further details see: Wyckoff, Wehinger, and Gehren (1981).

Reference:
Wyckoff, S., Wehinger, P. A., and Gehren, T., 1981, Astrophys. J., 247, 750.

DISCUSSION

ABELL: What are the redshifts of the two quasars where nebulosity was not detected optically?

WEHINGER: Out of 15 QSO's ($z = 0.158$ to 0.528), the two not resolved were among three of the highest redshift objects in our sample. One was at $z = 0.361$, the other at $z = 0.528$.

BL Lac OBJECTS AND THEIR ASSOCIATED GALAXIES

Donna Weistrop
Laboratory for Astronomy and Solar Physics
Goddard Space Flight Center
Greenbelt, Maryland

ABSTRACT

CCD photometry of five BL Lac objects indicates that at least three, and possibly four, are located at the centers of giant elliptical galaxies. The redshift for one of these objects, 1218+304, is estimated. A lower limit is placed on the redshift of 1219+28, for which no associated galaxy has been detected. Separation of the galaxy emission from the total observed flux makes possible comparison of the optical - far red flux from the point source alone with radio and X-ray data. This comparison suggests the emission from 1727+50 and 1218+304 can be interpreted as due to direct synchrotron emission. Observations of a small group of galaxies associated with the BL Lac object 1400+162 are also discussed.

I. INTRODUCTION

Investigations of the galaxies associated with BL Lacertae objects are important for several reasons. First, we wish to determine whether all BL Lac objects are embedded in galaxies. For those which are located in the centers of galaxies, determination of the galaxy redshifts will establish the distance. Comparison of the galaxy redshift with the redshift determined from the point source, where possible, will confirm or disprove the cosmological nature of the BL Lac redshifts. By separating the flux from the galaxy from the total optical emission, the optical flux arising in the point source alone can be compared to the compact radio source and X-ray observations. For example, the unreddened composite spectral index of PKS 0548-322 is 1.4 ± 0.2, $[F_\nu \propto \nu^{-\alpha}]$. The point source alone for this object has a spectral index $\alpha = 0.3 \pm 0.2$ (Weistrop, Smith and Reitsema 1979). Thus, eliminating the emission due to the associated galaxy can significantly effect the visible spectral index, and conclusions drawn concerning the nature of the underlying emission. Observations of the underlying galaxies help to define the role of the BL Lac phenomenon in galaxy evolution. Investigations of the nature of galaxies near BL Lac objects are useful to study the environment of the BL Lac's.

The observations consist of direct imaging of BL Lac objects with a charge-coupled device (CCD) in three or four broad passbands, centered at 0.45 μm (B), 0.53 μm (V), 0.75 μm, and 1.0 μm. The CCD is ideally suited to this project, since it has high quantum efficiency, especially in the red, spectral sensitivity from 0.4 - 1.1 μm, low noise characteristics, and is a linear device. Observations were obtained both with the 4.0 m telescope at CTIO and the University of Arizona's 2.3 m telescope. Where possible, an object known to be a star is included in the field, so that the point spread function can be defined. By comparison with the point spread function, we can determine whether the observed intensity distribution from a BL Lac is due to a point source alone or can be modelled using a point source plus elliptical galaxy intensity distribution. The exact procedure has been described elsewhere (Weistrop, Smith and Reitsema 1979 and Weistrop et al. 1981). The results are demonstrated by Fig. 1, which compares the point spread function, adopted models, and observations for the BL Lac object 1727+50.

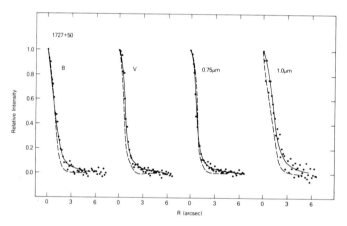

Figure 1. Comparison of the observed spatial intensity distribution of 1727+50 (filled circles) with models (solid line) and point spread function (dashed line).

II. RESULTS FOR INDIVIDUAL OBJECTS

A. 1727+50 and PKS 0548-322

Both of these objects are resolved and have known redshifts, so that by fitting the observed spatial intensity distribution and total fluxes with a point source plus giant elliptical galaxy model, the absolute magnitudes and broadband spectral distributions are obtained (Weistrop, Smith and Reitsema 1979 and Weistrop et al. 1981). In both cases, the absolute magnitudes determined for the associated galaxies are similar to those of bright galaxies, and the galaxies found associated with other BL Lac objects (Table 1). The broadband spectral energy distributions are consistent with the indentification of the

Table 1. Absolute Magnitudes for Associated Galaxies

Object	M_V	z
1727 + 50	-21.9 mag	0.0554
0548 - 322	-22.1	0.069
		H_0 = 50 km/sec/Mpc

associated galaxies as giant ellipticals (Fig. 2). It should be emphasized that the model fitting is done independently in each passband so that the broadband spectral energy distribution is a result of the analysis. The radio and X-ray data reveal a difference

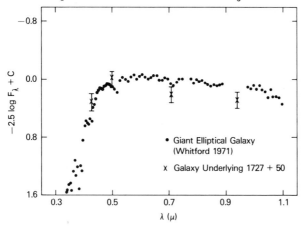

Figure 2. Comparison of the broadband spectral energy distribution of the galaxy associated with 1727+50 with the distribution for a known giant elliptical galaxy.

between the two point sources. Extrapolation of the optical data predicts the X-ray flux of 1727+50 (Fig. 3), while the high energy flux of 0548-322 is considerably fainter than would be expected from the extrapolation of the optical data (Fig. 4). The 0548-322 results can be understood in terms of a common synchrotron origin for the radio and optical data, with inverse Compton scattering the source of the high energy tail. The 1727+50 results can be interpreted as due solely to direct synchrotron emission.

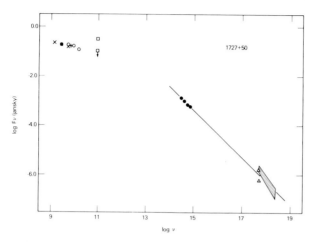

Figure 3. Comparison of the flux at several frequencies from the point source in 1727+50. The solid line indicates $\alpha = 0.97$. The symbols indicate the source of the data (see Weistrop et al. 1981).

B. 1218+304

The BL Lac 1218+304 was first brought to our attention when X-ray observations were obtained (Cooke et al. 1978). The spectrum may contain some very faint features, but no redshift has been determined (Wilson et al. 1979). The image is extended on the CCD frames. By assuming the associated nebulosity is actually a giant elliptical galaxy of absolute magnitude $M_V = -22.4$, a typical value for galaxies in which BL Lac objects have been found to be embedded, the redshift can be estimated to be $z = 0.13 \pm 0.03$ (Weistrop et al. 1981). Changes of 0.5 mag in the assumed absolute magnitude of the galaxy change the redshift by about 10%. The spectral energy distribution from radio to X-ray for the point source alone is similar to that for 1727+50, although the radio coverage is much sparser. As in the case of 1727+50, the radio, optical and X-ray emission can be interpreted as due solely to direct synchrotron emission with continuous particle injection. For all these sources, the X-ray optical and radio observations discussed were not obtained simultaneously, so care must be taken not to overinterpret the data.

C. 1219+28

The observations do not indicate any extended features associated with this BL Lac object. If it is in fact located in the center of a giant elliptical galaxy, a lower limit of $z = 0.10 \pm 0.03$ can be placed on its redshift. The spectral index is rather steep, $\alpha = 1.7$, independent of the presence of an elliptical galaxy (Weistrop et al. 1981). There is a diffuse object 10- 12 arcsec from 1219+28. Because of its low surface brightness and the relatively short integration times of the frames, it cannot be determined from these data whether the object is an elliptical galaxy or non-thermal source.

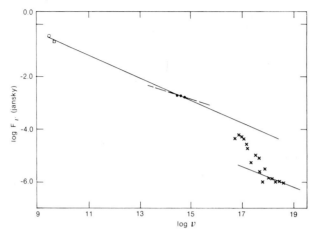

Figure 4. The flux at several frequencies for the point source in PKS 0548-322. Values of $\alpha = 0.3$ (dashed line) and 0.4 (solid line) are indicated. See Weistrop, Smith and Reitsema (1979) for data sources.

D. 1400+162

The analysis of these observations is still preliminary, and will be discussed only briefly. The BL Lac object is the only one for which the apparent association with a group of galaxies has been confirmed by redshift data (Baldwin et al. 1977). There are nine objects plus the BL Lac itself in the field. It is assumed that all nine are galaxies, although a statistical estimate indicates one or two of them are probably stars (Bahcall and Soneira 1980). From the data available, it is not possible to determine which objects may be stars. The broadband spectral energy distribution suggests two of the galaxies are late-type spirals or irregulars, and the remainder ellipticals or early spirals. Thus the fraction of late-type galaxies in this distant group is similar to that found for nearby rich clusters. The luminosity function of the galaxy group may be deficient in bright galaxies compared to the groups studied by Turner and Gott (1976). There is great temptation to speculate that the BL Lac is in the center of such a bright galaxy. The steep spectral index of the BL Lac (Table 2) lends some support to such a speculation, but further observations are needed to verify this hypothesis.

III. CONCLUSIONS

Visual and far red surface photometry has been obtained for five BL Lac objects. Three have extended images that can be fit by models consisting of a bright point source plus giant elliptical galaxy.

Table 2 presents the absolute magnitudes and optical-far red spectral indices for the point sources alone, assuming the indicated redshift. For at least two BL Lac objects, 1727+50 and 1218+304, the optical flux distribution predicts the flux in X-rays. This may also be the case in 1400+162, although further observations are necessary to determine whether an associated galaxy is influencing the observed optical spectral energy distribution and to define more precisely the X-ray energy distribution.

Table 2. Data for Point Sources

Object	M_V	α	z
1727+50	−20.7 mag	0.97 ± 0.15	0.0554
0548−322	−22.2	0.3 ± 0.2	0.069
1219+28	≤−23.7	1.7 ± 0.2	≥0.10*
1218+304	−23.3	0.9 ± 0.2	0.13*
1400+162	−24.0+	1.6 ± 0.2	0.244

$H_0 = 50$ km/sec/Mpc
$q_0 = +0.1$

*Absolute magnitude of galaxy assumed to derive redshift.
+Assumes no emission from associated galaxy.

The CCD camera used for part of this work was made available by the Space Telescope Wide Field/Planetary Camera Investigation Definition Team.

REFERENCES

Bahcall, J. N. and Soneira, R. M. 1980, Ap. J. (Suppl.) 44, pp. 73-110.
Baldwin, J. A., et al. 1977, Ap. J., 215, pp. 408-416.
Cooke, B. A., et al. 1978, Mon. Not. Roy. Astron. Soc., 182, pp. 489-515.
Turner, E. L. and Gott, III, J. R. 1976, Ap. J., 209, pp. 6-11.
Weistrop, D., Smith, B. A., and Reitsema, H. J. 1979, Ap. J., 233, pp. 504-509.
Weistrop, D., Shaffer, D. B., Mushotzky, R. F., Reitsema, H. J., and Smith, B. A. 1981, Ap. J., 249, (in press).
Wilson, A. S., Ward, M. J., Axon, D. J., Elvis, J., and Meurs, E. J. A. 1979, Mon. Not. Roy. Astron. Soc., 187, pp. 109-115.

X-RAY EMISSION FROM BL Lac OBJECTS: COMPARISON TO THE SYNCHROTRON SELF-COMPTON MODELS

Daniel A. Schwartz and Greg Madejski
Harvard/Smithsonian Center for Astrophysics

William H.-M. Ku
Columbia Astrophysics Laboratory

As one part of our joint study of the X-ray properties of BL Lac objects, we compare the measured X-ray flux densities with those predicted using the synchrotron self-Compton (SSC) formalism (Jones et al. 1974). Naive application of the formalism predicts X-ray fluxes from 10^{-3} to 10^5 those observed. We therefore ask what we can learn by simply assuming the SSC mechanism, and looking for ways to reconcile the observed and measured X-ray fluxes. This paper reports our investigation of beaming factors due to relativistic ejection of a radiation source which is isotropic in its own rest frame. We conclude that large Lorentz factors, $\Gamma \geq 10$, do not apply to BL Lac objects as a class.

Our present study (Table 1) is based on 16 sources for which a VLBI size has been measured by Weiler and Johnston (1980) and which have been observed (15 detected) in X-rays with the Einstein Imaging Proportional Counter. The X-ray data have been reduced in the manner described by Zamorani et al. (1981), except that we assume a steeper energy spectrum, of index 1.5. We use equation 4 of Jones et al. (1974) to predict the X-ray flux density S'_x(Jy) at 1 keV.

TABLE 1 Distribution of Beaming Factors

Object	S'_x(Jy)	S_x(Jy)	δ	Object	S'_x(Jy)	S_x(Jy)	δ
0048-097	1.3E-3	2.2E-7	8.8	1101+38	3.2E-7	1.4E-5	0.39
0219+438	8.8E-9	1.5E-7[1]	0.49	1219+28	3.2E-6	4.8E-7	1.6
0235+164	1.0E-1	1.6E-7	28.0	1308+326	2.1E-1	2.7E-7	30.0
0735+178	4.6E-5	2.4E-7	3.7	1400+162	1.9E-10	1.6E-7[1]	0.18
0754+100	3.6E-5	2.0E-7	3.6	1538+149	1.9E-5	1.3E-7	3.5
0818-128	3.9E-6	<1.4E-7	>2.3	1652+398	2.1E-7	1.2E-5	0.36
0829+046	1.8E-7	1.9E-7	0.98	2201+04	8.9E-8	2.6E-7	0.76
0851+202	3.4E-4	2.3E-6	3.5	2254+074	8.1E-9	1.2E-7[1]	0.51

[1]From Maccagni and Tarenghi.

We consider a model where the radio source is ejected with a Lorentz factor Γ, at an angle θ away from a line toward us. If we define the beaming factor $\delta = 1/\Gamma(1-\beta\cos\theta)$, then a Lorentz transformation gives $S_x = S_x' \delta^{-2(\alpha+2)}$. In Table 1 we have assumed $\nu_m = 1$ GHz and $\alpha = 0$. Calculating with $\nu_m = 0.3$ or 5 and $\alpha = 0.2$ or 0.5 changes S_x' by a large factor; however, since S_x depends on δ to a power similar to the dependence on the radio size, synchrotron absorption frequency ν_m and radio flux density S_m, δ is determined within a factor of a few.

If the predicted S_x' were all within a factor of 10 or 100 from the measured S_x, it could be evidence for a single component SSC model. Instead, we must find a way to reconcile at least the cases where the predicted X-ray flux density greatly exceeds that observed. The assumption of relativistic beaming is sufficient and furthermore is reasonable in the sense that a quasi-isotropic distribution of θ, and relatively small Γ are required.

We may predict the intrinsic distribution of δ for a set of objects if the Lorentz factors have a probability density function $\rho(\Gamma)$:

$$\rho(\delta) = \frac{1}{2\delta^2} \int_{\Gamma_{min}}^{\infty} \frac{\rho(\Gamma) \, d\Gamma}{\sqrt{\Gamma^2 - 1}} \qquad (1)$$

where the minimum $\Gamma_{min} = (1+\delta^2)/2\delta$. From this equation we can immediately see qualitatively that for a δ-function distribution of Γ, values of $\Gamma_0 = 2$ to 5 would suffice to span most of the range of observed δ factors, from 0.1 to 10. A power law $\rho(\Gamma) \propto \Gamma^{-2}$ would span this same range. On the other hand, for $\Gamma \geq 10$, the overwhelming majority of sources would have $\delta \ll 1$, which is not what we observe.

Suppose we only recognize an object as a "BL Lac" if it is beamed toward us within some angle θ_m. In this case we can only integrate equation (1) up to some Γmax at which $\delta \geq 1/\Gamma$max $(1-\cos\theta_m \sqrt{1-1/\Gamma_{max}^2})$, and the normalization factor 2 changes to $1-\cos\theta_m$. For the delta function distribution, $\theta_m = 1/\Gamma_0$, and it is easy to see that for $\Gamma_0 \geq 10$ we do not expect any $\delta \leq 10$. We conclude that although some extreme values of Γ may occur for specific sources, the general distribution is dominated by $\Gamma \lesssim 5$.

REFERENCES

Jones, T.W., O'Dell, S.L., and Stein, W.A. Astrophys. J. <u>192</u>, 261 (1974).

Maccagni, D. and Tarenghi, M. Astrophys. J. <u>243</u>, 42 (1981).

Weiler, K.W. and Johnston, K.J. M.N.R.A.S. <u>190</u>, 269 (1980).

Zamorani, G. et al. Astrophys. J. <u>245</u>, 357 (1981).

EVIDENCE FOR RELATIVISTIC MOTION IN THE MILLISECOND STRUCTURE OF BL Lac

R. L. Mutel, University of Iowa, and R. B. Phillips,
University of Kansas

After several years of relative quiescence, the flux of BL Lac has increased dramatically at centimeter wavelength, starting about epoch 1979.9 (Fig. 1). We have begun a series of VLBI observations to monitor the milliarcsecond structure at $\lambda 6$ and $\lambda 2.8$ cm wavelengths, using a five element VLBI array consisting of telescopes at Bonn, West Germany; Westford, MA; Green Bank, WV; Ft. Davis, TX; and Owens Valley, CA. The first two observations, in 1980 May and September, were at 5 GHz and were not of sufficient resolution to distinguish individual components in the source (Mutel, Phillips and Aller 1981). They did show, however, that the source was highly elongated along position angle $\sim 10°$ and was expanding that axis with a velocity of $\sim 4c$. The position angle is the same as several previous VLBI observations of this source, both during quiet periods and during previous flux outbursts (Pearson and Readhead 1981; Shaffer 1978 and references therein).

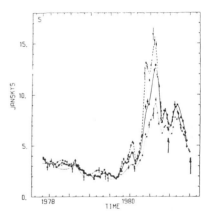

Fig. 1. Flux versus time at $\lambda 6$ cm (...), $\lambda 4$ cm (—), and $\lambda 2$ cm (---).

In order to increase the angular resolution of the observations, a frequency of 10.6 GHz was used during observations in 1980 December and 1981 June. The hybrid maps are shown in Fig. 2a. Again, the source was found to be elongated along a position angle $\sim 10°$, but with increased resolution separate components became apparent. Fig. 2b shows one-dimensional profiles of peak brightness along p.a. of $10°$ for both epochs. We arbitrarily aligned the brightest component to be coincident between epochs, but since absolute phase is not measured, the true registration between the maps is unknown.

A comparison of the source size at both epochs with the flux history of Fig. 1 shows immediately that the components are too far apart to have been causally related unless (1) there was a coincidental

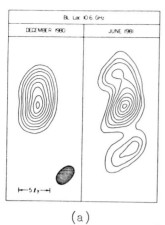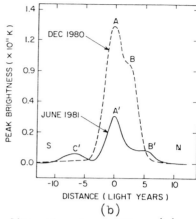

Fig. 2. Hybrid maps (a) and one dimensional profiles (b) through position angle 10° for epochs 1980.44 and 1980.93.

brightening of unrelated components ("Christmas tree" model) or (2) relativistic time dilation is responsible. The former explanation seems unlikely, especially considering the rather long period of quiescence prior to the current flux outburst.

For a relativistically moving source, the 'speed-up' of events occurring in the co-moving frame as seen by the distant observer scales as the Doppler factor γ. For the 1980 December map we derive a $\gamma \sim 4$ by comparing the apparent source size (~ 4 light-years) with ~ 1 year since the beginning of the flux outburst. A comparison of the 1980 December map with the 1981 June map shows that another component ('C') has appeared about 6.8 light-years south of component A. Furthermore, the northerly component ('B') has become more complex and has extended to ~ 5.5 light-years along a position angle of 10°. In both cases, if we assume expansion from the strong central A component, relativistic motions with a Doppler factor of $\gamma \sim 5\text{-}7$ are necessary. Alternatively, if we assume component B to be the central source, the southerly component (C) would require a $\gamma \sim 20$ to have been emitted from position 'B' ≤ 6 months prior to being seen (June 1981).

It is clear that further VLBI monitoring of the structure of BL Lac will be necessary to unambiguously determine the motion of individual components and hence the detailed dynamics of the source.

References

Mutel, R. L., Phillips, R. B., and Aller, H. 1981, submitted to Nature.
Pearson, T. J., and Readhead, A. C. S. 1981, Ap. J., in press.
Shaffer, D. B. in Pittsburgh Conference on BL Objects, U. of Pittsburgh, PA, 1978.

BACKER: Is the appearance of component "C" only in the June 1981 map a dynamic range effect (since the June 1981 flux was about a factor of two lower than December 1980)?

MUTEL: No, the lack of a third component in December 1980 is evident on the profile plot (Fig. 2b).

MARK III VLBI OBSERVATIONS OF THE NUCLEUS OF M81 AT 2.3 AND 8.3 GHz

N. Bartel[1], B. E. Corey[1], I. I. Shapiro[1],
A. E. E. Rogers[2], A. R. Whitney[2], D. A. Graham[3],
J. D. Romney[3], R. A. Preston[4]

[1] Massachusetts Institute of Technology
[2] Northeast Radio Observatory Corporation
[3] Max Planck Institut für Radioastronomie
[4] Jet Propulsion Laboratory

The normal spiral galaxy M81, which has some characteristics of a Seyfert (Peimbert, Torres-Peimbert, 1981), has a flat spectrum in the radio range (de Bruyn et al., 1976), variable on the time scale of days (Crane et al., 1976), and detectable radiation at infrared (Rieke, Lebofsky, 1978) and X-ray wavelengths (Elvis, van Speybroeck, 1981). At a distance of $\sim$3.3 Mpc, M81 is the nearest extragalactic object with a nucleus detectable with VLBI (Kellermann et al., 1976). We report here on simultaneous VLBI observations made with the Mark III system at 2.3 and 8.3 GHz. Observations on 14 and 16 March 1981 utilized the 100 m diameter telescope in Effelsberg, W. Germany (MPIR); the 43 m telescope at Green Bank, WV (NRAO); and the 40 m telescope near Big Pine, CA (OVRO).

Observations with the MPIR antenna yielded total flux densities per beam area of 135±15 mJy at 2.3 GHz and 82±7 mJy at 8.3 GHz. The visibility curve for the MPIR-OVRO baseline is a smooth function with a deep minimum at $\sim$14.5 and 16.1 hours GST at 2.3 and 8.3 GHz, respectively. The nucleus is unresolved at 2.3 GHz in the direction of its minor axis on the longest baseline, whereas it is partly resolved at 8.3 GHz. Parameters of elliptical-gaussian models of the brightness distribution at the two radio frequencies were estimated from the corresponding fringe amplitudes. The models are superimposed concentrically in Fig. 1, juxtaposed to a photograph of M81 taken from Sandage (1961). The position angle (PA) of the nucleus, though frequency dependent at the 4σ level, is closely aligned with the rotation axis of the galaxy (PA = $62\pm3°$; Rots, Shane, 1975).

Parameters of Elliptical-Gaussian Models of the Brightness Distribution of M81 (quoted errors are purely statistical)

Parameters	2.3 GHz	8.3 GHz
peak flux density (mJy)	58.2 ±2.6	76.0 ±4.9
peak brightness temp.(10^{10} K)	2.6 ±0.6	1.0 ±0.1
major axis (mas, FWHM)	1.3 ±0.1	0.53 ±0.03
minor axis (mas, FWHM)	0.6 ±0.1	0.36 ±0.03
pos. angle of major axis (°)	75.0 ±3.4	50.0 ±4.9

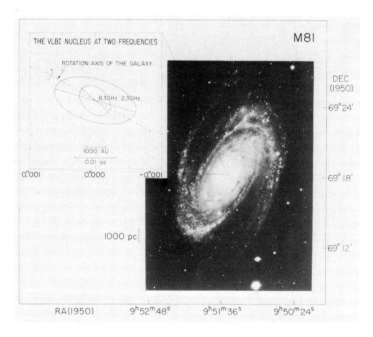

Figure 1: The galaxy M81 with its compact radio nucleus

We thank the Goddard-Haystack-MIT Mark III Development Team, and especially R. J. Cappallo, T. A. Clark, H. F. Hinteregger, N. R. Vandenberg, and C. A. Knight for their indispensable contributions.

REFERENCES

Bruyn, A. G. de, et al. 1976, Astr. Ap. 46, 243.
Crane, P. C., et al. 1976, Ap. J. Lett. 203, L113.
Elvis, M., van Speybroeck, L. 1981, BAAS 13, 550.
Kellermann, K. I., et al. 1976, Ap. J. Lett. 210, L121.
Peimbert, M., Torres-Peimbert, S. 1981, Ap. J. 245, 845.
Rieke, G. H., Lebofsky, M. J. 1978, Ap. J. Lett. 220, L37.
Rots, A. H., Shane, W. W. 1975, Astr. Ap. 45, 25.
Sandage, A. 1961, The Hubble Atlas of Galaxies, Carnegie Inst. Pub. No. 618.

RADIO OBSERVATIONS OF THE GALACTIC CENTER

D.C. Backer
Radio Astronomy Laboratory
University of California, Berkeley

Low-resolution, radio-intensity maps of the Galactic Center are dominated by two structures: the extended, nonthermal region between ℓ = 0°.0 and 0°.3 called Sagittarius A(East); and the bright thermal source at ℓ = -0°.05 and b = -0°.05 called Sagittarius A(West). SgrA(East) is probably a supernova remnant. SgrA(West) is a dense, gaseous region which is the very center of our Galaxy. Within SgrA(West) is a pc-sized cluster of infrared sources -- late-type stars and gas condensations -- swirling about with a dispersion of 250 km s^{-1}. The cluster contains a unique radio source with the following properties: a size not exceeding 10^{15} cm, a radio luminosity exceeding 10^{34} erg s^{-1}, and rather stable properties on a time scale of 10^8 s. This compact, nonthermal object (SgrAcn below) may be surrounded by a 0.1pc infrared source. Models for SgrAcn include both 10^6 M$_\odot$ black holes related to the central objects in radio galaxies and quasars (Oort 1977, van Buren 1978) and 1 M$_\odot$ degenerate stars related to pulsars and X-ray binaries (Reynolds and McKee 1980). We present below the radio intensity spectrum as a basic datum for further development of either class of models.

The radio spectrum of SgrAcn must be measured with interferometers whose resolution falls in the window between confusion from nearby HII regions (3") and the frequency-dependent angular broadening by interstellar plasma (1".0ν(GHz)$^{-2}$). The angular broadening effects have been observed at many frequencies between 1 and 10 GHz.

The flux density of SgrAcn was measured from 1.4 to 85.7 GHz with several radio interferometers. In March 1981 the VLA was used in A configuration to determine flux densities at 1.413, 4.885 and 15.035 GHz. Measurements were made in the visibility domain, and were referred to 3C286 via secondary standards of 1748-253, 1748-253 and NRAO 530, respectively. System temperature and sky absorption corrections were applied. At 1.4 GHz the source intensity was estimated assuming the ν^{-2} apparent size law. In April 1981 observations with the 35-km interferometer in Green Bank provided flux densities at 2.695 and 8.085 GHz referred to 1748-253. Absolute intensities were computed using the VLA spectrum for 1748-253.

In April 1981 the Hat Creek Interferometer was used at 85.7 GHz. The antennas were on an East-West baseline with projected lengths between 73,000 and 86,000 λ (2″.8 -2″.4). NRAO 530 was observed as a secondary calibrator. System temperature and sky absorption corrections were applied. Coherent and incoherent integrations gave consistent results. The flux-density of NRAO 530 was referred to planetary observations after the antennas were moved to a short baseline.

These flux-density observations (Fig.1) indicate a continuous spectral distribution up to 86 GHz with an index of +0.25, and an observed luminosity of 1.3×10^{34} erg s^{-1}. The turnover frequency must be at or above 100 GHz. As noted by many authors, spectra of the form presented below can be generated either by a superposition of many self-absorbed synchrotron components, or by a smooth distribution of synchrotron emissivity with internal absorption.

REFERENCES:

Oort, J.H.: 1977, Ann. Rev. Astron. Astrophys., 15, pp. 295-362.
Reynolds, S.P., and McKee, C.F.: 1980, Astrophys. J., 239, pp. 893-897.
van Buren, H.G.: 1978, Astron. Astrophys., 70, pp. 707-717.

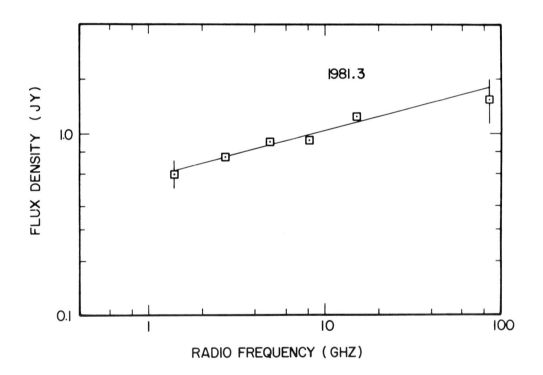

FIG.1 Flux density spectrum for SgrAcn between 1.4 GHz and 85.7 GHz measured in March-April 1981. Solid line corresponds to $\nu^{+0.25}$ spectrum.

EXTRAGALACTIC RADIO SUPERNOVAE IN NGC 4321 AND NGC 6946

R. A. Sramek
National Radio Astronomy Observatory, Socorro, NM
K. W. Weiler
National Science Foundation, Washington, DC
J. M. van der Hulst
University of Minnesota, Minneapolis, MN

ABSTRACT

The supernovae SN1979c in NGC 4321 and SN1980k in NGC 6946 have both been detected at centimeter wavelengths at the VLA. The radio emission turns on very rapidly, but may be delayed by as much as a year with respect to the optical outburst. In both supernovae, the 20 cm radiation peaks after the 6 cm, and the radio emission has a very slow post-maximum decay.

On April 6, 1980 the bright Type II supernova SN1979c in NGC 4321 was detected at the VLA at 6 cm (Weiler, et al, 1981). There was only a single prior observation, made a year earlier, about one week after the optical maximum, which showed no detection with a 0.3 mJy upper limit. Following the successful detection, regular observations of SN1979c were made at 20 cm, 6 cm, and occasionally at 2 cm to obtain the first radio light curve of a supernova. At 20 cm, SN1979c remained undetected until a December 4, 1980 observation showed that it had risen quickly to 2.1 mJy. Meanwhile, a second bright Type II supernova, SN1980k, was seen is NGC 6946. Radio monitoring was begun, and it was detected at 6 cm, 36 days after the optical maximum. Two and one half months later it was detected at 20 cm. The radio light curves up to May 6, 1981 for the two supernovae are shown in Figure 1.

The light curves are characterized by:
a) A very fast rise time, with time to the power 4 or 5 at 20 cm.
b) A spectrum that is steeply rising at early times then flattens out. The 2 cm data indicates that the spectrum between 6 cm and 2 cm slopes down with an index of about -0.8.
c) Peak flux densities that are comparable at 6 cm and 20 cm.
d) The rise occurs first at higher frequencies, and is delayed relative to the optical outburst.
e) The decay is much slower than the rise.
f) neither the rise nor the decay is simple monotonic; both show irregularities.

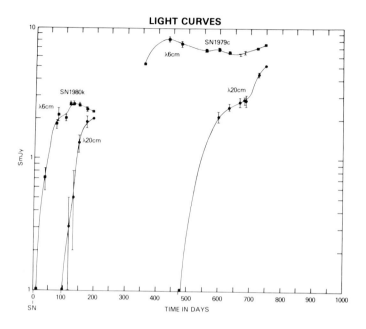

Figure 1. The radio light curves for two recent supernovae, with the times relative to their respective optical maxima.

These light curves are probably best explained by time varying absorption, either synchrotron self-absorption or free-free absorption, although the rise time in a) gives some problems for the former.

REFERENCE

Weiler, K.W., van der Hulst, J.M., Sramek, R.A., and Panagia, N. 1981, Ap. J. Lett., 243, L151.

THE ANGULAR SIZE DISTRIBUTION OF RADIO SOURCES AT LOW FLUX DENSITIES

Ann Downes
Mullard Radio Astronomy Observatory, Cavendish Laboratory,
Madingley Road, Cambridge CB3 OHE, U.K.

ABSTRACT Observations of complete samples of extragalactic radio sources at low and intermediate flux densities are described. Many types of source are found. The angular sizes form a smooth extrapolation from higher flux densities, and can be predicted from the known properties of samples at high flux density either with linear size evolution (for $\Omega=1$ or $\Omega=0$ Universes) or without linear size evolution (for $\Omega=0$). The question of whether such evolution is required therefore remains open.

1. INTRODUCTION

It is now possible to study extragalactic radio sources at low and intermediate flux densities in detail comparable to that which has hitherto been available only for the brightest sources. While the ultimate aim of the work described here is the definition of the radio source population down to flux densities of 55 mJy at 408 MHz and 20 mJy at 1407 MHz, the present paper concentrates on the angular size distribution down to these levels and, in particular, on the problem of whether linear size evolution with cosmological epoch is required. Such evolution was inferred from the small angular size θ of quasars at high redshifts z in bright samples. Subsequent studies indicated that it might also explain the small values of θ found at low flux densities S, and linear size evolution as $(1+z)^{-N}$ where N=1-1.5 has been suggested (Kapahi 1975, 1977). On the other hand, the θ-z result may simply reflect an inverse correlation between linear size and radio luminosity, and the radio luminosity function used by Kapahi provides a poor fit to the source counts at low S (Swarup 1977) so that conclusions based on this distribution of sources should be treated with caution.

Observations of complete samples of sources at low and intermediate flux densities are described here and compared with new predictions. These samples overlap those used by Kapahi as well as extending to significantly lower flux densities.

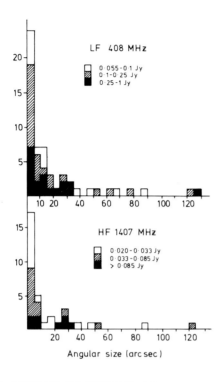

Figure 1 Angular size distributions for radio sources at low flux densities (Fielden et al.)

(a) Low frequency (LF) sample of sources in the flux density range .055-1 Jy (for 5C6 and 7) and 0.1-1 Jy (for 5C12) at 408 MHz.

(b) High frequency (HF) sample of sources in the flux density range above 20 mJy at 1407 MHz.

2. THE SAMPLES OF SOURCES AND THE OBSERVATIONS

At low flux densities the brightest sources were selected from the 5C6 and 7 (Pearson and Kus 1978) and 5C12 (Benn et al. 1982) deep radio surveys. These were chosen from areas well inside the primary beam to ensure accurate flux densities and to prevent discrimination against low brightness sources. Samples complete at 408 MHz to 55 mJy (5C6 and 7) and 100 mJy (5C12) and at 1407 MHz to 20 mJy (5C6, 7 and 12) were thus formed. In the original 5C surveys most of these sources were unresolved at 408 MHz (HPBW 80"x80"cosec δ) while about 30% were extended at 1407 MHz (HPBW 23"x23"cosecδ). High-resolution radio observations of the samples with the 5-km telescope and the VLA are described by Downes et al. (1981) and Fielden et al. (1982).

The distribution of angular sizes at both frequencies (Fig. 1) extends up to about 2'. A wide range of radio morphologies is found among the well-resolved sources, with clear examples of both classical doubles (indicating a high radio luminosity) and sources whose structure is characteristic of a low radio luminosity. At 408 MHz about 30% of the sample is unresolved, and 25% is both steep-spectrum ($\alpha > 0.5$) and unresolved; the corresponding figures at 1407 MHz are 46% and 34%. This reflects the increased number of flat-spectrum objects at the higher frequency, but also shows a substantial conribution from compact steep-spectrum sources in both samples.

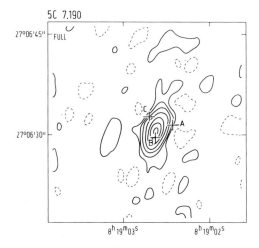

Figure 2 5C7.190: a flat-spectrum source in the 1407-MHz sample.
(a) 1.4-GHz map made at the VLA (Downes et al.). Contours are at ±2,5,10, 30,50,70,90% of the peak 30.2 mJy/beam. The positions of candidate identifications are marked with crosses. Further observations at 5 GHz (Fielden et al.) confirm B, a stellar object with $m_r=20.7$ as the identification.

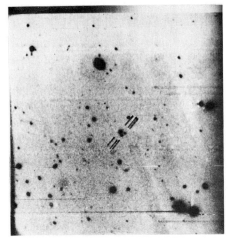

(b) CCD image of the 3.6'x3.6' field around 5C7.190 (Perryman et al.). The three candidates for the identification are marked.

The accurate positions and structures for sources in 5C6 and 7 have been used together with optical observations with a CCD detector in a continuation of a deep identification program begun by Perryman (1979a, b). The CCD observations reach a limiting magnitude $m_r=23$ and are described by Perryman et al. (1982). Fig. 2 shows, as an example, a radio map and CCD image of the field of 5C7.190. Approximately 30(-55)% of the sources are identified (the fraction is slightly higher for unresolved than for resolved sources). Only two sources (including 5C7.190; Fig. 2) are identified with stellar objects. It is therefore likely that, if the properties of the sources are similar to those at high flux density, a large fraction of the sample must lie at redshifts $z \gtrsim 0.6$.

At intermediate flux densities, Allington-Smith (1982) has observed a complete sample of 59 B2 radio sources in the range 1-2 Jy at 408 MHz. This corresponds to the region in which the source counts exceed non-

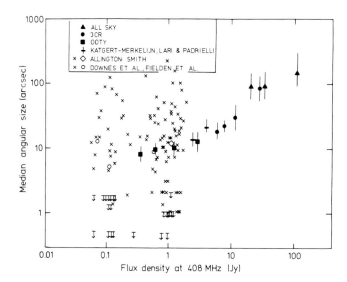

Figure 3 The angular size-flux density distribution at 408 MHz, adapted from the diagram by Swarup (1975). The medians are from Swarup, Katgert-Merkelijn et al. (1980) and the present observations (unfilled symbols), for which individual measurements are also shown (x). Allington-Smith's values are plotted using his measurements of the flux density at 408 MHz.

evolutionary predictions by the greatest amount, and should therefore represent the most strongly evolving sources. Radio structures have been determined with the One-mile and 5-km telescopes, and have been used to help identifications with CCD observations to magnitudes similar to those for 5C6 and 7 (Allington-Smith et al. 1982). Once again a large variety of source structures was found, with an angular size distribution ranging up to 4' and many compact sources. The fraction of identified sources is, however, considerably higher than at the 5C levels, with a preliminary estimate of 70-75% identified.

3. THE ANGULAR SIZE-FLUX DENSITY DIAGRAM

The angular sizes of individual sources in the 408-MHz samples are plotted in Fig. 3 together with the medians for both the present samples and previous measurements. The median is 7.5" in the range 0.1-1 Jy and 12" for 1-2 Jy, and the results form a smooth extrapolation from higher flux densities, with excellent agreement where the samples overlap with occultation measurements made at Ooty. At 1407 MHz the median angular size is 7". The angular sizes again agree with extrapolation from high flux densities and with those inferred from indirect measurements (Ekers and Miley 1977, and references therein).

The distribution of θ at low S can be synthesised from a complete "parent sample" at high S if the members of this sample are taken as representative of the overall population and all the redshifts are known. The track followed on the θ-S diagram by a single source of given size as the redshift is increased is shown in Fig. 4. It is harder to predict small angular sizes for $\Omega=1$ than for $\Omega=0$, as the angular size is larger for given flux density, and the redshift required for a given drop in flux density is higher. On the other hand, if linear size

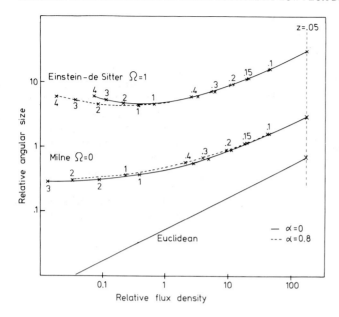

Figure 4   The angular size-flux density track obtained by moving a single source (with no linear size evolution) to higher redshifts. The curves are aligned at redshift z=0.05, and the increasing redshift is marked along each. Predictions for both a steep-spectrum ($\alpha=0.8$) and a flat-spectrum ($\alpha=0$) source are shown.

evolves as $(1+z)^{-N}$, this decrease will be more rapid with decreasing S for $\Omega=1$ because of the higher values of z.

The space density of each source is taken to follow an evolving radio luminosity function (RLF). The relative numbers of each type of source at given flux density vary because the strength of evolution is dependent on both radio luminosity and redshift, so the distribution of angular sizes changes with flux density (thus direct comparison of Fig. 4 with Fig. 3 is not sufficient proof of linear size evolution).

Previous attempts to model the distribution at low S using a subset of the 3CR "166 sample" (Jenkins et al. 1977) as parent and the RLF of Wall et al. (1980) suggested that although the median angular size could be predicted if linear size evolution was invoked, the distribution then became too compressed, so that another method of obtaining small angular sizes was required (Downes et al. 1981). The present modelling uses the evolving RLF of Peacock and Gull (1981), which is derived from the source counts, V/Vmax, local RLFs and known redshift distributions, and treats steep- and flat-spectrum sources separately. Two parent samples have been used:
(i)   the 3CR "166 sample" at 178 MHz (plus 5 giant sources whose low surface brightness prevented their inclusion in 3CR)
(ii) a complete sample of 168 bright sources at 2.7 GHz (Peacock and Wall 1981, 1982).
The higher number of flat-spectrum sources in the latter should not prejudice the result because of their separation in the RLF. On the other hand, the sample contains 36 steep-spectrum ($\alpha>0.5$) compact sources 10 of which are excluded from the "166 sample" only because of spectral curvature. Significant numbers of such sources have also been found by Kapahi (1981) at 5 GHz. As the 5C sample is likely to lie at substantial

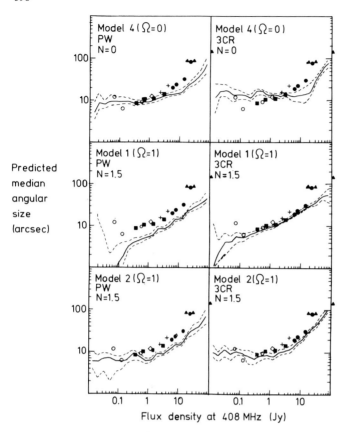

Figure 5 The predicted median angular size (solid line; errors are shown by dashed lines) as a function of flux density (Fielden et al.). The medians from Fig. 3 are shown for reference. Predictions are for a parent sample at 178 MHz (3CR "166 sample" plus 5 giant sources) on the right and at 2.7 GHz (Peacock and Wall 1981, 1982; PW) on the left. Evolving radio luminosity functions 4(top pair), 1 (middle pair) and 2 (lower pair) of Peacock and Gull (1981) can all provide a good fit. Linear size evolution as $(1+z)^{-N}$ was used and N is shown for each prediction.

redshifts, it is equivalent to selection at a higher frequency (possibly above the turnover of these steep-spectrum compact sources) so that the second parent sample may be more appropriate.

The modelling was both with linear size evolution as $(1+z)^{-N}$ where N=1.5 and with no linear size evolution (N=0). Fig. 5 shows the results of the most successful attempts to model the medians at low flux densities. The errors in Fig. 5 are determined from the number of sources in the parent sample which dominate the behaviour at any flux density (in the sense that the errors are increased if the distribution is defined by very few sources).

For $\Omega=0$, model 4 of Peacock and Gull (which has a redshift cutoff at z=5) provided a good fit to the observations at low flux densities with no linear size evolution and the parent sample at 2.7 GHz. This model was the most successful of those for $\Omega=0$; results for N=1.5 predicted angular sizes lower than observed, though N<1.5 for the 3CR parent sample should give a reasonable fit. For $\Omega=1$, no models with N=0 provided a satisfactory match to the data. Predictions with N=1.5 were too low, suggesting that less strong evolution (Kapahi, 1977, favoured N=1-1.1) may be more appropriate. The medians predicted at 5C

ANGULAR SIZE DISTRIBUTION OF RADIO SOURCES AT LOW FLUX DENSITIES 399

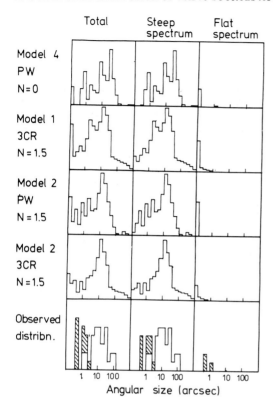

Figure 6 The angular size distribution at 408 MHz in the range 0.05-1 Jy for four of the models in Fig. 5 (Fielden et al.). The observed angular size distribution is plotted at the bottom, in which hatched areas represent upper limits. The distribution is shown separately for all sources (left column), for steep-spectrum ($\alpha > 0.5$) sources (centre) and for flat-spectrum sources (right column).

levels were lower for the 2.7-GHz sample than for the 3CR sample, by a factor up to ~ 2. In addition, the predictions for the same parent sample and form of linear size evolution can vary by a similar amount according to the choice of RLF.

The predicted distribution of angular sizes in the range 0.05-1 Jy at 408 MHz is plotted in Fig. 6 for the four most successful models in Fig. 5, together with the observed distribution. In all cases the predictions are consistent with the observations to within the errors. In particular, the number of flat-spectrum (compact) sources is correct, confirming that the small angular sizes are not a result of an increased fraction of these sources when the 2.7-GHz parent sample is used. Also, all the angular size distributions extend to sufficiently large values to match the data.

The observed angular size distributions can therefore be modelled for $\Omega=0$ either with or without linear size evolution, depending on the choice of parent sample. Models for $\Omega=1$ require linear size evolution (though less strong than $(1+z)^{-1.5}$) to match the observations. The predicted angular sizes are dependent on the selection of cosmological model, the assumed RLF and the parent sample. Refinement of the RLF by the inclusion of new data on the identification content (or redshift distribution) at low and intermediate flux densities may resolve these uncertainties.

4. CONCLUSIONS

On the basis of present knowledge of the radio source population, and its behaviour at different epochs, linear size evolution is sufficient but not necessary in order to explain the small observed angular sizes at low flux densities.

I thank Trinity Hall, Cambridge for a Research Fellowship and U.R.S.I. and the Royal Society, London for travel funds.

REFERENCES

Allington-Smith, J.R.: 1982, Mon. Not. Roy. Astron. Soc., in press.
Allington-Smith, J.R., Perryman, M.A.C., Longair, M.S., Gunn, J.E. and Westphal, J.A.: in preparation.
Benn, C.R., Grueff, G., Vigotti, M. and Wall, J.V.: in preparation.
Downes, A.J.B., Longair, M.S. and Perryman, M.A.C.: 1981, Mon. Not. Roy. Astron. Soc., 197, in press.
Ekers, R.D. and Miley, G.K.: 1977, in IAU 74 "Radio astronomy and cosmology", ed. Jauncey, D., p. 109.
Fielden, J., Allington-Smith, J.R., Benn, C.R., Downes, A.J.B., Longair, M.S. and Perryman, M.A.C.: in preparation.
Jenkins, C.J., Pooley, G.G. and Riley, J.M.: 1977, Mem. Roy. Astron. Soc., 84, 61.
Kapahi, V.K.: 1975, Mon. Not. Roy. Astron. Soc., 172, 513.
Kapahi, V.K.: 1977, in IAU 74 "Radio astronomy and cosmology", ed. Jauncey, D., p. 119.
Kapahi, V.K.: 1981, Astron. Astrophys. Suppl., 43, 381.
Katgert-Merkelijn, J., Lari, C. and Padrielli, L.: 1980, Astron. Astrophys. Suppl., 40, 91.
Peacock, J.A. and Gull, S.F.: 1981, Mon. Not. Roy. Astron. Soc., 196, 611.
Peacock, J.A. and Wall, J.V.: 1981, Mon. Not. Roy. Astron. Soc., 194, 331.
Peacock, J.A. and Wall, J.V.: 1982, Mon. Not. Roy. Astron. Soc., in press.
Pearson, T.J. and Kus, A.: 1978, Mon. Not. Roy. Astron. Soc., 182, 273.
Perryman, M.A.C.: 1979a, Mon. Not. Roy. Astron. Soc., 187, 223.
Perryman, M.A.C.: 1979b, Mon. Not. Roy. Astron. Soc., 187, 683.
Perryman, M.A.C., Longair, M.S., Fielden, J., Gunn, J.E. and Westphal, J.A.: in preparation.
Swarup, G.: 1975, Mon. Not. Roy. Astron. Soc., 172, 501.
Swarup, G.: 1977, Bull. Astron. Soc. India, 5, 36.
Wall, J.V., Pearson, T.J. and Longair, M.S.: 1980, Mon. Not. Roy. Astron. Soc., 193, 683.

DISCUSSION: see page 409.

THE EVOLUTION OF LINEAR SIZES

V.K. Kapahi and C.R. Subrahmanya
Tata Institute of Fundamental Research, P.B. No. 1234,
Bangalore 560012, India.

Possible evidence that the linear sizes of extragalactic radio sources were smaller at earlier epochs was first provided by the angular size redshift (Θ-z) relation for double radio quasars (Legg 1970, Miley 1971, Wardle and Miley 1974). But because of the strong correlation between redshift (z) and radio luminosity (P) in flux limited radio samples, it is hard to decide if the observed decrease in sizes with z is caused by an epoch dependence of linear sizes (ℓ) or by an inverse correlation between P and ℓ. Several authors (eg. Stannard and Neal 1977, Wardle and Potash 1977, Hooley et al. 1978, Wills 1979, Masson 1980) have attempted to separate the two effects by comparing the properties of quasars from the 3CR survey with those from the 4C and Parkes samples. Although most of these studies appear to marginally favour a P-ℓ correlation, none of them can rule out even a fairly strong evolution in ℓ with z. Apart from the small numbers involved, the difficulty is that 3C and 4C quasars do not differ a great deal in their redshifts or luminosities. A complete sample of quasars at much weaker flux levels would be quite valuable in this regard.

Evolution in ℓ can be investigated also by studying the angular size-flux density (Θ-S) relation (Swarup 1975, Kapahi 1975). Although S in this case has to be related to z through models of the evolving radio luminosity function that explain the observed source counts, there is the advantage that large unbiased samples of sources covering a wide range in S can be used. Earlier studies of the Θ-S relation, based generally on the median values of angular size (Θ_m), have indicated (Kapahi 1975, 1977; Swarup and Subrahmanya 1977; Katgert 1977) the presence of size evolution that can be simply represented by $\ell(z) \propto (1+z)^{-n}$ with n= 1 to 2. Recent measurements of the structures of several complete samples have not only extended the Θ_m-S relation to lower flux densities but have also led to improved distributions of angular sizes at all flux levels. Furthermore, there has been a significant advancement in the optical identification of sources in the 3CR sample (Gunn et al. 1981, and references therein) which has led to improved models of the evolutionary behaviour of the radio luminosity function. It is now worthwhile, therefore, to reexamine the Θ-S relation, and in particular to determine if an inverse correlation between P and ℓ can obviate the need for size evolution.

THE OBSERVED Θ-S RELATION

We shall confine ourselves to samples selected at low frequencies, in the range of 178 to 408 MHz. Such samples contain few sources of the compact 'flat spectrum' variety, so that we are dealing mainly with the extended 'steep spectrum' variety of sources. A new determination of the Θ_m-S relation from the best available data is shown in Figure 1. A few remarks about the samples used in constructing the relation. The highest flux density point is based on the 37 sources in the 'All Sky' catalogue (Robertson 1973) with 15 Jy< S_{408} < 68 Jy, and δ<10°; it is therefore independent of the 3CR sample with δ>10°. The 3CR points are derived from the '166 source' sample (Jenkins et al. 1977) with the addition of 7 other sources (see next section). The 4C point comes from a complete sample of 72 sources in the range of 4.8 to 9 Jy at 178 MHz, structures for which have been reported by Katgert-Merkelijn et al. (1980). The points labelled 'B2' are based largely on new observations of several complete samples (Kapahi 1981b, Padrielli et al. 1981) from the Bologna catalogues, made with the Westerbork synthesis telescope at 5 GHz. B2 sources in the upper two flux ranges (3.5 to 10 Jy at 408 MHz) have a considerable overlap with the 4C and 3CR sources. The observed distribu-

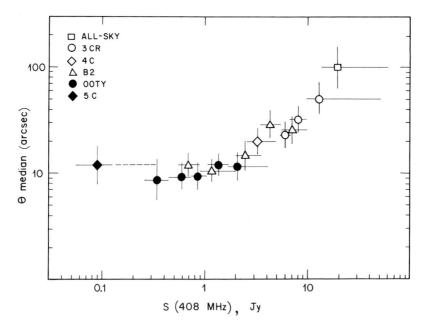

Fig. 1. The observed relation between median values of S and Θ. The estimated errors in Θ_m are indicated by vertical lines and the ranges in flux density by horizontal lines.

THE EVOLUTION OF LINEAR SIZES

tions of Θ in the other three flux ranges between 0.55 and 3.5 Jy are shown in Figure 3. The 5 Ooty points, in the range of 0.3 to 5 Jy at 327 MHz, have now been rederived from all the published occultation results to date (Joshi and Singal 1980, and references therein). The lowest flux density point is from recent VLA measurements (Downes et al. 1981) at 1465 MHz, of a complete sample of 24 sources in the 5C6 and 5C7 regions with $S_{408} > 0.055$ Jy. The Θ distribution for this sample is also shown in Figure 3. The flux densities of samples at 178 and 327 MHz have been translated to 408 MHz using a constant spectral index of $\alpha = 0.75$.

It is clear from Fig. 1 that the value of Θ_m decreases fairly rapidly down to about 1-2 Jy at 408 MHz and remains nearly constant at lower flux levels. It should also be noted that there is good agreement between the Ooty and Bologna points that cover a similar range in S but are based on almost entirely different sources. A comparison of the Θ-distributions in the two sets of data indicates that though a few large sources ($\Theta \gtrsim 60$"arc) could have been missed in the Ooty occultation surveys, their number cannot be large enough to affect the median values of Θ significantly.

COMPARISON WITH PREDICTIONS OF EVOLUTIONARY MODELS

In order to understand the observed Θ distributions we have calculated the distributions expected at lower flux levels from the known properties of sources in the 3CR sample by the method that has been described in detail by Downes et al. (1981). The method uses the observed values of S, α, z and Θ for each source in the 3CR parent sample to estimate its contribution to the number and Θ-distribution in any specified range of flux density. Apart from taking into consideration the evolution in number density of radio sources according to the successful evolutionary models, the method has the advantage that any correlation between P and ℓ in the 3CR sample is automatically taken into account.

For the parent sample we have used both, the complete '166 source' sample with $S_{178} \geq 10$ Jy and also a subsample with $S_{178} \geq 15$ Jy. As the 3CR sample is known to be incomplete (eg. Véron 1977), however, particularly with regard to sources of large Θ, we have added the following 7 sources to the sample, for which fairly complete information is now available; DA240 (S_{178}=19.4 Jy, Θ=2040"arc), 4C73.08 (15.0 Jy, 1100"), 4C64.19 (10.7 Jy, 120"), 4C35.40 (15.2 Jy, 840"), NGC6251 (10.0 Jy, 4320"), 4C11.71 (11.1 Jy, 720"), 3C296 (13.0 Jy, 208"). Furthermore, we take the total flux densities of 3C236 and 3C326 to be 19.8 and 16.7 Jy respectively, values that are considerably larger than those listed in Jenkins et al. (1977).

Redshifts are now known for about 75% of the sources in the 3CR sample. For most of the remaining these have been estimated from the magnitudes of the identified galaxies. For the unidentified sources and those identified with objects of m>22, we assume z=1. The uncertainty in these redshifts makes little difference to the results. The computations were done numerically for the 3 successful evolution models of Wall et al. (1980; WPL) (labelled as models '4a', '4b' and '5' by WPL), and for various amounts of linear size evolution assumed to be represented by $(1+z)^{-n}$. Values of $H_0 = 50$ km sec^{-1} Mpc^{-1} and $q_0 = 0.5$ were used; $q_0 = 0$ was also used for models '4b' and '5' (model '4a' does not give a satisfactory fit to the source counts for $q_0 = 0$; WPL and C.R. Subrahmanya, in preparation). The results of using the entire 3CR sample and the subsample with $S \geq 15$ Jy are quite similar but the errors in the predicted numbers should be smaller for the larger sample.

The predicted Θ_m-S relations (from the $S \geq 15$ Jy sample, $q_0 = 0.5$) for the three evolutionary models without size evolution (n=0) are shown in Figure 2. It can be seen that these predict significantly larger values of Θ_m than observed. But with size evolution included, satisfactory fits to the data can be obtained in all the models for the values of n indicated in Fig. 2. It is also interesting to note that rather strong size evolution (n∼3.5) is needed for model '5' of WPL. This is because the average redshift of sources in this model at all flux densities is considerably smaller than in the other models. It has been shown recently by Swarup et al. (1981) that model 5 is inconsistent with optical identification data because it predicts a much higher rate

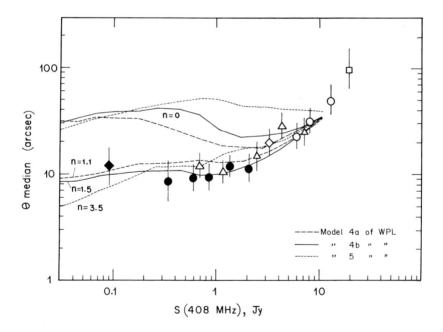

Fig. 2. The Θ_m-S relations predicted from the 3CR sample (S > 15 Jy, $q_0=0.5$) for three evolutionary models of WPL, without size evolution (upper curves) and with size evolution denoted by the values of index n as indicated (lower curves).

of percentage identification in the entire range of flux densities than is observed. The strong size evolution required in this model appears to be inconsistent with the observed Θ-z relation for quasars and thus provides additional evidence against the model.

The observed and predicted distributions of Θ in 4 complete samples (5C and B2) that have no overlap with the 3CR sources are shown in Fig. 3 for model '4b', with and without size evolution. While the agreement appears to be quite good throughout the observed range of Θ at different flux levels for the case with size evolution, too few sources with small angular sizes ($\Theta \lesssim 6"$arc) and too many with large sizes ($\Theta \gtrsim 40"$arc) are predicted in the absence of size evolution. A x^2 test of the observed and predicted distribution (from the entire 3CR sample) indicates (see Table 1) that these are indeed significantly different for the case of n=0. In performing the x^2 test the lowest Θ-bin in each flux range was chosen to include the upperlimits to Θ for the unresolved sources. At higher Θ values an interval of 0.2 in logΘ was used and neighbouring bins marged where necessary so as to have at least 5 sources for the comparison. There are thus 26 bins in logΘ, but since the total predicted number of sources in each flux range has been normalized to agree with the observed number (the differences are however within the expected standard errors), there are 22 degrees of freedom. It is also apparent from Table 1 that size evolution is needed even in the case of a q_0=0 Universe.

Table 1. x^2 analysis of Θ distributions (degrees of freedom = 22)

q_0	model	n for min. x^2	x^2 min	x^2 (n=0)
0.5	4a	1.1	19	38
	4b	1.4	12	49
	5	3.6	19	122
0.0	4b	1.1	14	33
	5	2.5	16	63

The results shown in Figs. 2 and 3 do not agree with those of Downes et al. who concluded on the basis of predictions from the S $\gtrsim$ 15 Jy 3CR sample that no single value of n provides even a tolerable fit to the observed Θ_m values over the entire flux density range in any of the WPL models, and that for n>1 insignificant numbers of sources with large sizes ($\Theta \gtrsim 25"$arc) are predicted. The Θ_m values predicted by us from the same 3CR sample with identical assumptions regarding the redshifts of those sources that do not have measured values and for the same values of n used by them (n=0 and n=1) appear infact to be considerably different throughout the flux density range for all three WPL models considered. The reason for the discrepancy is unclear. We are reasonably confident, however, that a computational oversight on our part is unlikely to be the cause.

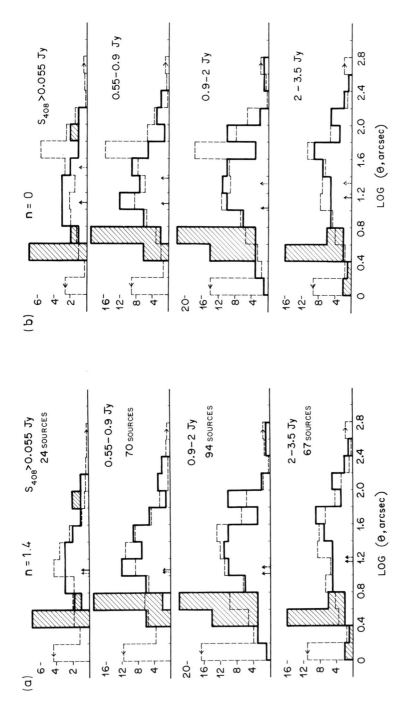

Fig. 3. Comparison of the observed (solid lines) and predicted (broken lines) distributions of Θ in four flux density ranges, (a) with size evolution, $n=1.4$ and (b) without size evolution, $n=0$. The shaded region refers to upper limits to unresolved sources. Predicted sizes $<1\rlap{.}''6$ have been grouped together in the first Θ bins. The solid and dotted vertical arrows indicate the observed and predicted values of Θ_m.

It has been suggested by Downes et al. that the fairly large number of unresolved sources ($\theta \lesssim 3$"arc) in the 5C samples could be similar to the compact steep spectrum sources that form a significant fraction of the strong source population at high frequencies (e.g. Kapahi 1981a, Peacock and Wall 1981) but are underrepresented in the 3CR sample because of low frequency turnovers in their radio spectra. Such sources are unlikely, however, to obviate the need for size evolution because such evolution appears to be needed to explain the observed θ-distributions at large angular sizes as well. Our analysis also shows that over 90% of the counts with $\theta < 3$"arc in the 5C sample are contributed by galaxies in the 3CR sample that have $z < 0.5$ and angular sizes of upto about 20"arc. Most of the compact steep-spectrum sources found in strong source surveys at frequencies of 2.7 or 5 GHz on the other hand are either quasars or have $m > 20$. It is therefore unlikely that such sources make a substantial contribution to the counts of unresolved sources at the 5C flux levels.

CONCLUSIONS

The available angular size data down to 55 mJy at 408 MHz appear to require evolution in linear sizes of the powerful radio sources. The likely value of n if the evolution is expressed as $(1+z)^{-n}$, appears to lie in the range of about 1 to 1.5. Size evolution is needed in addition to any P-ℓ correlation implicit in the 3CR data. Such a correlation alone is not sufficient because the best available evolutionary models of the radio luminosity function imply that the median luminosity of radio sources remains almost constant in the flux density range of about 10 to 1 Jy at 408 MHz where θ_m decreases by a factor of ~ 3. The median redshift of sources in this range of flux density increases typically from about 0.35 to $\gtrsim 1$, but since for a constant linear size, θ decreases rather slowly with increasing z in world models with $q_0 \geq 0$, a decrease in ℓ with increasing z is required to fit the observations.

It would be useful to measure the structures of unresolved sources in the present samples with much higher angular resolution in order to improve the distributions at small angular sizes and to determine the nature of the compact steep-spectrum sources.

REFERENCES

Downes, A.J.B., Longair, M.S. and Perryman, M.A.C. 1981, Mon. Not. R. astr. Soc., in press.
Gunn, J.E., Hoessel, J.G., Westphal, J.A., Perryman, M.A.C. and Longair, M.S. 1981, Mon. Not. R. astr. Soc., 194, 111.
Hooley, T.A., Longair, M.S. and Riley, J.M. 1978, Mon. Not. R. astr. Soc., 182, 127.
Jenkins, C.J., Pooley, G.G. and Riley, J.M. 1977, Mem. R. astr. Soc., 84, 61.
Joshi, M.N. and Singal, A.K. 1980, Mem. Astron. Soc. India, 1, 49.

Kapahi, V.K. 1975, Mon. Not. R. astr. Soc., 172, 513.
Kapahi, V.K. 1977, in IAU Symposium 74 "Radio Astronomy and Cosmology", ed. D. Jauncey, p. 119.
Kapahi, V.K. 1981a, Astron. Astrophys. Suppl., 43, 381.
Kapahi, V.K. 1981b, in preparation.
Katgert, P. 1977, Ph. D. Thesis, University of Leiden.
Katgert-Merkelijn, J., Lari, C. and Padrielli, L. 1980, Astron. Astrophys. Suppl., 40, 91.
Legg, T.H. 1970, Nature, 226, 65.
Masson, C.R. 1980, Astrophys. J., 242, 8.
Miley, G.K. 1971, Mon. Not. R. astr. Soc., 152, 477.
Padrielli, L., Kapahi, V.K. and Katgert-Merkelijn, J.K. 1981, Astron. Astrophys. Suppl., in press.
Peacock, J.A. and Wall, J.V. 1981, Mon. Not. R. astr. Soc., in press.
Robertson, J.G. 1973, Austr. J. Phys., 26, 403.
Stannard, D. and Neal, D.S. 1977, Mon. Not. R. astr. Soc., 179, 719.
Swarup, G. 1975, Mon. Not. R. astr. Soc., 172, 501.
Swarup, G. and Subrahmanya, C.R. 1977, in IAU Symposium 74 "Radio Astronomy and Cosmology", ed. D. Jauncey, p. 125.
Swarup, G., Subrahmanya, C.R. and Venkatakrishna, K.L. 1981, Astron. Astrophys., submitted.
Véron, P. 1977, Astron. Astrophys. Suppl., 30, 131.
Wall, J.V., Pearson, T.J. and Longair, M.S. 1980, Mon. Not. R. astr. Soc., 193, 683.
Wardle, J.F.C. and Miley, G.K. 1974, Astron. Astrophys., 30, 305.
Wardle, J.F.C. and Potash, R. 1977, Ann. NY Acad. Sci., 302, 605.
Wills, D. 1979, Astrophys. J. Suppl., 39, 291.

DISCUSSION, of previous two papers, by Downes, and by Kapahi and Subrahmanya

SWARUP: How do you justify the use of parent sample at 2.7 GHz, for which the emitted frequency will be ~ 5 GHz at z = 1, for comparing with the θ − S data at 408 MHz for the 56 sources?

DOWNES: The method of estimation of angular sizes gives a local density to each source in the parent sample proportional to $1/V_{max}$, the maximum volume in which the source would have been included in the sample. This means that high weight is given to low-luminosity, low-redshift sources, for which the emitted frequency $(1 + z)2.7$ GHz would be close to 2.7 GHz. Contributions from high-redshift sources are small both because V_{max} is large and because it is hard to move them far enough away to reach the required low-flux densities without involving unreasonably large redshifts. The 5C sources, however, are likely to lie at $z > 0.6$ so would have an emitted frequency ~ 1 GHz. 2.7 GHz is the closest frequency to this for which almost complete θ and z data exist.

REES: Presumably the θ(z) relation is, to a large extent, probing the z-dependence of the intergalactic environment of extended radio lobes. But this environment evolves in a complex way: between clusters, the gas was probably denser in the past; within clusters, the density may be higher now (and indeed may be higher than the intergalactic density was anywhere at z ~ 2). For this reason, and others, one would not expect any simple power-law to be a good fit. Obviously these data will eventually tell us a great deal, when we have AXAF-type maps showing the properties of intergalactic gas out to large z. But how much do you really think these studies can now tell us about the physics of the sources?

DOWNES: It is not clear whether one would expect sources at high z to have different linear sizes. Comparisons of linear sizes outside and inside clusters (i.e., high-density environments) indicate no significant differences in size. On the other hand, high inversion Compton losses may extinguish high-z sources when they are smaller. Any indication of whether high-z sources are indeed smaller would therefore add to our understanding of the source population.

WILLS: In future studies, do you consider it more profitable to restrict the observed samples to one optical class, with measured redshifts, or to continue using much larger samples that are a mixture of all optical classes?

DOWNES: Measurements of redshifts of sources in the present samples, particularly at the lowest flux densities, would place considerable contraint on radio luminosity functions, and would seem the most profitable course of action to resolve the situation.

BALDWIN: Both speakers recognize the problem of attaining completeness for radio sources of the largest angular size. In comparing data with the models is some attempt made to use some limiting rest-frame surface brightness? If not, how much does it matter?

DOWNES: (1) No. (2) It probably doesn't matter a great deal--at 80" resolution (in the 5C surveys) very few of the sources are resolved so that we do not expect a significant contribution from low-brightness extended objects. Low-brightness compact objects would simply be below the sample limits.

LAING: (1) Could the different definitions of angular size for Class I and Class II sources cause any problems? (2) Many flat-spectrum "compact" sources have weak, extended outer components which are difficult to detect. If these are neglected, then the angular sizes are considerably underestimated. (3) In the 3CR sample at least, the "steep-spectrum compact sources" are of two types: (a) small symmetrical doubles, and (b) steep-spectrum radio cores. These may have different evolutionary properties.

KAPAHI: (1) The proportion of Class I radio galaxies ($P < 10^{25}$ W Hz^{-1} $ster^{-1}$) does not increase (probably decreases) with decreasing S in the flux range of interest because they are not subject to strong density evolution. Such sources can in any case be fairly easily recognized from the magnitudes of their optical counterparts at least down to about 1 Jy. Leaving them out from the samples (including 3CR) seems to make little difference to the conclusions. (2) Most of the sources of this type have probably been considered "unresolved" in our data. The number of flat-spectrum sources are anyway not large enough in low frequency samples to affect the results significantly.
(3) These could well be of two kinds, but little is known about their evolutionary behaviour.

HOT-SPOTS AND RADIO LOBES OF QUASARS

G. Swarup[1,4], R.P. Sinha[2] & C.J. Salter[3,4]

1. Sterrewacht, Leiden, The Netherlands
2. Systems & Applied Science Corp., Hyttasville, Md., U.S.A.
3. Istituto di Radioastronomia, Bologna, Italy
4. Radio Astronomy Center, Ootacamund, India

A number of investigations have been made of the dependence of largest angular size (LAS) of quasars on their redshifts Z (Miley 1971; Wardle and Miley 1974; Hooley et al. 1978; Wills 1979; Masson 1980). The earlier workers found a relation LAS $\propto Z^{-1}$ indicating a decrease in linear size of quasars as $(1+Z)^{-n}$, where $1 < n < 3$. The latter workers have argued that the apparent decrease is perhaps caused by an inverse correlation between LAS and radio luminosity. It is also interesting to investigate the dependence of size of hot-spots on luminosity and Z.

We report here preliminary results for 28 quasars out of a sample of 35 observed in the full 35-km configuration of the VLA at a wavelength of 6 cm in March 1981. Each source was observed for a few minutes at 2 or 3 hour angles. The sources were selected as those having the largest angular sizes in 6 redshift ranges from the complete 3CR and Schmidt's 4C sample (Hooley et al. 1978), the 4C sample of Wills (1979) and the B2 sample of Fanti et al. (1977).

In Fig.1 are plotted the largest angular size against redshift for the 28 quasars. The sources are divided into 3 ranges of radio luminosity at 178 MHz (H_0=50 kms^{-1} Mpc^{-1}; q_0=1) and are indicated by different symbols as labelled in the figure. The horizontal bar on a symbol indicates that no prominent hot-spots were observed. Even though the radio luminosity of these sources without hot-spots is relatively high, they show a relaxed morphology indicating that they may be 'dying away'. This would be expected for some of the sources of largest linear sizes at a given Z. Such sources comprise about 30% of the sample. No such sources were observed in the highest luminosity range. If we exclude the 3 sources of lowest radio luminosity in Fig.1, the data is consistent with the absence of linear size evolution with Z. This question, however, needs further investigation.

About 40% of the observed quasars have hot-spots of size < 2.5 to $\sim$ 4 kpc which contain typically 10-30% of the flux-density of the outer lobes. The sizes of hot-spots in the other 30% of the quasars lie in the range $\sim$ 4-10 kpc. Hot-spots are generally more compact and prominent

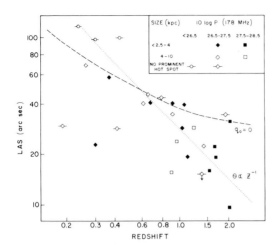

Fig.1. LAS versus redshift for 3 radio luminosity ranges and different hot-spot characteristics (see text)

in one compared to the other outer lobe of the 'double' radio sources. This probably indicates that the relativistic electrons are supplied intermittently to the two outer radio lobes. The apparent increase of the average angular size of hot-spots with redshift and decreasing flux density, as indicated by interplanetary scintillation observations (Readhead & Hewish 1976; Duffett-Smith et al. 1980), is likely to be the result of blending of the hot-spots in the two outer radio lobes.

Another interesting result is that at-least 4 out of these 28 quasars selected for their relatively large angular sizes, and hence likely to have radio axes at large angles to the line of sight, show prominent radio jets (3C9, 0932+02, 1244+32 and 3C280.1). Therefore, these one-sided jets are unlikely to be caused by doppler enhancement for the approaching, and attenuation for the receeding, side of relativistic beams. The radio jets in 3C9 (Z=2.012) and 3C280.1 (Z=1.659) have the highest redshifts and radio luminosities of all jets reported so far. The presence of radio jets in many high luminosity quasars, in contrast to their absence in high luminosity radio galaxies, indicates that the beams supplying energy to the outer lobes are more turbulent in the former than in the latter.

REFERENCES
Duffett-Smith,P.S., Purvis,A. & Hewish,A., 1980, M.N.R.A.S., 190, 891.
Fanti,C., Fanti,R., Formiggini,L., Lari,C. & Padrielli,L., 1977, Astron. Astrophys. Suppl., 28, 351.
Hooley,A., Longair,M.S. & Riley,J.M., 1978, M.N.R.A.S., 182, 127.
Masson,C.R., 1980, Astrophys. J., 242, 8.
Miley,G.K., 1971, M.N.R.A.S., 152, 477.
Readhead,A.C.S. & Hewish,A., 1976, M.N.R.A.S., 176, 571.
Wardle,J.F.C. & Miley,G.K., 1974, Astron. Astrophys., 30, 305.
Wills,D., 1979, Astrophys. J. Suppl., 39, 291.

THE OPTICAL AND INFRARED PROPERTIES OF 3CR RADIO GALAXIES

S.J. Lilly and M.S. Longair
Department of Astronomy, University of Edinburgh
Blackford Hill, Edinburgh, Scotland

1. INTRODUCTION

For the last 20 years, a huge observational effort has been devoted to the identification of all 3CR radio sources lying in directions away from the Galactic plane. The project is now more-or-less complete thanks to high-resolution radio observations with aperture synthesis radio telescopes, in particular the Cambridge 5-km telescope, and the use of a CCD camera on the Palomar 5-metre telescope (see e.g. Gunn et al 1980).

The aim of the present paper is to assess the reliability of the identifications of distant radio galaxies and to explore their properties using recent infrared observations.

2. THE IDENTIFICATION STATUS OF THE STATISTICAL SAMPLE OF 166 3CR RADIO SOURCES

Most of the effort to complete the optical identifications of 3CR radio sources has been devoted to a statistical sub-sample of 166 sources defined in Jenkins, Pooley and Riley (1977). This sample was selected by C.C. Jenkins, M.S. Longair and J.M. Riley prior to the last major IAU Symposium devoted to extragalactic radio astronomy, IAU Symposium No. 74 "Radio Astronomy and Cosmology". All 3CR sources lying in the declination range $\delta \geqslant 10°$ and $|b| \geqslant 10°$ were re-investigated individually. The flux densities and structures of confused sources were re-assessed by inspecting aperture synthesis maps made with different angular resolutions. All corrected flux densities were reduced to the scale of Kellermann, Pauliny-Toth and Williams (1969) and all those with $S_{178} \geqslant 10$ Jy were included in the statistical sample which amounted to 166 sources. The sources and their adopted flux densities are given by Jenkins, Pooley and Riley (1977).

Besides missing out 3C296 by mistake, the main uncertainty in the sample is discrimination against sources of large angular size and low surface brightness. These "giant radio sources" would have been resolved by the pencil beam of the cylindrical-paraboloid of the radio telescope used to produce the 3CR catalogue. A significant number of these have now been identified in surveys of confused 4C sources and in the 6C radio surveys. There is no definitive figure for how many sources should be added to the sample. It is known that at least 6 giant sources should be added and as many as 20 might have to be included. From the point of view of identifications, most of these sources are associated with relatively bright nearby galaxies and so the inclusion of the giant sources makes little difference to the identification content of the sample.

We wish to estimate how reliable the identifications of the faintest objects in the 166 sample are. Identifications have been claimed for almost all the sources in the sample. Most of those with objects fainter than 20th magnitude are believed to be galaxies on the basis of an extended optical image. The deepest identification surveys using the CCD camera have extended to about $V = 24$ at which the surface density of galaxies exceeds that of foreground stars.

Laing, Riley and Longair (1982) have re-investigated this problem for the 166 sample. The sources were divided into two groups according to whether they were identified with objects having $V \lesssim 19$ or $V > 19$. The sources in each group were then divided according to their radio structures into (i) compact sources with $\theta \lesssim 2$ arcsec, (ii) "classical" double or Fanaroff-Riley class II sources and (iii) the rest, which are Fanaroff-Riley class I sources.

The best criterion for claiming a certain identification results if the source is compact or possesses a compact central component (or radio core) which is coincident within 1 arcsec of an object. Even at $V = 24$, the probability of such a random coincidence is less than 0.2%. Thus, all the compact sources fall in the category of certain identifications. The next best criterion is the observation of a strong emission line spectrum. Only about 15% of elliptical galaxies possess a weak emission line spectrum and only for 1-2% is there a strong emission line spectrum.

For the third class of source, the Fanaroff-Riley class I sources, all of which are brighter than $V = 19$, there is only one source which does not have a radio-core or an emission line spectrum, 3C 314.1, for which the probability of a chance coincidence is 3%. Thus, we can be reasonably confident that these identifications are all correct.

This leaves the double sources which have been considered in three parts. First, sources with radio cores and identifications having $V \lesssim 19$ within 1 arcsec of this component have been selected and the position of the identification with respect to the outer maxima of the double source plotted. For these certain identifications, it was found that all but one lay within a circle of radios 0.2θ centred on the mid point of the double, where θ is the separation of the double. The analysis was repeated for the 25 double sources with $V \lesssim 19$ which do not possess radio cores and it was found that only 3 of the proposed identifications lay outside the search area. In no case did the probability of a random coincidence within this search area exceed 0.01 and all but 6 have strong emission line spectra. Of those objects which lie outside the error circle, 3C254 is perhaps most remarkable in that the identification, a quasar, is almost, but not quite, coincident with one of the double source components. Laing (private communication) has found a weak radio core associated with the quasar in a recent observation with the VLA. Another problem source is 3C340 which lies outside the 0.2θ circle (although on the axis of the source) and there are no emission lines in the optical spectrum. The chance of a random coincidence with the radio source is only 0.6%. Thus, for all doubles with $V \lesssim 19$, 55 out of 59 lie within the 0.2θ circle indicating that it is a very good criterion for assessing the reliability of the identification.

Of the remaining 68 double sources with $V > 19$, 2 proposed identifications are well outside the 0.2θ circle, 3C68.2 and 3C470 and, unless they are anomalous doubles like 3C254, are probably not correct identifications. Two fields are obscured by bright stars, 3C268.1 and 3C294. There is no identification for 3C437. Of the 68 sources, the chances of a random coincidence within the 0.2θ error circle exceeds 0.01 in 24 cases and is greater than 0.1 in 8 cases. Thus, it is likely that at most one or two of the identifications which satisfy the identification criterion are chance associations with the radio source. An obvious programme of importance is the search for weak radio cores in all sources in which they have not yet been found.

In summary, of the fields which are not obscured, there are no optical identifications for 3C68.2 and 437. The identifications of 3C340 and 470 are uncertain. Others require further spectroscopic and radio observations to resolve which of a number of galaxies is the correct identification, in particular 3C61.1 and 469.1. Thus, the sample is almost certainly more than 95% complete.

From the point of view of studying a homogeneous sample of distant radio galaxies, those in the 3CR catalogue thus provides a more or less complete sample to the faintest magnitudes. This is the sample we have selected for study at infrared wavelengths.

3. INFRARED OBSERVATIONS OF 3CR RADIO GALAXIES

The infrared waveband holds great potential for many different types of extragalactic and cosmological research. The advantages of observing very distant galaxies in the near infrared have been discussed by Grasdalen (1980) and Lebofsky (1980), and arise because the K-correction for the K waveband (2.2μm) is increasingly negative for redshifts up to and greater than 1. We describe below the first results of a systematic survey of the near infrared properties of 3CR radio galaxies which we have carried out with the UKIRT 3.8m infrared telescope.

The galaxies selected for study are from the 166-source sub-sample. Additional constraints are imposed that limit the declination to less than 55° and the redshift to greater than 0.03. These limit the number of galaxies to 87 and includes the 2 empty fields. The apparent optical magnitudes of these galaxies range from 14 to 23, the magnitude distribution being shown in Fig. 1. Spectroscopic redshifts are available for about 66% of the galaxies, and range from 0.03 to in excess of 1.0. The optical morphologies of these identifications are very similar to those of first ranked elliptical galaxies, including some cD systems and N-galaxies, the latter being associated with broad-line radio galaxies (BLRGs).

The bulk of the observations were carried out during six nights of excellent observing conditions on Mauna Kea during March 1981. 35 3CR galaxies, selected from throughout the magnitude range of the 87-source sub-sample have now been studied, and their magnitude distribution is shown in Figure 1. In addition, we have confirmed the detection of 3C68.2 reported by Grasdalen (1980), and have obtained a marginal detection of 3C437.

All sources were observed at K, and for all but the faintest at H(1.65μm) and J(1.2μm) as well. For most of the observations a 10.8" aperture was used, except for three cases when a 7.2" aperture was used to avoid nearby objects visible on the best plate material.

4. THE COLOUR-REDSHIFT DIAGRAMS

From our observations it is straightforward to construct the infrared colour-redshift diagrams. In Figures 2 and 3 we show the observed (H-K) and (J-K) colours for the galaxies. The redshifts are taken from the compilation of Smith and Spinrad (1980) or from recent work by Spinrad (private communications).

Considering first the low redshift galaxies (z < 0.4), it is clear that with the exception of the four galaxies classified as BLRG by Grandi and Osterbrock (1977), and represented by crosses on the diagram, the infrared colours of the radio galaxies occupy a well defined locus on the colour-redshift planes. This implies that they may all be represented, with small cosmic scatter, by a single energy distribution. This was derived using the observed infrared colours of the galaxies with z < 0.4 and is shown by the solid line on the diagrams.

The zero redshift intercepts of the colours predicted from this energy distribution are very similar to those found in a large sample of nearby elliptical galaxies by Frogel et al (1978), (hereafter FPAM). We can therefore conclude that the infrared energy distributions of these galaxies are essentially the same as those of normal elliptical galaxies, and that any additional component associated with the active nucleus must be small. There is no obvious relation between emission line strength and infrared properties amongst the NLRGs.

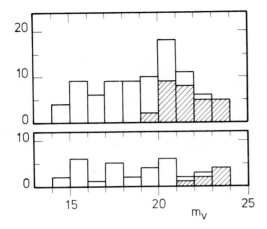

Figure 1. The apparent magnitude distribution for 87 radio galaxies from the statistical samples of 3CR radio sources and that for the 35 radio galaxies studied here. Galaxies without redshifts are shown by hatched boxes.

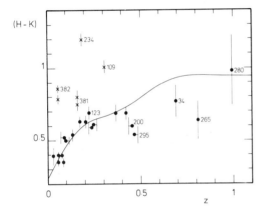

Figure 2. The (H-K) - redshift relation for 3CR radio galaxies. The crosses are N-galaxies with strong non-thermal components.

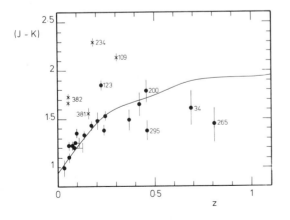

Figure 3. The (J-K) - redshift relation for 3CR radio galaxies. The crosses are N-galaxies with strong non-thermal components.

At higher redshifts, the galaxies broadly follow the predicted relations computed on the basis of the infrared spectrum constructed earlier and the energy distribution shortward of 1μm of Coleman et al (1980). There is no strong evidence for colour evolution in the infrared, and this is as predicted by conventional evolutionary models of elliptical galaxies, for example those of Bruzual (1981). Lebofsky (1981) has recently published (H-K) colours of gE galaxies over a similar redshift range. We remark that the two high z 3CR galaxies considered by her to have anomalous (H-K) colours are not anomalous on our diagram, because of our improved determination of the mean galaxy spectrum, and that we do not find evidence for the large scatter in low z galaxies.

The BLGRs are clearly red in both (J-K) and (H-K), and from the (K,z) relation it is deduced that these are brighter than the typical galaxies in our study, at all the wavelengths observed. Sandage (1973a) has shown that these systems may be successfully decomposed into a central nuclear component situated in a normal galaxy, and we have followed a similar procedure, for 3C 109, 234 and 382, using our observations of the other galaxies to define the colours and magnitudes of the underlying galaxy. To within the uncertainties of this subtraction procedure, we find that the additional component has a power-law spectrum, and that in the case of 3C382 this extends to 3.5μm. In addition, the variability of this latter source which we observed over a 6 month timebase is compatible with a nuclear component of approximately constant spectral index ($\alpha \approx 1.5$) but of varying intensity. In the small sub-sample for which there exists good spectrophotometry we find that the BLRGs displaying an infrared excess have more than 10 times the Hβ flux as compared with those other galaxies which do not.

The optical-infrared colours may be similarly constructed from published optical photometry (e.g. Sandage (1972b, 1973b), Kristian et al (1978), Smith et al (1979), Spinrad et al (1981)), although greater uncertainty will be introduced. The (R-K) and (V-K) colours as a function of redshift are plotted in Figures 4 and 5. At high redshift the CCD r magnitudes were transformed to the R system using an extension of the colour equation given by Wade et al (1979). In the diagrams, the solid lines are the predicted colour-redshift relations based upon the infrared energy distribution derived earlier and the optical spectrum from Coleman et al (1980). The other lines are various evolutionary models from Bruzual (1981). The reddest model assumes no evolution, the others representing different histories of the star formation rate (SFR), being either a constant burst of duration 1 Gyr (the C model) or an exponential decay of the SFR, with 0.7, and 0.5 respectively, of the mass of the galaxy in stars at the end of the first 1 Gyr. At high redshifts it is clear on both diagrams that there are large deviations from the relations predicted by the non-evolving models. We have included the colours derived from the K magnitudes of the high z 3CR galaxies observed by Lebofsky (1981), and these are represented by open circles. We cannot exclude the possibility that part of the blue colours of 3C299, 318 and 265 have a non-thermal origin, since the first two of these do not have "classical double" radio structure, and

THE OPTICAL AND INFRARED PROPERTIES OF 3CR RADIO GALAXIES

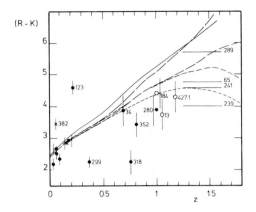

Figure 4. The (R-K) redshift relation for radio galaxies. The straight lines show the (R-K) values for galaxies of unknown redshift. Open circles are data from Lebofsky (1981).

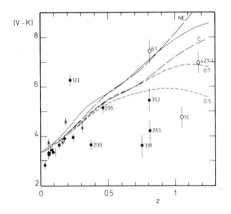

Figure 5. The (V-K) redshift relation for radio galaxies. The uppermost lines show the predicted relation if the galaxies undergo no colour evolution.

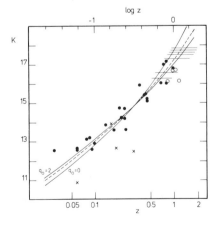

Figure 6. The redshift - K magnitude relation for radio galaxies. Crosses are N-galaxies and the open circles are data due to Lebofsky (1981).

3C265 does not have stellar absorption features in its optical spectrum (Smith et al (1979)). For the remainder of the galaxies, however, the V-K colours can be accounted for by the evolving galaxy models, with relatively slowly decaying SFR. We have also plotted on the (R-K) diagram the colours of 4 very faint galaxies of unknown redshift, and it may be seen that these galaxies have colours that are consistent with this conclusion. This result has implications for attempts to derive the redshifts of distant galaxies from their colours alone.

The red colours of 3C123 are almost certainly due to galactic absorption, the colours being consistent with an Av of 2 magnitudes.

At low redshift, it may be seen that there is a systematic trend for the (V-K) colours as observed to be some 0.2 magnitudes too blue, as compared with the predictions which themselves intercept the axis at the value of 3.3 found by FPAM in their study of nearby giant ellipticals. The reasons for this deviation, which is probably also present in (R-K) is not known at the present time. This discrepancy does however mean that it would be imprudent to use the K magnitudes of these very low redshift giant ellipticals to anchor the low redshift end of the (K,z) Hubble diagram considered in the next section.

5. THE INFRARED HUBBLE DIAGRAM FOR RADIO GALAXIES

Because essentially all the observations were made through the same aperture, we plot on the Hubble diagram, figure 6, the magnitudes as observed, and incorporate the K-corrections (which differ in the evolving and non-evolving models) and aperture corrections (which are different for different world models) into the predicted magnitude-redshift relations for different values of q_o. The aperture corrections were derived from the beam profiles and from the curve of growth given by Sandage (1972a). The best fit model may then be determined by a Chi-squared test, with the absolute magnitude/Hubble constant combination a free parameter in each world model. This method also gives the certainty with which other models may be rejected. Also shown on the diagram are the sources with unknown redshifts, and the high z data from Lebofsky (1981), as horizontal lines and open circles respectively.

With the exclusion of the BLRGs, the radio galaxies form a well defined Hubble relation, the cosmic scatter about the best fit models being 0.40 mag, which is very similar to that found at optical wavelengths by Smith (1977). For the unevolving energy distribution, the apparent, uncorrected, value of q_o is considerably in excess of 1. The best fit value occurs at 2.8, and the $q_o = 1$ model may be rejected with 70% confidence. The relations for $q_o = 0$ and 2 models are shown in Figure 6. However, the lack of infrared <u>colour</u> evolution does not imply a lack of <u>luminosity</u> evolution in the infrared. The effect of the $\mu = 0.5$ model discussed in connection with the optical-infrared colour diagrams is to increase the predicted K luminosity at a redshift of 1 by 0.8 mag. Application of the K-corrections appropriate to this model result in a best fit value of $q_o = 0.5$. Therefore, if lower values of q_o are preferred for other reasons, then the Hubble diagram provides further evidence of substantial evolution over lookback times of half the Hubble time.

REFERENCES

Bruzual, G.; Thesis, UCa, Berkeley.
Coleman, G.D., Wu, C.C., Weedman, D.W., 1980, Ap.J.(Supp.), 43, 393.
Frogel, J.A., Persson, S.E., Aaronson, M., Matthews, K.; 1978 Ap.J., 220, 75.
Grandi, S.A., Osterbrock, D.E.; 1977, Ap.J., 195, 255.
Grasdalen, G.L.,; 1980, IAU Symp., 92, 269.
Gunn, J.E., Hoessel, J.G., Westphal, J.A., Perryman, M.A.C., Longair, M.S., 1981, MNRAS, 194, 111.
Jenkins, C.J., Pooley, G.G., Riley, J.M., 1977, Mem.R.astr.Soc., 84, 61.
Kellermann, K.I., Pauliny-Toth, I.I.K., Williams, P.J.S.; 1969, Ap.J., 157, 1.
Kristian, J., Sandage, A., Westphal, J.A., 1978, Ap.J., 221, 383.
Laing, R.A., Riley, J.M., Longair, M.S.; 1982 (in preparation).
Lebofsky, M.; 1980, IAU Symp. 92, 257.
Lebofsky, M.; 1981, Ap.J. (Letters), 245, L59.
Sandage, A., 1972a, Ap.J., 173, 485.
Sandage, A., 1972b, Ap.J., 178, 25.
Sandage, A., 1973a, Ap.J., 180, 687.
Sandage, A., 1973b, Ap.J., 183, 711.
Smith, H.E., 1977, IAU Symps. 74, 279.
Smith, H.E., Junkkarinen, V.T., Spinrad, H., Grueff, V., Vigotti, M.; 1979, Ap.J., 231, 307.
Smith, H.E., Spinrad, H., 1980, Pub.astr.Soc.Pac., 92, 553.
Spinrad, H., Smith, H.E., 1979, Ap.J., 206, 355.
Wade, R.A., Hoessel, J.G., Elias, J.H., Huchra, J.P.; 1979, Pub.astr.Soc.Pac., 91, 35.

DISCUSSION;

WEISTROP: Can you tell from the CCD frames whether those galaxies with blue excess in the color-redshift diagrams have a bright, blue nucleus?

LONGAIR: The CCD images were taken in r and i and so it is difficult to say whether there is a blue excess or not. Our analysis of the images indicates that most of the faint identifications are radio galaxies, and we could set limits to the contribution of a point-like object at their nuclei.

ROBERTSON: At high redshifts, most galaxies with known redshifts have emission lines, because these are the ones that can be measured. How does this affect your results?

LONGAIR: We find that the broad-line radio galaxies are clearly separable on the infrared color diagrams. For the others, the continuum radiation is much weaker than the stellar radiation from the galaxy, and we can find no correlations between the infrared colors and the properties of the narrow line region.

WINDHORST: What is the typical value of the IR aperture correction applied to the Hubble diagram in k? Is it very uncertain due to the fact that the Hubble profile of ellipticals is not well known in the k band?

LILLY: The aperture correction incorporated into the predicted (k,z) relations changes by 1.2 magnitudes over the entire redshift range, 0.03 to 1.0, with half of this change occurring below a redshift of 0.1. Unfortunately, the k growth curve of Frogel et al. (1978) does not extend to sufficiently large metric diameters to be used here. In the region of overlap, their growth curve is similar to the v growth curve of Sandage used in this paper.

REDSHIFT ESTIMATES FOR DISTANT RADIO GALAXIES BASED ON BROADBAND
PHOTOMETRY

J. J. Puschell
University of California, San Diego

F. N. Owen and R. Laing
National Radio Astronomy Observatory

Calculations modeling the effects of stellar evolution on elliptical galaxies (e.g., Bruzual and Kron 1980, Bruzual 1981) suggest that the shape of the spectral flux distribution should remain almost constant in the red and near-infrared out to at least $z \sim 2$. Thus, it should be possible to derive the redshift of a distant elliptical galaxy by fitting a model galaxy spectrum to broadband near-infrared (RIJHK) photometry.

In order to test this idea, we have begun a program of JHK photometry of elliptical galaxies with high measured redshifts and optically identified and unidentified faint steep-spectrum radio sources believed to be elliptical galaxies, using the NASA 3-m IRTF at Mauna Kea. Details of the observations and interpretation will be discussed by Puschell, Owen and Laing (1981). Eight of nine optically unidentified sources have been detected in at least one infrared band. The infrared data are consistent with these objects being elliptical galaxies, although observations shortward of 1 μm are needed before this is certain. By combining our measurements with published visual and infrared photometry (e.g., Gunn et al. 1981, Lebofsky 1981, Kristian, Sandage and Westphal 1978), we have derived the results shown in the table. Spectroscopic redshifts were taken from Spinrad, Kron and Hunstead (1979), Minkowski (1960), Spinrad (private communication) and Spinrad, Stauffer and Butcher (1981). The model galaxy spectral energy distribution consisted of the M31 nuclear bulge spectrum of Coleman, Wu and Weedman (1980) from 0.14-1.00 μm and a synthesized spectrum from 1-4 μm. The agreement between photometric and spectroscopic redshifts is good for the two galaxies with $z < 0.5$. The redshifts of the more distant objects, which are bluer than expected in the optical band, are systematically underestimated. This is in accord with the predictions of Bruzual (1981). A surprising result is that both of the galaxies with substantial discrepancies between photometric and spectroscopic redshifts, 3C 13 and 3C 368, have anomalous infrared colors of J - K = 0.9 and 1.4, respectively (note that 3C 13 has an alternative spectroscopic redshift of 0.4). Both our non-evolving s.e.d. and the evolving models of Bruzual (1981) predict that

$J - K \simeq 2$ at $z = 1$: either the effects of stellar evolution extend to the infrared colors or there is some non-stellar source of radiation. Objects like 3C 368, whose spectra cannot be fitted at the correct redshift by our model, should be recognizable from their positions on a plot of $J - K$ against K.

Galaxy	z_{SPECT}	z_{PHOT}	χ^2
0442 − 18	0.28	0.31 + 0.08 − 0.05	0.4
3C 295	0.46	0.43 + 0.06 − 0.07	2.7
JB 1647 + 43	−	0.56 + 0.18 − 0.27	0.8
3C 34	0.69	0.48 + 0.07 − 0.12	0.8
3C 184	0.99	0.64 + 0.04 − 0.09	1.1
3C 13	1.05 (Alt: 0.4)	0.55 + 0.12 − 0.12	5.5
3C 368	1.13	0.22 + 0.10 − 0.04	0.04
3C 427.1	1.18	0.91 + 0.16 − 0.10	1.9
3C 289	−	1.11 + 0.12 − 0.27	2.7
3C 65	−	1.08 + 0.36 − 0.14	2.6

References

Bruzual A., G. 1981, Ph.D. dissertation, Univ. of California, Berkeley.
Bruzual A., G., and Kron, R. G. 1980, Ap. J. 241, 25.
Coleman, G. D., Wu, C., and Weedman, D. W. 1980, Ap. J. Suppl. 43, 393.
Gunn, J. E., Hoessel, J. G., Westphal, J. A., Perryman, M.A.C., and Longair, M. S. 1981, Mon. Not. R. astr. Soc. 194, 111.
Kristian, J., Sandage, A., and Westphal, J. A. 1978, Ap. J. 221, 383.
Lebofsky, M. 1981, Ap. J. (Lett.) 245, L59.
Minkowski, R. 1960, Ap. J. 132, 908.
Puschell, J. J., Owen, F. N., and Laing, R. 1981, submitted to Ap. J. (Lett.).
Spinrad, H., Kron, R. G., and Hunstead, R. W. 1979, Ap. J. Suppl. 41, 701.
Spinrad, H., Stauffer, J., and Butcher, H. 1981, Ap. J. 244, 382.

RADIO EVOLUTION IN HIGH REDSHIFT CLUSTERS

Walter Jaffe
National Radio Astronomy Observatory
Edgemont Road
Charlottesville, Virginia 22901

In this project we try to follow the evolution of the Radio Luminosity Function (RLF) with redshift by observing a set of galaxy clusters with known spectroscopic redshifts. This method has the advantage over previous studies, based on general radio surveys, that most of the redshifts are accurately known, that intrinsically weaker sources are well represented, and that the parent sample of galaxies is well defined. The chief disadvantages are that, because the sample is optically selected, there are few high luminosity radio sources, and that we cannot examine differences between cluster and field populations.

With the VLA we observed 55 clusters with redshifts of $0.25 \rightarrow 0.90$ obtained by J. Gunn, B. Oke and J. Hoessel, R. Kron, and H. Butcher. In 3 of the clusters, the spectra showed only a blue continuum and we classified the redshifts for these clusters as large (~ 0.6). We observed each cluster for ± 1 hr at 1460 MHz. We could typically detect radio sources with $S(1460) \gtrsim 4$ mJy. We count as cluster sources those within 0.3 Abell radii of the cluster centers (using $q_o=0.0$). There were 13 such detections, of which less than 1 is expected by chance. Local, $z=0$, RLFs (Owen 1975, Auriemma et al. 1977) indicated a differential detection fraction $dN/d \ln P(1400)$ which is more or less constant for $P(1400) \lesssim P^* = 10^{24.4}$ WHz^{-1}, and which varies as P^{-m}, with $m=1.5$, for $P > P^*$. In Table I we compare in 3 luminosity decades the number of detections based on the Owen RLF with those actually found, and find the maximum likelihood value of the slope of the RLF, m, in the decade. We calculated our luminosities by summing the fluxes of all detected sources and converting to intrinsic power using $H_o=100$km s^{-1} Mpc^{-1}, $q_o=0$, $\alpha = -0.7$. While the detections in the middle decade don't differ much from the $z=0$ expectation, the slopes have flattened considerably from the local value of -1.5. It is surprising, in view of the supposed rapid evolution at high z (e.g., Windhorst et al. 1982), that there are fewer detectors at $z \simeq 0.6$ than at $z \simeq 0.35$. This suggests that the rate of evolution has considerable structure as a function of P and z, and the RLF may evolve as shown in the sketch below.

Our results differ from Katgert et al., and Windhorst et al. in that, for these low luminosity sources, we see a more rapid evolution at low z, and a slow, or negative evolution at higher z. The difference could derive either from the inaccuracy of the non-spectroscopic distances used by the Westerbork groups, or from a true/cluster field difference in the RLF evolution.

TABLE I

Redshift bin <z>	0.25 - 0.40 0.35			0.40 - 0.50 0.45			0.50 - 0.90 0.60		
	Ns	Ne	Nd	Ns	Ne	Nd	Ns	Ne	Nd
log P = 23.4 - 24.4	16 $m = 0.9\pm0.7$	5	4	5 ---	1	0	0 ---	0	0
log P = 24.4 - 25.4	20 $m = -0.2\pm0.7$	4	5	16 $m = +0.3\pm0.9$	3	3	14 ---	2	1
log P = 25.4 - 26.4	20 ---	0.1	0	18 ---	0.1	0	17 ---	0.1	0

Ns = Number of clusters surveyed in this luminosity-redshift bin
Ne = Number of detections expected from Owen (z=0) RLF
Nd = Number of actual detections in this survey

When more than one source was found, values for the slope m of the RLF in the luminosity-redshift bin are given.

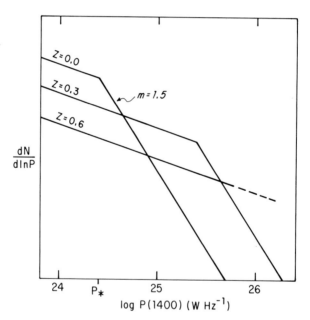

Fig. 1. Suggested form of evolution of the cluster RLF.

References:
  Auriemma, C., Perola, C. G., Ekers, R., Fanti, R., Lari, C., Jaffe, W. J., Ulrich, M. H. 1977, A & A, 57, 41.
  Katgert, P., de Ruiter, H. R., vander Laan, H. 1979, Nature, 280, 20.
  Owen, F., 1975, Ap. J., 195, 593.
  Windhorst, R. A., Kron, R. G., Koo, D., Katgert, P. 1982, This Symposium, p. 427.

COLORS OF RADIO GALAXIES AT HIGH REDSHIFTS

Rogier A. Windhorst[+], Richard G. Kron[++], David Koo[+++], Peter Katgert[+]
[+] Sterrewacht, Leiden
[++] Yerkes Observatory, Chicago
[+++] Astronomy Department, Berkeley

In this paper we present the first results of the Westerbork-Berkeley Deep Survey, the purpose of which is to derive the epoch dependence of the radio luminosity function (RLF) and the optical spectral energy distribution (SED) of elliptical radio galaxies. From calibrated photographic photometry we conclude that no spectral evolution is seen for $z \lesssim 0.4$, but that colors of radio galaxies are $0\overset{m}{.}5-1\overset{m}{.}5$ bluer for $0.4 \lesssim z \lesssim 0.9$ than predicted from classical model spectra for ellipticals (with star formation only during the first 10^9 yrs after formation). First-ranked cluster radio galaxies may be slightly bluer than the average radio galaxy for $z \gtrsim 0.6$.
The epoch dependence of the 1.4 GHz RLF is determined in a direct way. We find almost no population evolution for $z \lesssim 0.3$, but strong evolution for $0.3 \lesssim z \lesssim 0.9$ for all radio powers above the break in the RLF, with enhancement factors of ~ 100 at $z \sim 0.8$.

THE OBSERVATIONS

The Westerbork-Berkeley Deep Survey consists of several deep radio surveys of areas in SA 57, SA 68, Hercules and Lynx. For these areas multicolor Mayall 4^m plates are available, taken by two of us for studies of faint galaxies. The 21 cm radio observations were done with the 3 km Westerbork array, with 12" resolution and ~ 100 µJy rms noise in 12^h. About 500 radio sources with $S_{1.4} > 0.6$ mJy were found, of which 297 form a complete sample within 5.3 deg^2. Observations and analysis are described by Windhorst et al (1981).
For most areas we have several high quality prime-focus plates in four passbands, half of them with sub-arcsecond seeing. The plate limits (for stellar objects) are $24\overset{m}{.}0$ in U(3600 Å) and J(4650 Å), $23\overset{m}{.}0$ in F(6100 Å) and $22\overset{m}{.}0$ in N(8000 Å). Absolute photographic photometry was derived, using accurately known photoelectric standards, with magnitude errors of $0\overset{m}{.}05$ for the brightest objects and errors in colors of $\sim 0\overset{m}{.}3$ for objects $\sim 0\overset{m}{.}5$ above the plate limits (see Kron, 1980, for details). Star/galaxy discrimination is done by a -2nd order moment of the object profile.
A complete subsample studied so far contains 150 radio sources, for which

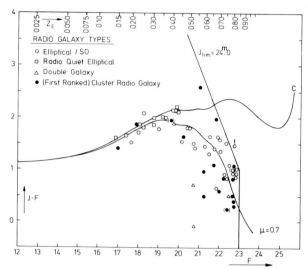

Fig. 1. The observed color magnitude distribution (J-F) vs F of radio galaxies in the Westerbork-Berkeley Deep Survey.

we found 45% reliable identifications (after correction for contamination) In general the faint identifications were seen on both J and F plates. The U plates gave no new identifications, but the N plates added a few very red radio galaxies. About 85% of all identifications are radio galaxies, predominantly fainter than J = $22^m.0$. About a third of all radio galaxies was found in clearly visible clusters, defined as having, within a 40" diaphragm, more than 3 times the general background density.

COLORS OF RADIO GALAXIES

The observed color magnitude distribution (J-F) vs F of elliptical radio galaxies is shown in fig. 1, together with some data for radio-quiet ellipticals (Koo, Kron and Spinrad, unpublished). The radio galaxies appear to become progressively bluer towards fainter magnitudes. This cannot be explained by the shape of the error distribution close to the plate limits (denoted by the lines J = $24^m.0$, F = $23^m.0$). Note that not many upper limits for J exist for objects visible on F, nor the other way around. Instead, most plate limit objects are visible on both J and F plates, which essentially means J-F $\sim 1^m.0$. Curve c corresponds with the theoretical colors of an evolving elliptical SED, with constant star formation rate (SFR) during the first 10^9 yrs after formation, and after that SFR = 0 (Bruzual and Kron, 1980, Bruzual, 1981). The μ = 0.7 curve corresponds to an evolving elliptical SED with an SFR that declines exponentially with cosmic time, with 70% of the mass already in stars after the first 10^9 yrs. We conclude that any model with star formation occurring only during the first 10^9 yrs (let alone non-evolving models) do not reproduce the observed color-magnitude diagram of elliptical radio galaxies. The μ = 0.7 model gives a reasonable fit to the data,

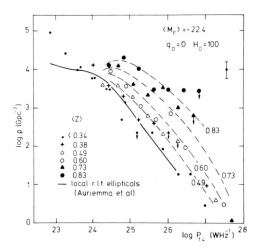

Fig. 2. The 1.4 GHz RLF of elliptical radio galaxies at various epochs. Data from different Westerbork Surveys have been taken together.

but is probably not at all unique. For most radio galaxy identifications we observe, or infer, $U-J \geq 0\overset{m}{.}5$. For spirals at all redshifts, as well as for non-thermal nuclei, one expects $U-J \lesssim 0\overset{m}{.}0$. This is indeed what we observe for our quasars. So it is likely that all our radio galaxies are ellipticals without a major non-thermal continuum contribution, or rather, that the colors of the radio galaxies are dominated by the stellar population.
Beyond $z \sim 0.6$ there may be a tendency for the first-ranked cluster radio galaxies to be slightly bluer than the other elliptical radio galaxies. A similar trend was observed by Kristian et al. (1978). If this were due to a dependence of star formation history on (cluster) environment one might expect the radio-quiet cluster members to be similarly blue. Radio selected clusters at high redshifts are suitable objects to further study such an effect, which could be similar to that found by Butcher and Oemler (1978).

EPOCH DEPENDENCE OF THE RLF OF ELLIPTICAL RADIO GALAXIES

The observed color-magnitude distribution of elliptical radio galaxies effectively samples their evolving SED as a function of z. Assuming that the SED evolution manifests itself predominantly in the UV part of the spectrum (as a spectral upturn increasing with redshift) we have derived an evolving SED from the observed colors. This allows us to calculate K-corrections and infer redshifts from apparent magnitudes. Using F magnitudes this procedure should be safe out to $z \sim 0.8 - 0.9$. Of course, one has to assume that the radio galaxy population has the, locally observed, small dispersion in absolute magnitude at all relevant redshifts. Further, $<M_F>$ may depend on z, i.e. there may be appreciable evolution in the red part of the spectrum as well. Presently there is however no evidence for such evolution. Using the SED determined from the observed colors we predicted the Hubble diagram in V. This turned out to be completely consistent with the observed diagram, based on spectroscopically measured redshifts of dozens of 3CR and 4C radio galaxies, out to $z \sim 1$ (Laing et al 1978, Gunn et al. 1981, Peacock and Wall 1981).

Given the bivariate flux density distribution for the identified radio galaxies one can derive directly the RLF of elliptical radio galaxies as a function of redshift. Rather than presenting the results of the Westerbork-Berkeley Deep Survey separately, we have combined the present data with those of previous Westerbork surveys (Katgert et al. 1979). The result is presented in fig. 2, which is based on a total of ~ 200 objects. There appears to be considerable evolution above the break in the RLF at $\log P_{1.4} = 24.4$ W Hz^{-1}. The magnitude of the increase of the RLF with redshift does not seem to depend strongly on radio luminosity. Little or no evolution is seen for $z \lesssim 0.3$; beyond that redshift there is a strong increase of the radio galaxy space density, with at $z \sim 0.8$ enhancement over the local values by factors of 100 or more. It is interesting to note that the redshift dependence of the RLF as derived indirectly by Robertson (1980) using a free-form evolution function to fit source counts and luminosity distributions, also shows the initial flat part and the steep rise beyond $z \sim 0.3$ apparent in our direct determination.

We thank Harry van der Laan for his continuous support of this project, and also Geoffrey van Heerde and Marc Oort for their help with the data reduction.

REFERENCES

Bruzual, A.G.: 1981, Ph.D. Thesis University of California, Berkeley
Bruzual, A.G., Kron, R.G.: 1980, Astrophys. J. 241, p. 25
Butcher, H., Oemler, A.: 1978, Astrophys. J. 219, p.18
Gunn, J.E. et al.: 1981, M.N.R.A.S. 194, p. 111
Katgert, P. et al.: 1979, Nature 280, p. 20
Kristian, J. et al.: 1978, Astrophys. J. 221, p. 383
Kron, R.G.: 1980, Astrophys. J. Suppl. 43, p. 305
Laing, R.A. et al.: 1978, M.N.R.A.S. 184, p. 149
Peacock, J.A., Wall, J.V.:1981, M.N.R.A.S. 194, p. 331
Robertson, J.G.: 1980, M.N.R.A.S. 190, p. 143
Windhorst, R.A. et al.: 1981, in preparation

DISCUSSION

ROBERTSON: What percentage of the radio sample do you estimate are quasars, and given that quasars have a wide dispersion in absolute magnitude, what have you done about them in your analysis?

WINDHORST: About 8% of all radio sources down to 2 mJy are quasars, or 15% of all identifications. Below that level, these functions seem to increase somewhat. We did not include the quasars in the RLF for the moment, but their fraction is small anyway. Unlike the radio galaxies, most quasars are brighter than $J \simeq 21.5^m$, so there is good hope to get spectra for them somehow.

WALL: I believe that the colors of faint radio galaxies which you present indicate that the drastic blueing of such galaxies suggested by Katgert et al. (Nature 280, 20, 1979) is incorrect. Would you like to comment on this?

WINDHORST: Sure, as Katgert et al. already suggest, the weak point in their data was, that no photoelectrically calibrated photographic photometry was available for surveys covering such a large area over many 48" plates. But still, the blueing is present in their data. The pure fact that their radio galaxies near the red plate limit did show up in the blue at all--and often even very clearly--shows inconsistency with a non-evolving SED. However, with our WBDS colors and Longair and Lilly's 3CR infrared colors, the controversy is solved convincingly.

JAFFE: My results differ considerably from yours where they overlap. This could result from the crudeness of the analysis or it may show a real cluster/non-cluster difference in evolution.

WINDHORST: Two comments about the differences between our samples:
(1) Our sample is distance limited in the radio out to a redshift of unity and in the optical out to z = 0.6, which is the redshift at which we still can see those ellipticals above the break in the bivariate RLF with the faintest M_F. Our sample in Fig. 2 contains 303 identifications of 1300 radio sources in various surveys, so I think we have to take the quantitative measure of the evolution serious out to z = 0.6. Beyond that we are magnitude limited in the optical, and I am the first one to admit that we really need the redshifts to see whether we are dealing with a parent population with the same M_F as locally. It is at the moment impossible to say to what extent the Malmquist bias or optical luminosity evolution, or both, play a role for z > 0.6. Your sample might be distance limited in the optical, but it is not clear to me that this is also true in the radio. A larger, deeper sample is needed. Note that we find radio selected faint clusters even below 1 mJy.
(2) But still, the differences could be real indeed. We also found that several known distant, often extreme rich and red clusters did not show up as radio sources, while blue ones did show up.

A STUDY OF SMALL ANGULAR SIZE OOTY SOURCES

T.K. Menon
Department of Geophysics and Astronomy
University of British Columbia
Vancouver, B.C. V6T 1W5 Canada

I. INTRODUCTION

Recent theoretical models of radio radiation from central cores of radio sources invoke axial relativistic beaming of the radiation and predict significant enhancement of the radiation from the central cores when viewed at small angles to the direction of the beam. Investigations of the angular diameter distributions of radio sources by Swarup (1975) has shown that the median angular size of radio sources below 4 Jy at 327 MHz is about 10 arc seconds. Hence a sample of sources, with angular sizes substantially less than the above median, chosen from the Ooty occultation survey may be expected to contain a large fraction of double sources viewed along the axis of the double structure.

II. CONCLUSIONS

I have selected an unbiased sample of 73 sources with angular size $\leq 4"$ from the Ooty survey and measured their flux densities at 5 GHz and 2.7 GHz using the 300 ft telescope of NRAO at Green Bank. In this paper I shall confine myself to the discussion of the spectral indices α computed between 327 MHz and 2.7 GHz. The statistics of the spectral index distribution and identification content of the sample are given below.

	QSO or BSO	G	NSO	Crowded or Uncertain	E.F.	Total
$\alpha \leq 0.5$	8	3	0	1	5	17 (24%)
$\alpha > 0.5$	10	2	3	9	32	56 (76%)

The percentage of flat spectra sources in the present sample is in direct contrast to the very small percentage (3.7%) of flat spectra sources found in a complete low-frequency sample of 4C sources discussed by Veron and Veron (1980). Furthermore for the 56 sources with $\alpha > 0.5$

the median spectral index is only 0.75. This is to be contrasted with the median spectral index of 0.92 for a sample of 46 well resolved double sources from the Ooty survey studied earlier by Menon (1980). The comparatively flat spectral index of 0.75 for the present angular size limited sample implies that half of the sources with $\alpha > 0.5$ have substantial contribution from a flat spectrum central component at high frequencies. Since these sources have angular sizes 2.5 times smaller than the median angular size of sources in their flux density range they are most likely to be double sources inclined at a small angle to the line of sight. The enhancement of the intensity of the central components in such sources may be related to the above circumstance as is suggested in the models of Blandford et al. (1977) and Scheuer and Readhead (1979). VLBI observations of a similar sample of 30 few arc-second sources by Gopal-Krishna et al. (1980) also suggest that radio cores are more prominent in such a sample as compared with the cores found typically in extended doubles. Other possible interpretations are discussed by Gopal-Krishna et al. (1980) and Kus et al. (1981).

Of the 25 sources with α between 0.5 and 0.75 only 8 have definite identifications while 15 are in empty fields. This contrasts with the result that practically all extended sources with central components have optical identifications. Hence it would appear that the radio to optical luminosity ratio of the inferred compact components of the present sample are anomalous and this anomaly may be related to the question of relativistic beaming in such sources. Detailed structural studies can provide important information regarding the evolution of such sources.

I am grateful to the Director of NRAO for observing facilities at Green Bank. NRAO is operated by Associated Universities, Inc., under contract with the National Science Foundation. This work has been supported by a grant from the Natural Sciences and Engineering Research Council of Canada.

References

Blandford, R.D., McKee, C.F. and Rees, M.J.: 1977, Nature 267, pp. 211-216.
Gopal-Krishna, Preuss, E. and Schilizzi, R.T.: 1980, Nature 288, pp. 344-347.
Kus, A.J., Wilkinson, P.N. and Booth, R.S.: 1981, Mon. Not. R. Astr. Soc. 194, pp. 527-535.
Scheuer, P.A.G. and Readhead, A.C.S.: 1979, Nature 277, pp. 182-185.
Swarup, G.: 1975, Mon. Not. R. Astr. Soc. 172, pp. 501-512.
Veron, M.P. and Veron, P.: 1980, Astron. Astrophys. Suppl. 40, pp. 191-198.

A COMPARISON OF THE STRUCTURES OF 3CR QUASARS AND BLANK FIELD RADIO SOURCES

F. N. Owen, J. J. Puschell and R. A. Laing
National Radio Astronomy Observatory

The purpose of this communication is to update our knowledge of the radio structural properties of quasars and blank field radio sources (blank field ≡ any radio source without an identification on the Palomar Sky Survey prints). The quasar sample consists of all sources (25) with angular sizes greater than 10 arcsec in the list of Jodrell Bank quasars observed by Owen, Porcas and Neff (1978). The blank fields consist of 16 3CR sources also with structures >10 arcsec based on Cambridge 5 km telescope observations. The sources were selected in low-frequency surveys; their emission at $\nu < 1$ GHz is dominated by extended components with steep spectra. Thus, both samples should be oriented randomly in space except for a slight bias to be in the plane of the sky.

Before this study we knew that blank field sources were almost never found to have central components at the 10 mJy level at 6 cm (e.g. Longair, 1975). On the other hand most, but not all, quasars were known to have central components at similar levels (e.g. Owen, Porcas, Neff, 1978). A few quasars were known to have jets (e.g. Potash and Wardle, 1980) but no such structures were known in blank field sources.

Our observations were made in the "A" array of the VLA at 6 cm. Typically, the blank fields were observed for 40 minutes with a 50 MHz bandwidth while the quasars were observed for 20 minutes with a 25 MHz bandwidth. At the time of this report we have self-calibrated all the blank field sources obtaining a 200-2000 dynamic range on each map. The quasars have not been self-calibrated and have a dynamic range of 100 to 400.

A summary of the results for the 3CR blank fields is as follows:
1) We detect central components in 7 out of 15 blank field sources with one questionable case. The central components range from 13 to 0.8 mJy with limits for the rest of 0.5 mJy; (see Table I)
2) No jets are found;
3) The central components agree in position with a previously sugges-

ted optical identification in all but the questionable case, 3C280. In this case there is a 39 mJy component very near one of the lobes with no obvious counterpart.

Table I
3CR Blank Field Sources

	Central Components
3C13	<0.5 mJy
3C226	7.5
3C228	13.3
3C247	3.5
3C252	1.1
3C267	<0.5
3C280	(39.2)??
3C289	<0.5
3C322	<0.5
3C324	<0.5
3C325	1.2
3C337	<0.5
3C340	<0.5
3C356	<0.5
3C427.1	0.8
3C469.1	2.6

For the quasars we find that:
1) We detect all the central components at flux densities ranging from 6 to 500 mJy;
2) At least 20 percent of the quasars have radio jets;
3) No obvious differences exist between the lobes of the quasars and blank fields.

References

Longair, M.S. (1975) M.N.R.A.S. 173, 309.
Owen, F. N., Porcas, R. W., and Neff, S. G. (1978) A.J. 83, 1009.
Potash, R. I. and Wardle, J. F. C. (1980) Ap.J. Letters 239, L42.

SPACE DISTRIBUTION OF QUASARS BASED ON OPTICALLY SELECTED SAMPLES

Maarten Schmidt
Palomar Observatory, California Institute of Technology, and
Richard F. Green,
Steward Observatory, University of Arizona

Even though the number of known quasars is approaching 2000, relatively few belong to well defined complete samples which are needed to derive their space distribution. For many years, the Braccesi survey (Braccesi, Formiggini, and Gandolfi 1970) was the only published one — its limited spectroscopic coverage allowed construction of only a small complete sample of 19 quasars brighter than B = 18, over an area of 36 sq. deg.

A large-scale search for bright quasars was started in 1972. This survey for ultraviolet-excess stellar objects covers some 10,700 sq. deg. and is aimed to be complete, on the average, to B = 16.2. We undertook spectroscopic observations of all objects and produced a sample of over 100 quasars with redshifts and magnitudes that constitute the Palomar Bright Quasar Survey. We have used the results of this Survey, together with the Braccesi sample, two objective-prism quasar surveys discussed by Osmer and Smith (1980) and Osmer (1980), and a small, deep sample in SA 57 (Kron and Chiu 1981) to study the space distribution and optical luminosity function of quasars. This study will soon be submitted for publication, and we will summarize here some of the main results.

1. QUASAR DISTANCE SCALE

The slope of the cumulative counts with magnitude is very steep: d log N/dB = 0.93±0.06, or in the radio equivalent: d log N/d log S = -2.3±0.15. These slopes are significantly steeper than those (0.60, and -1.5, respectively) corresponding to a uniform distribution in Euclidean space. Hence, if quasars were local (and their redshifts non-cosmological) their space density would have to increase with distance (approximately as $r^{1.6}$).

In this case, we would be located in a unique position in the universe, namely in the central deep density minimum of the quasar cloud — a conclusion to be rejected on Copernican grounds. This argument only breaks down if quasar distances are so large that their travel time is a substantial fraction of the age of the universe.

This is precisely the distance scale corresponding to the cosmological interpretation of the redshifts.

Since the arguments given above are statistical in nature, we cannot use them to assert that every quasar must have a cosmological redshift. It may be shown that up to 2% of quasars at magnitude B = 20 can be local without violating the above arguments.

2. LUMINOSITY EVOLUTION

Pure luminosity evolution of quasars has been discussed recently by Mathez, by Braccesi and others. In this case it is assumed that the luminosity function of quasars at all cosmic epochs is the same, except for a time-dependent shift in luminosity. As a consequence, the total number of quasars is the same at all times.

We find that the observed number of quasars in the various surveys at redshifts less than around 2 can be well fitted by a shift in absolute magnitude $\Delta M_B = 7\tau$ mag. ($q_o = 0$), or $\Delta M_B = 5.5\tau$ mag. ($q_o = 1/2$), where τ is the light travel time expressed as a fraction of the age of the universe. It should be noted that this fit is achieved only if we assume that the nuclei of Seyfert galaxies are also subject to this luminosity evolution and hence, at larger redshifts, are all identified as quasars.

At redshifts larger than 2, however, the luminosity evolution models predict too few quasars. In particular, the $q_o = 1/2$ luminosity evolution model predicts only one third of the observed number of quasars brighter than B = 19.5 with redshifts in the range 1.8-3.1.

If luminosity evolution represents the evolution of individual objects all born at an early cosmic epoch, then their total radiated energy will be 10^{63}-10^{64} ergs. This excessive energy requirement is alleviated if quasars live less than 10^{10} years. If this is the case, then luminosity evolution does not represent the evolution of the luminosity of each individual object, but instead the statistical result of the births, light history and deaths of quasars on the luminosity function. In this case, there is no obvious reason why the luminosity function should have the same shape at all epochs and we discuss in the next section evolution models where this requirement is relinquished.

3. LUMINOSITY-DEPENDENT DENSITY EVOLUTION

The mean V/V_{max} of quasars in the Palomar Bright Quasar Survey shows a strong dependence on absolute optical luminosity. For quasars of highest luminosity it is close to 0.8, while for those of lowest luminosity it is not much larger than 0.5. This suggests that the increase in space density toward larger redshifts is luminosity dependent. This is confirmed by detailed comparison of the observed numbers in the different complete samples, as a function of absolute magnitude. Assuming that the space density in co-moving coordinates varies as $\exp(k\tau)$, we find that k is as large as 30 for the highest luminosities, and less than 10 for the lower luminosities.

For a tentative model based on $q_o = 0$ and $H_o = 100$ km s^{-1} Mpc^{-1} we find the following results. The space density of quasars rises from 360 Gpc^{-3} at $z = 0$ to 30,000 Gpc^{-3} at $z = 3$. The space density of Seyfert nuclei, which we assume not to evolve, is around 13,000 Gpc^{-3}. The model predicts that there should be some 20 quasars brighter than B = 20.5 with z = 3.7 - 4.7 in an area of 5 sq. deg. Osmer (1981) found none in a survey designed to detect such quasars, suggesting strongly that there is a significant deficiency of quasars with such redshifts. This may reflect the turn-on time of the earliest quasars, or the effect of dust as suggested by Ostriker and Cowie (1981).

The total number of quasars ever formed can be derived once an assumption is made about the lifetime of quasars. Assuming that the lifetime is inversely proportional to the quasar luminosity, we find a space density of dead quasars of

$$10^{-3} \left(\frac{10^{61} \text{ergs}}{E_{rad}} \right) \text{Mpc}^{-3}$$

where E_{rad} is the total energy radiated by a quasar during its lifetime. Depending on E_{rad} it seems that the space density of dead quasars could be a substantial fraction of that of galaxies.

The apparent birth rate of quasars out to a redshift of 3.4 is

$$10^{-1} \left(\frac{10^{61} \text{ergs}}{E_{rad}} \right) \text{yr}^{-1}.$$

If the birth of a quasar is signaled by a gravitational collapse, then the detection of such an event within a century, would require $E_{rad} < 10^{62}$ ergs.

This research was supported in part by the National Science Foundation under grant AST 77-22615.

REFERENCES

Braccesi, A., Formiggini, L., and Gandolfi, E.: 1970, Astron. Astrophys. 5, pp. 264-279.
Kron, R. G., and Chiu, L. G.: 1981, preprint.
Osmer, P. S.: 1980, Astrophys. J. Suppl., 42, pp. 523-540.
Osmer, P. S.: 1981, preprint.
Osmer, P. S., and Smith, M. G.: 1980, Astrophys. J. Suppl., 42, pp. 333-349.
Ostriker, J. P., and Cowie, L. L.: 1981, Astrophys. J. (Letters) 243, pp. L127-131.

DISCUSSION

SHAPIRO: The absence of quasars with z > 3.5 found by Osmer in his search for redshifted Ly-α emission lines may be the effect of Ly-α absorption by intergalactic neutral hydrogen. The absence of such absorption in the spectra of quasars with z < 3.5 then requires that the H atoms be ionized by roughly this redshift. In order to explain the absence of <u>partially</u> absorbed quasars in the redshift range corresponding to the transition of the absorbing gas from neutral to ionized, it may be necessary that the space distribution of absorbing gas be patchy on large scales during this epoch (V. Trimble, private communication). Since the absorbing gas is likely to be transparent to X-rays, at least those above a few keV, these high z Ly-α-obscured quasars may be identifiable as X-ray sources whose optical spectra show intergalactic absorption troughs.

BARNOTHY: Do you think that the space distribution of quasars and the possibility of quasar evolution would be affected if a significant percentage of quasars were gravitational lens intensified nuclei of Seyfert galaxies? (J. M. Barnothy 1965, <u>A. J.</u> 70, 666). Due to the circumstance that not only the number of objects, but also the number of potential gravitational lenses, and the intensification through the lens increases with increasing distance of the object, the number of lensed quasars should, in a first approximation, increase as the fifth power of z, and simulate a rapid increase in the number of observable quasars per unit volume. (J. M. Barnothy, 1966, <u>Observatory</u> 86, 115). The V/V_m ratio is about the same for a luminosity evolution, or gravitational lens without evolution. (J. M. Barnothy 1975 <u>Ap. J.</u> 201, 287). A further modification in the explanation of quasar distribution could occur, should the universe correspond to the modified static solution of the Friedmann equations, in which latter the luminosity distance has the form sin ln (1+z). (J. M. Barnothy, IAU Symp. No. 44).

SCHMIDT: J. Ostriker (private communication) has stated that lensing of quasars would yield a quasar luminosity function proportional to L^{-3} dL at the bright end. Observations show $L^{-3.5}$ dL or L^{-4} dL, suggesting that lensing has no major effect on the luminosity function. An independent argument is provided by the space density increase of steep radio spectrum 3CR quasars, which can be derived from radio statistical evidence only. In this case, most of the radio sources are so large that they should be little affected by galaxies acting as gravitational lenses.

TERRELL: It seems to me that your lnN-lnS argument against local quasars clearly does not apply to very local quasars, ejected by our own galaxy.

SCHMIDT: You are entirely correct. Since the ejection you postulate is precisely centered on our own galaxy, no Copernican arguments can be used in this case.

COSMOLOGICAL EVOLUTION OF QSOs AND RADIO GALAXIES FROM RADIO - SELECTED SAMPLES

J.V. Wall
Royal Greenwich Observatory, Herstmonceux Castle, Hailsham, Sussex, U.K.

C.R. Benn
Mullard Radio Astronomy Observatory, Cavendish Laboratory, Madingley Road, Cambridge, U.K.

We describe recent advances in observation and analysis which lead to improved understanding of the spatial distribution of QSOs and radio galaxies.

1. REVIEW

A severe bias exists in the distribution of powerful radio sources; on a cosmological interpretation of redshifts, they favour the more distant regions of their observable volumes, their space density at earlier epochs of the Universe thus exceeding that at the present day by factors in excess of 1000. The physical basis for this evolution remains elusive, although hints emerge as observations continue to establish the details.

The more recent investigations (e.g. Robertson 1980; Machalski 1981; Wall, Pearson & Longair (WPL) 1980) have used improved statistical techniques and much improved data bases - larger samples of sources with near-complete identification and redshift information, and accurate source counts defined over large flux-density ranges (Fig. 1). The analyses generally assume that extragalactic radio sources belong to one of two sub-populations, the steep-spectrum sources of extended (usually double-lobed) structure identified predominantly with galaxies, and the flat-spectrum sources of compact (VLBI-scale) structure identified predominantly with QSOs. The steep-spectrum sources constitute most sources catalogued at frequencies of 408 MHz and below, and for these, the analyses confirm the degree of evolution and its 'differential' or luminosity-dependent nature found by Longair (1966). The most luminous show drastic evolution, the weaker ones little or none. The flat-spectrum sources occur in similar numbers to steep-spectrum sources in surveys at frequencies of 2.7 GHz and above, and for these, recent analyses suggest mild evolution, with increases in space densities at $z = 1$ of ~ 100 rather than ~ 1000 as found for the powerful steep-spectrum sources (Machalski 1981; WPL 1981).

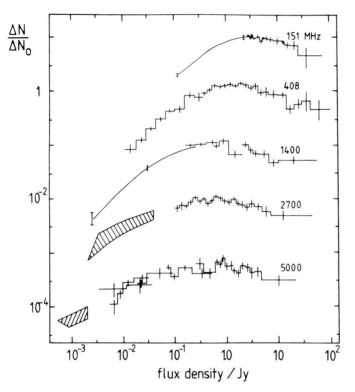

Fig. 1 Source counts in relative differential form, from surveys and P(D) analyses (hatched areas) at 151 to 5000 MHz. References appear in Wall (1980), and further discussion in §2.3 below.

Perhaps the most important result of these analyses was to define new observations crucial to furthering our understanding of the evolution for each population. Three such programmes are described in the next section; the final section discusses an improved technique of evolution-model synthesis which is suitable to incorporate the new data.

2. NEW OBSERVATIONS

2.1 The 5C12 Survey

WPL (1980) showed that powerful constraints on permissible models of evolution for steep-spectrum sources could be provided by a faint steep-spectrum sample for which optical identifications and redshifts are obtained. The conclusion has led to 5C12.

The 5C (Fifth Cambridge) surveys are carried out with the 3-element One-Mile Telescope, 64 days of observing yielding concentric maps at 408 and 1407 MHz. Respectively, these maps have diameters of 4 deg and

1 deg, reach apparent flux densities of 10 mJy and 1 mJy, and contain ~250 sources and ~100 sources. The 5C12 survey is centered at 13^h, $+35°$, close to the North Galactic Pole to minimize optical obscuration and chance coincidences with foreground stars. It coincides with the first Westerbork deep survey at 1415 MHz (Katgert et al. 1973), and with one of the original Sandage-Véron-Braccesi fields searched for radio-quiet QSOs. The catalogue from the survey (Benn et al. 1981) contains 299 sources with $S_{408} > 10$ mJy. Improved phase-calibration and super-cleaning techniques have resulted in positions accurate to 3 arcsec and a noise level of 1.8 mJy, parameters significantly better than for previous 5C surveys.

The complementary optical survey for 5C12 consists of repeated, sky-limited plates, some 5 in each of 3 broad-band colours, taken with the 48-in Schmidt telescope of Hale Observatories. Magnitude calibration was achieved with a Racine wedge ($\Delta m = 5.0$ mag), and by taking deep plates of SA57, plates which overlap the 5C12 area plates by a strip 1.5 deg wide. Most of the 299 sources have been observed with the VLA, and the final identifications will be made using the VLA accurate positions. But a preliminary examination of all fields has yielded the following two results:

(i) A first estimate of 33 percent for total identification rate down to $m_b = 22.5$ mag has been obtained using a new statistical procedure (Benn 1981). This rate is higher than that found by Perryman (1979a,b) for the 5C6 and 5C7 surveys, and the difference may be due to some underestimation of position errors in these surveys. The 5C12 identification rate is in good agreement with predictions from evolution models in which 'transition power' between the evolving and non-evolving sources is not a function of epoch (e.g. model 4, WPL 1980). Models which ascribe a significant amount of evolution to sources of lower powers (e.g. model 5, WPL 1980) predict many more identifications at the 10-mJy level than appear in 5C12, and such models are therefore rejected.

(ii) An initial investigation of the colours of 5C12 radio galaxies is shown in Fig. 2. The fainter galaxies have magnitudes

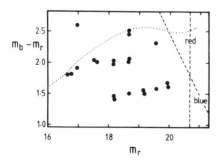

Fig. 2 The colours of 21 5C12 radio galaxies; $m_r = 20$ corresponds to $z \simeq 0.4$. Dashed lines represent red and blue plate limits, while the dotted line is the variation in colour with magnitude predicted from a redshifted E-galaxy spectrum.

corresponding to $z \simeq 0.4$, and these show some blue excess. But the effect does not exceed $\Delta(m_b - m_r) = 0.7$ mag, in disagreement with the suggestion by Katgert et al. (1979) that at $z \simeq 0.3$ radio galaxies become drastically bluer by some 2 magnitudes in $(m_b - m_r)$. Problems with magnitude calibration near their plate limit may be responsible.

2.2 A Large-area Sample at 2.7 GHz

The technique advocated by WPL (1980) to derive cosmological information from radio-source statistics demands a well-defined luminosity distribution for each class of object. In a preliminary analysis of 2.7-GHz data to determine evolution for flat-spectrum sources (WPL 1981), the luminosity distribution used contained only 20 objects. To provide a more satisfactory sample, Peacock & Wall (1981) compiled a Northern Hemisphere catalogue of the brightest sources at 2.7 GHz. The sample is complete to limits $S_{2.7} \geqslant 1.5$ Jy, $\delta \geqslant +10°$, and $|b| \geqslant 10°$; it is intended as a high-frequency counterpart to the 3CR '166 sample' (Jenkins, Pooley & Riley 1977). The 2.7-GHz sample contains 168 sources of which 161 (96 per cent) are identified, and 108 (64 per cent) have measured redshifts. All sources in the sample have been mapped with the Cambridge 5-km telescope at 2.7 or 5.0 GHz.

There are 51 flat-spectrum sources in the sample, of which 49 are optically identified and 41 have measured redshifts. As a result, an improved luminosity distribution (Fig. 5 of Peacock & Wall 1981) is now available. Other points of interest have emerged from the study:

(i) The sample contains 33 sources with relatively steep spectra and yet with compact morphologies in the sense that they were not resolved by the 2 arcsec beam of the 5-km telescope. The sources show some spectral curvature in the sense of flattening towards lower frequencies, enough to keep them all out of the 3CR sample. Their nature is unknown. Are they powerful, very distant doubles, extreme Cyg A's, in which self-absorption in the hot-spots bends the spectra? Or are they 'semi-compact', either single component or very small double-component sources, some kind of missing link between extended and compact sources? High-resolution observations are necessary.

(ii) There are 39 flat-spectrum QSOs in the sample, and for these, Peacock et al. (1981) found $\langle V/V_{max} \rangle = 0.68 \pm .04$. This is higher than the early values found for such objects (Schmidt 1976; Masson & Wall 1977); contrary to these results and to the results from preliminary source-count analysis (WPL 1981), it suggests that the flat-spectrum QSOs do partake in the strong cosmological evolution exhibited by the most powerful steep-spectrum sources. We return to the point in §3.

2.3 Construction and Dissection of the 5-GHz Source Count; 5C12 Again

Surveys at 5 GHz give a still greater yield of flat-spectrum sources. To extract information on their spatial distribution, the WPL (1980, 1981) analyses indicate that prime reqirements are (i) a luminosity distribution from a large-area sample, the 5-GHz analogue of the 2.7-GHz sample of §2.2, (ii) definition of the total source count down to mJy levels, and (iii) samples at different intensity levels with complete radio spectral data, so that sub-counts for both steep- and flat-spectrum populations can be constructed.

With regard to (i), Kühr (1980) has compiled a large-area sample at 5 GHz for which identifications approach completeness. Requirement (ii) has now been met via recent deep surveys and P(D) analyses at 5 GHz carried out on the NRAO 91-m and the MPIfR 100-m telescopes. The deep surveys (Ledden et al. 1980; Pauliny-Toth, Steppe & Witzel 1980; Maslowski et al. 1981) provide direct counts down to 15 mJy, while the P(D) analyses (Ledden et al. 1980; Maslowski et al. 1981; Wall et al. 1981) yield source surface densities near 1 mJy. The new results are consistent, and demonstrate that below 0.1 Jy, the source surface density decreases monotonically from the Euclidean law (Fig. 1; hatched area, 5 GHz, from Wall et al. 1981). In particular, they confirm that the estimate of 5-GHz surface density by Wall (1978) is too high for reasons discussed by Wall et al. (1981).

Several of the new surveys contribute to (iii), the construction of sub-counts, or the question of population mix as a function of 5-GHz flux density. A further deep survey at 5 GHz with the MPIfR 100-m telescope (Wall, Benn, G. Grueff & M. Vigotti, in preparation) covers the 16 deg^2 area of 5C12 and is complete to 18 mJy. It contains a wealth of information relevant to requirement (iii), because the 408-MHz survey (5C12 itself) provides radio spectral data over a long frequency baseline, while the 5C12 deep Schmidt plates permit faint optical identifications to be made. Fig. 3b depicts the two-point spectral-index distribution from the 5-GHz survey of 5C12. It shows that as flux density at 5 GHz is decreased, an astonishing change in the population content takes place. The spectral-index distribution for bright sources

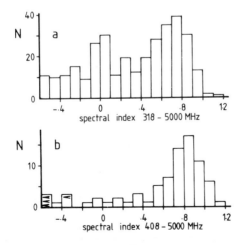

Fig. 3 Spectral-index ($S \propto \nu^{-\alpha}$) distributions for samples selected at 5 GHz: (a) $S_5 > 0.6$ Jy (Condon & Jauncey 1974), and (b) $S_5 > 18$ mJy (5C12 area).

(Fig. 3a) shows the well-known double peak, with similar contributions from steep- and flat-spectrum sources. At low flux densities, only a single peak remains; the flat-spectrum sources have almost disappeared, the remnants of this population spilling out in a tail to negative spectral indices. In parallel, the proportion of QSO identifications has fallen drastically; of the 99 sources in the 5C12 5-GHz sample, a preliminary search (Benn 1981) finds 25 with galaxies as optical counterparts, and only 6 with QSOs. Thus the count for flat-spectrum sources - which are QSOs for the most part - drops off below 0.1 Jy at a rate very much faster than that for steep-spectrum sources. Indeed, spectral-index distributions for 5-GHz samples at intermediate flux-density levels confirm the trend and provide some further detail of the form of the flat-spectrum source count (Kellermann 1980).

A further point of interest is the shift to higher indices of the steep-spectrum peak in Fig. 3b. Via the P - α correlation, this suggests that the steep-spectrum sources dominating the 5-GHz count at faint intensity levels are of relatively high luminosity.

We thus have the following picture for the 5-GHz source count. Of the two sub-counts comprising it, that of the flat-spectrum sources has its plateau centered on $\sim$1 Jy and drops rapidly from Euclidean values below 0.1 Jy, while that of the steep-spectrum sources has its plateau centered on $\sim$0.2 Jy and its decrease with respect to Euclidean is mapped essentially by that of the total count, Fig. 1. The increasing width of plateau of the total counts, so apparent (Fig. 1) as survey frequency is raised, is thus due to the increasing prominence of the flat-spectrum count, whose plateau centre is displaced to higher flux densities than that for the count of steep-spectrum sources.

3. MULTI-FREQUENCY / DUAL-POPULATION MODELS OF EVOLUTION

How best to synthesize these new data into descriptions of the spatial distribution of radio sources? The analyses mentioned in §1 are all flawed in some sense. The WPL (1980) technique, for instance, requires guessed analytical functions to describe the evolution of steep-spectrum sources. Robertson's (1980) generalization of the method is 'free-form' only in a restricted sense, in that it assumes a particular type of transition between evolving (high-luminosity) and non-evolving components of the steep-spectrum population. None of the analyses explores the question of differential evolution for the flat-spectrum sources.

A new procedure (Peacock & Gull (PG) 1981) considers both steep- and flat-spectrum populations simultaneously, and represents the closest possible approach to truly free-form analysis. PG write the luminosity functions for both populations as power series expansions over the (radio-power - redshift) or (P - z) plane. The coefficients of expansion are then determined by optimization using an algorithm in which the χ^2 statistic tests goodness-of-fit between model prediction and data. As input data, PG used (i) published observations of bright galaxies to establish the local radio-luminosity function at low luminosities, (ii) source counts at 408, 1400, 2700, and 5000 MHz, with sub-counts constructed at the latter two frequencies from radio spectral data to subdivide steep- and flat-spectrum populations, (iii) luminosity or redshift distributions from complete samples at several frequencies and flux-density levels, (iv) V/V_{max} results from complete samples, again at several different frequencies and flux-density levels, and (v) a P - α correlation to relate the steep-spectrum populations at low and high frequencies.

The initial achievements are impressive. The first of these is to delineate quantitatively the areas of the (P - z) planes in which our knowledge of space density is deficient. These uncertainty maps for steep- and flat-spectrum populations can be used to reflect the impact of each succesive new data set, and are therefore invaluable in defining the next generation of observational programmes.

The second achievement is to demonstrate conclusively that strong cosmic evolution is a feature of the flat-spectrum source population as well as the steep. As noted in §2.2, the result is in apparent disagreement with earlier indications from source counts at cm-wavelengths and V/V_{max} data on flat-spectrum QSOs. But a strength of the PG scheme is that it examines compatibility of all data sets via the χ^2 statistic; and in this instance PG show that all observations are reconcilable on a picture of differential evolution for the flat-spectrum sources - evolution identical in style to that of the steep-spectrum sources in which only the most powerful have had dramatic enhancement of their space densities at the earlier epochs.

Fig. 4 indicates the qualitative similarity in evolution of the luminosity functions for one of PG's models. This similarity suggests that factors peculiar to one population or the other, e.g. confinement

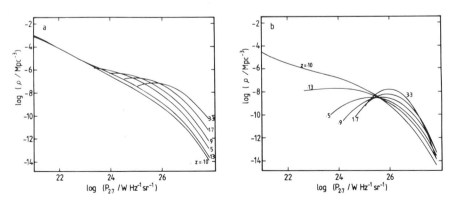

Fig. 4 Luminosity functions and their epoch dependence as given by one of the free-form evolution models of Peacock & Gull (1981), (a) for steep-spectrum sources, and (b) for flat-spectrum sources.

of extended (steep-spectrum) sources by the IGM, or preferential 'Compton snuffing' of the weaker such sources by the MWBG, cannot play a major role in determining the spatial distribution. Something common to both steep- and flat-spectrum objects such as fuel supply must dominate. On the other hand, Fig. 4 suggests that there are <u>quantitative</u> differences between the evolutions. The question is complex, requiring consideration of the uncertainty maps for both populations together with K-corrections and the frequency at which the evolution enhancements are specified.

Should such considerations and/or further analyses show that the evolutions <u>are</u> very similar, it might suggest that the steep- and flat-spectrum sources comprise a single population. Moreover, optically-selected QSOs are now known to exhibit differential evolution (M. Schmidt, this volume) although again quantitative differences may exist in that the rate of evolution appears to exceed that of the powerful radio sources. Possibly gravitational lensing (Turner 1980) has boosted the <u>apparent</u> evolution over that of the luminous radio sources, on which it has little effect (J.A. Peacock, this volume). Similar spatial distributions for all these populations would point to a common origin, and would lend support to scenarios in which they are related perhaps by orientation and a range in jet speed from non-relativistic to extreme-relativistic (Scheuer & Readhead 1979; I.W.A. Browne, this volume).

Speculation aside, we conclude that the sequence (evolution-model synthesis) → (improved data) → (improved synthesis) yields systematic advances in our understanding of the gross properties of radio-source populations.

REFERENCES

Benn, C.R.: 1981, PhD Thesis, University of Cambridge.
Benn, C.R., Grueff, G., Vigotti, M. & Wall, J.V.: 1981, Mon. Not. R. astr. Soc., in press.
Condon, J.J. & Jauncey, D.L.: 1974, Astron. J., 79, 1220.
Jenkins, C.J., Pooley, G.G. & Riley, J.M.: 1977, Mem. R. astr. Soc., 84, 61.
Katgert, P., Katgert-Merkelijn, J.K., Le Poole, R.S. & van der Laan, H.: Astron. Astrophys., 23, 171.
Katgert, P., de Ruiter, H.R. & van der Laan, H.: 1979, Nature, 280, 20.
Kellermann, K.I.: 1980, Phys. Scripta, 21, 664.
Kuhr, H.: 1980, PhD Thesis, University of Bonn.
Ledden, J.E., Broderick, J.J., Condon, J.J. & Brown, R.L.: 1980, Astron. J., 85, 780.
Longair, M.S.: 1966, Mon. Not. R. astr. Soc., 133, 421.
Machalski, J.: 1981, Astron. Astrophys. Suppl., 43, 91.
Maslowski, J., Pauliny-Toth, I.I.K., Witzel, A. & Kuhr, H.: 1981, Astron. Astrophys., 95, 285.
Masson. C.R. & Wall, J.V.: 1977, Mon. Not. R. astr. Soc., 180, 193.

Pauliny-Toth, I.I.K., Steppe, H. & Witzel, A.: 1980, Astron. Astrophys., 85, 329.
Peacock, J.A. & Wall, J.V.: 1981, Mon. Not. R. astr. Soc., 194, 331.
Peacock, J.A. & Gull, S.F.: 1981, Mon. Not. R. astr. Soc., 196, 611.
Peacock, J.A., Perryman, M.A.C., Longair, M.S., Gunn, J.E. & Westphal, J.A.: 1981, Mon. Not. R. astr. Soc., 194, 601.
Perryman, M.A.C.: 1979a, Mon. Not. R. astr. Soc., 187, 223.
Perryman, M.A.C.: 1979b, Mon. Not. R. astr. Soc., 187, 683.
Robertson, J.G.: 1980, Mon. Not. R. astr. Soc., 190, 143.
Scheuer, P.A.G. & Readhead, A.C.S.: 1979, Nature, 277, 182.
Schmidt, M.: 1976, Astrophys. J., 209, L55.
Turner, E.L.: 1980, Astrophys. J., 242, L135.
Wall, J.V.: 1978, Mon. Not. R. astr. Soc., 182, 381.
Wall, J.V.: 1980, Phil. Trans. R. Soc., A296, 367.
Wall, J.V., Pearson, T.J. & Longair, M.S.: 1980, Mon. Not. R. astr. Soc., 193, 683.
Wall, J.V., Pearson, T.J. & Longair, M.S.: 1981, Mon. Not. R. astr. Soc., 196, 597.
Wall, J.V., Scheuer, P.A.G., Pauliny-Toth, I.I.K. & Witzel, A.: 1981, Mon. Not. R. astr. Soc., in press.

GRAVITATIONAL LENSES AND COSMOLOGICAL EVOLUTION

J.A. Peacock
Royal Observatory, Edinburgh

1. LENSING AND SELECTION EFFECTS

Empirical descriptions of cosmological evolution rely on an assumed relation between luminosity distance and redshift in order to derive the luminosity function and its epoch dependence. This relation is usually taken to be that of an idealised Friedmann cosmology, despite the obvious abundance of small-scale structure in the Universe. However, the assumption that the probability of a lensing event is negligible is now made less tenable with the discovery of the double and triple QSO's, both of which appear to have been magnified by factors of 10-15. With at least two events of this amplification in 1500 known quasars, it may be that a large fraction of observed QSO's have been magnified by significant amounts. Turner (1980) suggests that all evolutionary statistics (e.g. V/Vmax) may thus be misleading, although he considers only an illustrative model of the effect. The problem is that, even if the intrinsic probability of a lensing event is very low, the effect becomes important if the background density of faint sources rises sufficiently quickly. We must therefore calculate not only the intrinsic probability of a given amplification, but incorporate this with the luminosity function to see if lensing could be made dominant by selection effects.

2. THE LENS POPULATION

The first step is to identify the likely lenses: these are simply galaxies or groups/clusters of galaxies; in each case, we know the form of the optical luminosity function (Felten 1977; Bahcall 1979). We now model the mass distribution as that of a singular isothermal sphere, in which case the lensing effects depend only on the velocity dispersion, V. Using the relation $L \propto V^4$ (Faber & Jackson 1976), we can derive the space density of lenses of a given strength.

The probability distribution of amplification, f(A), is derived as follows: for a given A, we can calculate N, the expected number of interactions leading to an amplification greater than A. If $N \ll 1$, then differentiation yields f(A) directly. From this, and the condition

$\langle A \rangle = 1$ (flux conservation implies no amplification on average) we can construct realistic forms for $f(A)$. The most important features of these distributions are:
  i) For $A \gtrsim 2$ we have $f(A) = a/A^3$, where the constant a depends on redshift.
  ii) $f(A)$ has a cut-off at high A which depends both on redshift and on the physical size of the object being lensed.

Representative values for a 10-pc flat-spectrum radio source at $z=1$ are $a=0.03$, $A_{max}=2500$. A_{max} scales linearly with physical size, so lensing effects cannot be important for extended radio sources on scales $\gtrsim$ kpc.

3. RESULTS

To find the observational effects of lensing we should take the luminosity function inferred in the standard manner and calculate the lensing correction (assuming this to be small). However, the luminosity function is not always uniquely defined (see Peacock & Gull 1981) and we shall be content for now with some simple illustrative calculations. If we take $z=1$ to be a representative redshift for QSO's and flat-spectrum sources (the constant a varies slowly with z in this range), then the importance of lensing depends on the slope of the source counts. For differential counts $dN \propto S^{-\beta}$, the distribution of amplifications at a given observed flux denisty is $f(A).A^{\beta-1}$; the probability of significant amplification can thus become very large if $\beta \gtrsim 3$. In practice, both radio and optical counts flatten at low flux densities, tending to slopes of $\beta = 1.5$ and 2.2 respectively; this implies that the probabilities of $A>2$ in the brightest observed flat-spectrum sources and optically selected quasars are about a and $3a$ respectively.

Lensing thus has a small effect on statistical conclusions based on samples of bright radio sources. For optically selected quasars, the effects are likely to be more important, since much smaller objects than galaxies may act as lenses. At $z=1$, a quasar continuum source of size $\sim 10^{-3}$ pc can be lensed by masses as low as $\sim 10^{-4} M_\odot$ whereas for radio emission of scale ~ 10 pc, the critical mass is $\sim 10^5 M_\odot$. In principle, therefore, optically selected quasars could be magnified by large factors; this would help explain the high ratio of radio-quiet to radio-loud quasars.

We conclude that, while the effects of lensing do not invalidate cosmological conclusions based on radio samples, the phenomenon is not so rare as had been supposed: Arp's excess quasars near to bright galaxies and the extreme superluminal motions in 3C279 are two cases where lensing may well be dominant.

REFERENCES

Bahcall, N., 1979. Astrophys. J., 232, 689.
Felten, J.E., 1977. Astron. J., 82, 861.
Peacock, J.A. & Gull, S.F., 1981. Mon. Not. R. astr. Soc., 196, 611.
Turner, E.L., 1980. Astrophys. J., 242, L135.

THE INTERGALACTIC MEDIUM

A.C. Fabian & A.K. Kembhavi
Institute of Astronomy
Madingley Road
Cambridge CB3 OHA, England

ABSTRACT

The density of intergalactic gas may be an important parameter in the formation of extended radio sources. It may range from ~ 0.1 particle cm^{-3} in the centres of some rich clusters of galaxies down to 10^{-8} cm^{-3} or less in intercluster space. The possible influence of the intracluster gas surrounding NGC 1275 on its radio emission is discussed, and the possibility that a significant fraction of the X-ray background is due to a hot intergalactic medium is explored in some detail.

1. INTRODUCTION

The density of intergalactic matter has only been measured in the cores of clusters of galaxies. This gas emits X-radiation primarily by thermal bremsstrahlung and its density often exceeds 10^{-3} particles cm^{-3} in rich clusters. Little is known about the gas density at the edges of clusters (i.e. several Mpc from the centre), near relatively isolated galaxies or in intercluster space. All galaxies produce gas from stellar mass loss, and many are assumed to lose gas in the form of a wind. This, together with an expectation that galaxy formation is not 100 percent efficient, suggests that an intergalactic medium pervades most of space, with a density that probably depends fairly strongly on the local galaxy environment.

Circum- and possibly inter-galactic gas is an important ingredient in most theories of extended radio sources, both as a means of confining the radio-emitting plasma and in providing a 'working surface' for stimulating particle acceleration and emission (see e.g. De Young 1977). Unfortunately, these theories are not yet developed to the point where unambiguous particle densities may be inferred from properties of the radio emission, nor are the observed properties clearly related to the local environment (see Stocke 1979 and references therein). We shall concentrate here on those regions where we can estimate the particle densities with some accuracy, i.e. the clusters of galaxies, and on the possibility that a diffuse hot intercluster medium is detected as the X-ray background. The likely evolution of these examples of intergalactic gas may relate to the evolution of radio sources.

2. LOCALISED INTERGALACTIC GAS

a) The Intracluster medium

The X-ray emission from intracluster gas has been reviewed here by Jones (1982). The iron line emission features in the X-ray spectra of rich clusters suggests that much of this gas has been expelled or stripped from the member galaxies. This gas has a well observed effect on the radio sources associated with moving galaxies, producing headtail sources such as discussed here by Harris (1982).

The gas density in some of the rich clusters such as A1656 in Coma does not exceed $\sim 5 \times 10^{-3} cm^{-3}$ and so radiative cooling is not of importance on a Hubble time. In a number of other clusters, however, the density is much greater. Cooling flows (Fabian & Nulsen 1977, Cowie & Binney 1977) occur in the cores of these clusters, with densities surrounding the central galaxy approaching ~ 0.1 cm^{-3}. The region around NGC 1275 (3C84, Per A) in the Perseus cluster is a good example, in which ~ 300 M$_\odot$ yr^{-1} is being accreted by the pressure of the outer hot material (Fabian et al. 1981). The pressure of the surrounding gas is ~ 100 times the pressure in the local interstellar medium in our Galaxy. It seems possible that this could frustrate a nascent extended radio source. The inertia of the high-density cooling flow would be a severe impediment to a supersonic outflow. Nevertheless, extended radio emission is observed around NGC 1275 on most scales out to ~ 3 arcmin (Ryle & Windram 1968, Miley & Perola 1975, Gisler & Miley 1979, Reich et al. 1980, Noordam & de Bruyn 1982). The largescale radio blob, lying $\sim 2\frac{1}{2}$ arcmin SW of the nucleus, may have been produced by some jet from NGC 1275, or it may be some relic of the slow motion of the galaxy in the cluster core. Studies of this source should provide information on extended radio emission in high-pressure environment. Stewart et al. (1981) find traces of weak X-ray emission leading from the nucleus to this blob. An inverse-Compton interpretation, similar to that used for the extended M87 structures found by Schreier (1982), suggests a weak magnetic field (B $\sim 10^{-7}$G; $B^2/8\pi <<$ thermal pressure) in that region. The cosmic ray electrons may be in pressure equilibrium. Much more precise measurements of the radio (and preferably X-ray) spectral index are necessary before any firm conclusion may be reached.

It seems possible that much of the amorphous radio structure observed on a scale of ~ 30 arcsec could be directly associated with the accretion flow. The thermal instability causes rapid changes of density (and shocks) to occur in the cooling intracluster gas, much of which was presumably interstellar gas at some stage. It may thus contain magnetic fields and cosmic ray electrons from this earlier phase, which will then be amplified in the flow. It is not therefore surprising that an unstructured radio source is found in the centre of such a flow, although the energetics are admittedly uncertain. Similar amorphous sources should occur in most such cooling flows.

b) The Edges of Clusters

The X-ray emission from clusters is only well studied by current imaging techniques out to 1 or 2 Mpc from the centre. Beyond the core we can expect that conditions are mostly determined from the early phases of the formation of the cluster and that cooling is totally unimportant. The accretion of sub-clusters and the dynamical evolution of the cluster galaxies may, however, lead to some resettlement of the gas. It is possible that radio sources heat the gas significantly (cf. Lea & Holman 1978), but if this is a common occurrence we might wonder why any gas at all has remained. For the present we can only make crude estimates of the density falloff at large distances from the cluster centre. At radii approaching and exceeding 10 Mpc, the assumption of hydrostatic pressure balance probably breaks down as this is the distance that sound waves (in a gas at $\sim 10^8$K) can even out any imbalance. A slow net inflow, or outflow, of intracluster gas may be established at intermediate radii.

c) Isolated Galaxies

Little is known of the intergalactic environment surrounding relatively isolated galaxies. As already mentioned, galactic winds may transport some of the interstellar medium of such galaxies up to several Mpc in a vacuum. However, even a very-low density uniform intercluster gas confines such a wind to within a few 100 kpc. Perhaps the best evidence for matter at large (~ 100 kpc) distance around isolated galaxies is to be found in the shells discovered by Malin & Carter (1980). (For an interpretation in terms of galactic winds see Fabian *et al.* 1980).

3. A DIFFUSE INTERCLUSTER GAS

Observations of enormous radio sources such as DA 240, NGC 6251 and 3C236 (e.g. Strom *et al.* 1980) provide some support for a tenuous intercluster medium. The most tantalising evidence is the X-ray background. The spectral shape of the $\sim 3 - 40$ keV background was noted to be consistent with thermal bremsstrahlung from a diffuse hot ($kT \sim 40$ keV) gas by Cowsik & Kobetich and by Field in 1972. This provides either a measurement or a limit on a hot intercluster gas. Constraints on any lower temperature ($T < 10^7$K) medium rely on ultraviolet (see Parasce *et al.* 1980) or optical (see Young *et al.* 1980) measurements and are not discussed here. (Nor are implications for the survival of cold clouds, see Cowie & McKee 1975). It is possible that the X-ray background provides us with a direct measurement of the intercluster medium.

The 3 - 400 keV spectrum of the X-ray background (Fig. 1) has been accurately measured by the HEAO-1 satellite (3 - 50 keV, Marshall *et al.* 1980; ~ 100 - 400 keV, Matteson *et al.* 1979) and by balloon (~ 25 - 150 keV, Kinzer *et al.* 1978). (A composite spectrum from 1 keV to ~ 100 MeV due to R. Kinzer is shown in Fabian 1981). A bump observed in the spectrum at ~ 1 MeV is plausibly attributed to unevolved active galaxies. The composite spectrum deduced from a luminosity function compiled from observed Seyfert galaxies (Piccinotti *et al.* 1981, and mean photon index of 1.7) contributes ~ 20 percent at 3 keV and extrapolates to give all of the observed background at ~ 1 MeV (see Boldt 1981 and references therein). We show the resultant 3 - 300 keV background spectrum after subtraction of the unevolved Seyfert component in Fig. 1.

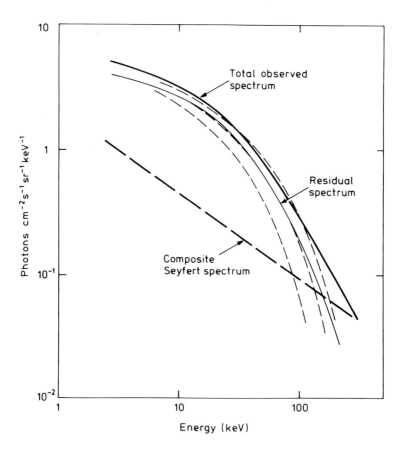

Figure 1. The total X-ray background spectrum is shown as the heavy solid line. The residual spectrum after subtraction of the composite Seyfert spectrum (heavy dashed line) is represented by the fine continuous line. Predicted spectra from intercluster gas adiabatically cooling from a redshift, z_m, of 5 to current temperatures of 6, 8 and 10 keV are shown as fine dashed lines.

The contribution of quasars to the X-ray background remains uncertain (see Tananbaum 1981). We find that at least 15 percent of the extrapolated background at 2 keV must be due to quasars (Kembhavi & Fabian 1981). It remains possible that most of the remainder could result from very faint quasars, although it should be immediately noted that most of the X-ray measurements of quasars are made at $\sim$ 2 keV, at which energy there may be an excess (Garmire & Nousik 1981) over the extrapolated background (i.e. from higher energies), and that little is known of the X-ray spectrum of most quasars. Their contribution to the 3 - 300 keV background will be ignored in the rest of this discussion. A large ($\gtrsim$ 20 percent) quasar contribution at much above 3 keV clearly changes the conclusions. The Seyfert-subtracted X-ray background spectrum (Fig. 1) is not compatible with that from a single-temperature bremsstrahlung but is consistent with a range of models involving a hot gas expanding and cooling with the Universe. Such models were first considered by Field & Perrenod (1977), who compared the total observed spectrum with that from electron-ion bremsstrahlung from a gas cooling from $z_m \simeq 3$. The particle density, n, falls as $(1+z)^3$, and thus the temperature, T, as $(1+z)^2$.

We have considered here a range of maximum redshift, z_m, and owing to the mildly relativistic nature of the electrons, electron-electron bremsstrahlung and further e-i corrections due to Gould (1980) are included. (These formulae may overestimate the e-e contribution for kT > 100 keV.) Spectra were computed for a range of initial temperatures, and the appropriate one selected by visual comparison with the residual (total-Seyfert) spectrum. This was considered reasonable considering the observational errors and possible normalization problems with three data sets. A typical fit is shown in Fig. 1. We find that the present (z=0) density, n_o, temperature, T_o, and pressure, P_o, scale as

$$n_o \simeq 2.10^{-6} (1+z_m)^{-1.5} (1+z)^3 \text{ cm}^{-3}$$

$$T_o \simeq 6.10^8 (1+z_m)^{-1} (1+z)^2 \text{ K}$$

and $P_o \simeq 3.2.10^{-13} (1+z_m)^{-2.5} (1+z)^5$ dyne cm^{-2}.

H_o was taken as 50 km s^{-1}Mpc^{-1}, and $q_o = \frac{1}{2}$. Generally $H_o^3 \Omega_{gas}^2$ = constant. We see that if $z_m \simeq 5$, the present intercluster density would be $\sim 10^{-7}$cm^{-3}($\Omega_{gas} \sim 0.04$) and $kT_o \sim 8$ keV.

The major problems with this explanation for the X-ray background are its energetic extravagance and inefficiency. (This may be a philosophical problem, for it also applies to many large radio sources.) Radiative cooling could never have tapped more than a few percent of the thermal energy of the gas. The total energy requirement amounts to $\sim 10^{63} (1+z_m)^{-\frac{1}{2}} N_{-2}^{-\frac{1}{2}}$ erg/galaxy, where the number density of relevant galaxies = 10^{-2} N$_{-2}$ galaxies Mpc^{-3}. (N$_{-2}$ = 1 corresponds to luminous spirals). This may be compared with the energy required to make the observed metals (see Bookbinder et al. 1980), the energy content of some radio galaxies and of massive spinning black holes, and with the output of quasars. How the gas might have been heated is a mystery (for suggestions see Field & Perrenod 1977, Sherman 1979, Bookbinder et al. 1980).

We note that the electron-ion coupling time generally exceeds the age of the hot gas and that then the assumption of an underlying Maxwellian distribution may fail. On the other hand, Compton cooling by the microwave background becomes a serious problem if $z_m \gg 5$. The microwave background spectrum might also become noticeably distorted (see Wright 1979). Clumping of the gas to increase the radiative efficiency and ease the energy problem cannot work if z_m is small (McKee 1980, Fabian 1981) because of the strong observed isotropy limits on any origin of the X-ray background. It may yet be possible to have small-scale clumping (on an observed scale $< 5°$) if $z_m \gtrsim 3$ or so. (This requires $\gtrsim 100$ clumps per square degree.)

4. THE INFLUENCE OF THE INTERGALACTIC MEDIUM ON RADIO SOURCES

This section is hampered by a lack of any clear understanding of the evolution of radio sources. Intracluster gas has observable consequences on radio sources in clusters. Galaxy motions (and possibly buoyancy and cluster winds) can shape extended radio emission which is also confined by thermal and ram pressure. Amorphous sources may form from cooling gas in the cores of some clusters. Henry et al. (1979) find little evidence for evolution in the X-ray properties of intracluster gas out to $z \sim 1$. This may be relevant to the evolution of many radio sources. A diffuse intercluster medium should evolve with $P \propto (1+z)^5$. If any radio sources are in equipartition and thermal pressure balance with such a gas, we may expect that $L \propto (1+z)^{15/4} U$, where U is the total energy. Size should scale as $R \propto (1+z)^{-3}$. Ram pressure balance gives $R \propto (1+z)^{-3/5}$. The high pressure in the past could also affect the spectral index of radio sources. However since most radio sources seem to be associated with galaxies, it could be that few such sources ever sample the true intercluster medium.

ACKNOWLEDGEMENTS

ACF thanks the Radcliffe Trust for financial support.

REFERENCES

Boldt, E.A.: 1981, Comments on Astrophys. 9, pp.97.
Bookbinder, J., Cowie, L.L., Krolik, J.H., Ostriker, J.P., and Rees, M.J.: 1980, Astrophys.J. 237, pp.647.
Cowie, L.L., and Binney, J.: 1977, Astrophys.J. 215, pp. 723.
Cowie, L.L., and McKee, C.F.: 1975, Astrophys.J. 195, pp.715.
Cowsik, R., and Kobetich, E.J.: 1972, Astrophys.J. 177, pp.585.

De Young, D.S.: 1977, Ann.N.Y.Acad.Sci. 302,pp.669.
Fabian, A.C.: 1981, Ann.N.Y.Acad.Sci. (in press).
Fabian, A.C., and Nulsen, P.E.J.: 1977, Monthly Notices Roy.Astron.Soc. 18,pp.479.
Fabian, A.C., Cowie, L.L., Hu, E., and Grindlay, J.: 1981, Astrophys. J. 248,pp.47.
Fabian, A.C., Nulsen, P.E.J., and Stewart, G.C.: 1980, Nature, 287,pp.613.
Field, G.B.: 1972, Ann.Rev.Astron.Astrophys. 10,pp.227.
Field, G.B., and Perrenod, S.C.: 1977, Astrophys.J. 215,pp.717.
Garmire, G., and Nousek, J.: 1981, Bull.Am.Astron.Soc. 12,pp.853.
Gisler, G.R., and Miley, G.K.: 1979, Astron.Astrophys. 76,pp.109.
Gould, R.J.: 1980, Astrophys.J. 238,pp.1026.
Harris, D.: 1982, These proceedings, p. 77.
Henry, J.P. *et al.*: 1979, Astrophys.J. (Lett.) 234, L15.
Jones, C.: 1982, These proceedings.
Kembhavi, A.K., and Fabian, A.C.: 1981, Monthly Notices Roy.Astron.Soc. in press.
Kinzer, R.L., Johnson, W.N., and Kurfess, J.D.: 1978, Astrophys.J. 222, pp.370.
Lea, S.M., and Holman, M.D. 1978, Astrophys.J. 222,pp.29.
Marshall, F.E., Boldt, E.A., Holt, S.S., Miller, R.B., Mushotsky, R.F., Rose, L.A., Rothschild, R.E.,and Serlemitsos, P.J.: 1980, Astrophys. J. 234,pp.4.
Malin, D.F., and Carter, D.: 1980, Nature 285,pp.643.
Matteson, J.L., Gruber, D.E., Nolan, P., Peterson, L.E., Kinzer, R.L.: 1979, Bull.Am.Astron.Soc. 11,pp.653.
McKee, C.: 1980, Physica Scripta 21,pp.738.
Miley, G.K., and Perola, G.C.: 1975, Astron.Astrophys. 45,pp.223.
Noordam, W., and deBruyn, G.: 1982, These proceedings.
Paresce, F., McKee, C.F., and Bowyer, S.: 1980, Astrophys.J. 240,pp.387.
Piccinotti, G., Mushotzky, R.F., Boldt, E.A., Holt, S.S., Marshall, F.E., and Serlemitsos, P.J.: 1981 preprint.
Reich, W., Stute, U., and Wielebinski, R.: 1980, Astron.Astrophys. 89,pp.204.
Ryle, M., and Windram, M.D.: 1968, Monthly Notices Roy.Astron.Soc. 138,pp.1.
Schreier, E.: 1982, These proceedings.
Sherman, R.D.: 1979, Astrophys.J. 232,pp.1.
Stewart, G.C., Fabian, A.C., Nulsen, P.E.J., and Phinney, S.: 1981, in preparation.
Stocke, J.: 1979, Astrophys.J. 230,pp.40.
Strom, R.G., and Willis, A.G.: 1980, Astron.Astrophys. 85,pp.36.
Tananbaum, H.: 1982, These proceedings.
Wright, E.L.: 1979, Astrophys.J. 232,pp.348.
Young, P.J. *et al.*: 1980, Astrophys.J.(Suppl.) 42,pp.41.

MODELLING THE GRAVITATIONAL LENS OF THE DOUBLE QUASAR

P.K. Moore & S.M. Harding
Nuffield Radio Astronomy Laboratories,
Jodrell Bank, Macclesfield, Cheshire

The discovery of the double quasar (Walsh et al. 1979) provides an opportunity to study the mass distribution of elliptical galaxies and clusters of galaxies. This has been done initially by Young et al. (1981) who produced a model to account for the image positions and intensities. Since then VLBI observations have been made of 0957+561A and B (Porcas et al. 1981) which show very similar core and jet structures in the nuclei of both images. In addition to providing further evidence in favour of the gravitational lens hypothesis, these new observations provide additional constraints on the mass distribution of the lensing galaxy and cluster. We have attempted to produce a model in the light of these new results.

Due to the low flux density of this source it has not been possible to make VLBI maps of the two images. Instead a process of model fitting was used. The most reliable quantities from this process are the relative positions of the core and jet in each image. Individual flux densities and sizes for the core and jet are prone to uncertainty due to poor u-v coverage. Consequently we only tried to reproduce the observed jet-core separations of

$$\underline{\Delta \theta_A} = (0\overset{''}{.}016, 0\overset{''}{.}043)$$

and $\underline{\Delta \theta_{B1}} = (0\overset{''}{.}016, 0\overset{''}{.}054)$

with our model of the mass distribution.

We have investigated many different models using a method similar to that described by Young et al. (1981). Our most successful ones consist of two elliptical King distributions; one for the dominant galaxy (G1) and one for the rest of the cluster material. The following is the best model we have found so far.

	Galaxy (G1)	Cluster
Position of centre	(0″.0, 0″.0)	(-2″.0, -5″.0)
Structural length	0″.13	13″
Velocity dispersion	280 km s^{-1}	920 km s^{-1}
Axial ratio	0.76	0.4
Position Angle	48°	73°

QSO core position   (-0″.1676, -2″.2865)
QSO jet-core separation   (0″.0089, 0″.0252)

The coordinate system has x along the direction of increasing right ascension and y along the direction of increasing declination, structural lengths are defined as core radius/3, and position angles are defined from North through East. We have used values of H_0 = 60 km s^{-1} Mpc^{-1} and q_0 = 0. This model produces three images with the following positions, amplification factors and jet-core separations:

	Position	Amplification	Jet-core separation
A	(-1″.435, 5″.066)	3.573	(0″.015, 0″.047)
B1	(-0″.205, -1″.007)	-2.384	(0″.014, 0″.049)
B2	(-0″.021, -0″.298)	0.209	(-0″.006, -0″.017)

This results in the following image flux ratios:

B1/A	-0.667
B2/B1	-0.088

This model is in reasonable agreement with observation except perhaps for the amplification factor for image B2 which is rather large. In the absence of a measured value for the flux density of B2 we must ensure that our model does not produce too bright a third image. Our present model requires that all of the flux from component G on the VLA map of Greenfield et al. (1980) be due to the image B2 and none to the galaxy G1. We are currently trying to modify our model to produce a smaller amplification factor for image B2.

It thus seems likely that all of the observations of the double quasar can be explained by a gravitational lens consisting of an elliptical galaxy in a cluster. However, in order to accommodate the VLBI results we have to modify some of the galaxy or cluster parameters originally proposed by Young et al. (1981).

References

Greenfield, P.E., Burke, B.F. & Roberts, D.H., Nature 286, 865 (1980)
Porcas, R.W., Booth, R.S., Browne, I.W.A., Walsh, D. & Wilkinson, P.N., Nature 289, 758 (1981)
Young, P., Gunn, J.E., Kristian, J., Oke, J.B. & Westphal, J.A., Ap.J. 244, 736 (1981).
Walsh, D., Carswell, R.F. & Weymann, R.J., Nature 279, 381 (1979)

SUPERLUMINAL VELOCITIES OF COMPACT RADIO SOURCES: A GRAVITATIONAL LENS EFFECT

Jeno M. Barnothy
Evanston, Illinois

An intervening galaxy acting as a gravitational lens produces usually from a compact radio source two or four enlarged "crescent" shaped virtual images, which have the same surface brightness as the object, but look brighter, due to the lens caused enlargement of the object area in the image. Velocities between elements of a source being vector quantities, will also be seen enlarged, occasionally to superluminal velocities (Fig.1). It is as simple as that. (Barnothy and Barnothy 1971; Barnothy 1976.)

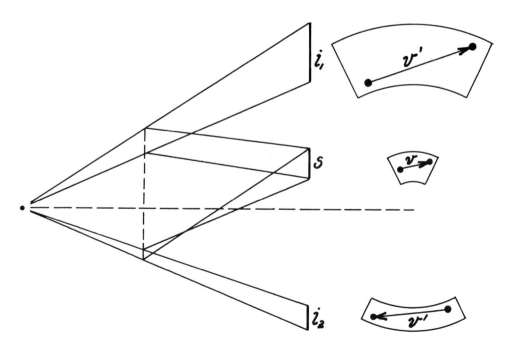

Figure 1. Path of light rays in a distributed mass lens; in a flat space. (Sketch is not to scale.)

The tangential enlargement of the image, the scale factor s_t, is inversely proportional to the distance of the optical axis of the lens from the object; its value may be very large (10^8). The radial enlargement, the scale factor s_r depends from the mass distribution of the lens galaxy and from the impact parameter x, at which the focused rays pass from the gravitational center of the lens. Its value is usually small (<5), albeit in some mass distributions and within a short range of x, its value may be very large. (Barnothy & Barnothy 1973).

The total intensification of the brightness of a source is for one crescent $I = s_t \times s_r$. The probability of a proper alignment of object and lens is inversely proportional to the square of s_t, while the chance to observe the radio source increases as the 6th power of its brightness; hence compact radio sources will be mostly observed when s_t is not too large and s_r not too small; meaning that $s_t \sim s_r$. Since the value of x is different for each crescent (image), it is likely that some of the crescents are so faint that due to the small dynamical range of VLBI's, only one of the images will be visible.

The virtual images will faithfully reproduce the details of the source. Should the object, say, the nucleus of a Seyfert galaxy contain several compact radio sources, aligned in a string like manner in the radial direction, then the virtual image may look like a one-sided, or two-sided jet. It is not a true physical phenomenon, merely a fata morgana.

It is noteworthy that superluminal velocities were so far observed only in quasars. The lens which produces the radio structure and the superluminal phenomenon, will also intensify the optical object, the Seyfert galaxy nucleus. At a lens mass of $10^{11} M_\odot$, and the lens axis passing at a distance of 0.1 kpc from the source, the intensification should be 5 - 6 magnitudes, sufficient to raise the brightness of a Seyfert galaxy nucleus to that of a quasar, as I have proposed this 16 years ago (1965).

REFERENCES

Barnothy,J.M. and Barnothy,M.F., 1971, B.A.A.S. 3,472
Barnothy,J.M. 1976, IAU Colloquium No.37; ed.C.Balkowski and B.E. Westerlund. C.N.R.S., Paris, 361-363
Barnothy,J.M. and Barnothy,M.F. 1973, B.A.A.S. 5, 448
Barnothy,J.M. 1965, A.J. 70, 666

SYMMETRY IN RADIO GALAXIES

R.D. Ekers
N.R.A.O., Socorro, New Mexico, U.S.A.

I. Introduction

One of the most striking properties of the radio galaxies is the predominance of a symmetrical double lobed structure. Any model to explain the energy release and collimation in radio galaxies must be able to produce this symmetry as the normal morphology. Although this large scale symmetry in radio galaxies is a well known phenomenon, I feel it is worth reemphasising since our current attention is more sharply focussed on the small scale and often much less symmetric structures which are seen with the higher resolution radio telescopes (VLB, VLA etc).

Throughout this paper I will argue that these simple morphological considerations can already put many constraints on radio galaxy theories.

II. 1-D Symmetry

(a) Component Intensity and Separation

Fig. 1a shows the distribution of the ratio of the flux densities of the components of double radio sources and fig. 1b shows the distribution of the ratio of the separation of the components from the central galaxy. Given the very wide range in power ($\gtrsim 10^5:1$) and separation ($\gtrsim 10^4:1$) of radio galaxies this symmetry is quite remarkable. The centroids are 1.5 for the intensity ratio and 1.25 for the separation ratio distribution. The degree of symmetry in either the intensity ratio (Mackay 1973) or the separation ratio (Longair and Riley 1979) can be used to put a limit $\lesssim 0.2c$ on the component ejection velocity. With somewhat dubious assumptions about the component evolution this can be further reduced to $\lesssim 0.03c$ (Mackay 1973).

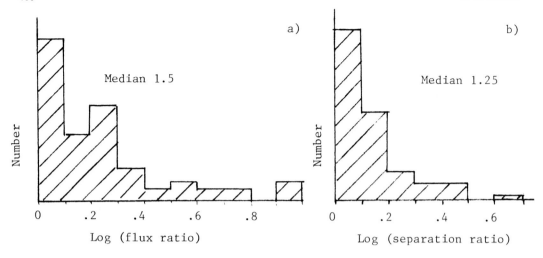

Figure 1. The distribution of the logarithm of the ratio of a) the stronger to the fainter component of 3C radio sources (from Mackay 1971) and b) the furthest to the closest component (from Ingham and Morrison 1975).

When asymmetry does occur it is well correlated between intensity and separation in the sense that the parent galaxy is closer to the centroid than the center (Fomalont 1969). In analysis of a magnitude limited sample of 93 Southern radio galaxies (Shaver et al, this meeting) there were no exceptions to the rule that the fainter component is further from the parent galaxy than the brighter component. The result for this sample is especially significant because the identifications are with bright galaxies coincident with a radio core component and consequently there is no significant selection bias towards centroid identifications. Ryle and Longair (1967) assumed that this asymmetry resulted from the age difference between the front and back components of an expanding double source. This correctly predicts the sense of this effect (the back component is seen at an earlier epoch when it is closer and stronger), however this model requires uncomfortably high values of v/c and it fails for the one radio galaxy (NGC612) for which we can tell front from back (Ekers et al 1978b). A more straightforward explanation of this result which is more consistent with current ideas on radio galaxy formation is to postulate gradients or irregularities in the external medium which cause a lobe encountering a higher density to be both closer and brighter.

(b) Small scale symmetry

The high degree of symmetry seen in the large scale structure is usually reduced as we go to higher resolution (e.g. compare the 3C maps made with the 1 mile telescope and the 5 km telescopes). In the closest radio galaxy, Centaurus A (see papers in this symposium), we find a strange mixture of symmetry and asymmetry. The very large outer

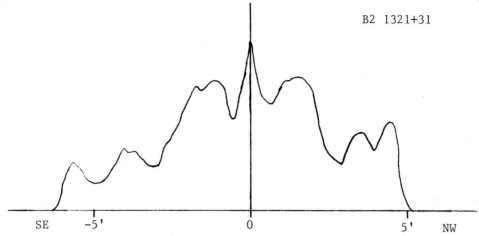

Figure 2. Brightness of the jet in B2 1321+31 as a function of the distance from the nucleus (from Ekers et al 1981).

lobes (1 - .1 Mpc scale) have almost identical integrated flux density (Cooper, Price and Cole 1965), the middle structure (10 kpc scale) is completely one sided, the inner double (1 kpc scale) is very symmetric, while the jet from the nucleus (100 pc scale) is again one sided on the same side!

A general conclusion which can be drawn from this is that the energy output over long time scales (ie integrated in the extended lobes) is divided equally between the two sides. The asymmetry on smaller scales could result from many effects such as short time scale fluctuations in the energy supply (e.g. Rudnick, this symposium), local variations causing the energy beams to become visible in a patchy manner, relativistic effects due to bulk motion of the plasma supplying energy to the lobes, or anisotropic radio emission. Simple observations of morphology can already be used to distinguish between some of these types of explanation. For example, relativistic effects can only explain one sided asymmetry, while very short time scale switchings from side to side are already excluded because most long asymmetric jets are completely one sided.

The jets in radio galaxies show a transition from mainly symmetrical structures in low luminosity sources to one sided structures at higher luminosity (Bridle, this symposium). Some of the symmetric low luminosity jets have an even higher degree of symmetry. Fig. 2 shows the brightness of the jet in B2 1321+31 as a function of distance from the nucleus of the galaxy (Bridle et al in preparation). In addition to the equality of total flux there is a striking one to one correspondence in the brightness changes along the jets. This kind of behaviour will be very hard to model in theories using local instabilities to change the appearance of a jet, but is natural in models using large scale symmetry e.g. a spherically symmetric density

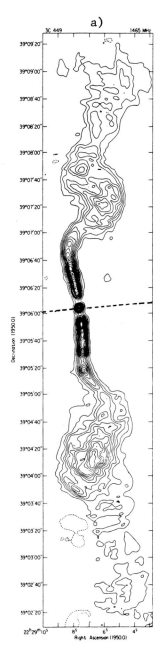
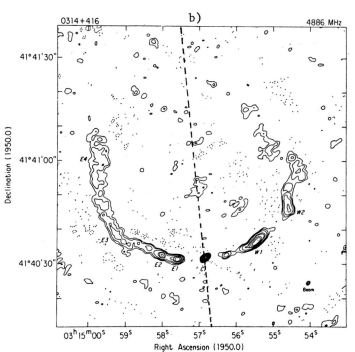

Figure 3. Examples of sources with mirror symmetry. Dashed line indicates line of symmetry. a) VLA observation of NGC1265 at 5GHz (from Owen et al 1978), b) VLA observation of 3C449 at 1.4 GHz (from Perley et al 1979).

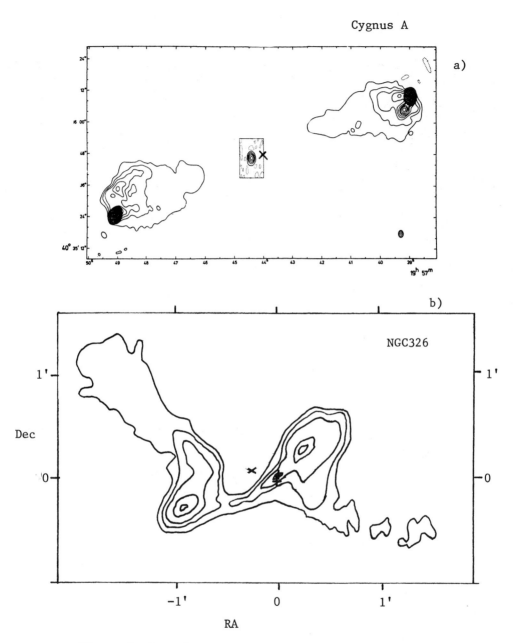

Figure 4. Examples of inversion symmetric sources. The X indicates the center of symmetry. a) Map of Cygnus A at 5 GHz (from Hargrave and Ryle 1974), b) VLA observation of NGC326 at 5GHz and 1.4 GHz (lowest contour). The pair of galaxies are indicated by + (Ekers, Fanti, Fomalont, Lari and Parma in preparation).

distribution in the surrounding medium or a time varying ejection maintaining two sided symmetry over long time scales.

III 2-D Symmetry

(a) Mirror and Inversion Symmetry

Most of the radio galaxies mapped in two dimensions have a well aligned linear structure but some, especially those of lower luminosity, have spectacular two dimensional symmetry. The two dimensional symmetries are of two types. The examples NGC1265 and 3C449 (fig. 3) have a strong symmetry when reflected about a line. This is reflection or mirror symmetry (also called C type). Cygnus A and NGC326 (fig. 4) have strong symmetry when reflected through a point. This is inversion or rotational symmetry (also called S or Z type).

If you have any doubts about the reality of these symmetries try comparing the figure with a transparent copy - turned over for the mirror symmetrical source and rotated by 180° for the inversion symmetric source. In the mirror symmetric sources the symmetry line usually changes in angle a little between the inner and outer structure. It should also be noted that the point of inversion symmetry is often significantly displaced from the nucleus of the radio galaxy. In the case of NGC326 (fig. 4b) this displacement is considerable and the jets from the nucleus to the lobes do not have the same symmetry as the lobes.

(b) Statistical Results

It is useful to define a quantitative measure of the degree of deviation from colinearity for use in statistical studies. The small inserts in fig. 5 show one such measure. This figure also shows the distribution of the distortions for three classes of objects. The B2 radio galaxies (fig. 5a) are a complete sample of bright galaxies ($M_p <$ 15.7). This sample has a large fraction of distorted structures almost equally distributed between the mirror and inversion symmetric classes. The 3CR sample (fig. 5b) has much higher average radio luminosity because it is not a magnitude limited sample. It is much more strongly peaked near $\chi = 0°$ (more colinear sources) and again the number of distorted sources are equally distributed between the mirror and inversion symmetric classes. Finally, a sample of radio galaxies from Abell clusters shows a very strong preference for strong mirror symmetric distortions (the head tail radio sources) and a complete absence of inversion symmetric sources. An analysis of the distortions in a sample of Southern radio galaxies by Shaver et al (this symposium) shows that the degree of distortion is larger when the radio galaxy has a nearby companion.

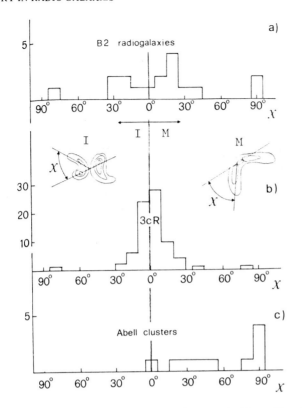

Figure 5. The definition and distribution of the distortion angle χ for a) B2 radio galaxies, b) 3CR radio galaxies, and c) radio galaxies in the Abell clusters. (Ekers et al 1981).

Summarizing these results we can conclude that i) the amount of distortion increases with decreasing radio luminosity, ii) the inversion symmetry is most likely to occur in isolated multiple systems, and that iii) rich clusters convert all morphologies to strong mirror symmetric distortions.

(c) Projection Effects

We can never measure more than the two dimensional projection of the three dimensional radio galaxies. Some take the point of view that the third dimension should be ignored because it is unknown and frown on models that invoke a specific three dimenisonal structure. With everyday objects we can rely on experience to suggest the appropriate three dimensional shapes. We immediately visualize three dimensional houses or trees or light bulbs when we see two dimensional images of them because our experience reminds us what they look like from other directions. Since we never experience radio galaxies from any other direction, our two dimensional maps remain stuck as two dimensional images in our minds.

Two additional points should be kept in mind when considering the effects of projection on morphology: i) intrinsic distortions from linear structures can be amplified by projection, and ii) projection can break mirror symmetry but not inversion symmetry.

Perhaps the only good way we have to get a feel for the three dimensional structures is by looking at a large number of distorted objects. By assuming these are viewed at random angles some constraints can be put on the three dimensional structure. For example an analysis of the 3C sample by Ingham and Morrison (1975) shows that both the source bending and most of the asymmetry in component separation ratio (15-30%) must be intrinsic to the source.

(d) Models

Relative translation between the parent radio galaxy and the surrounding medium can give a natural explanation of the mirror symmetric sources. Large relative velocities are expected in clusters and seem to give a very plausible explanation of the extreme mirror symmetric distortions such as in NGC1265 (Miley et al 1972). The more complex mirror symmetric distortions such as seen in 3C449 (fig. 3b) can be explained in an an analagous way by the slower orbital motion of the radio galaxy about its companion (Blandford and Icke 1978).

Possible explanations of the inversion symmetric distortions involve rotation of the radio ejection axis with respect to the extragalactic medium. Rotation (or shearing) of the medium itself is unlikely because of the large scales (>1 Mpc) involved in inversion symmetric sources like 3C315 (e.g. Miley 1980). A simple rotation of the engine is also inadequate to explain the kind of inversion symmetry seen in sources like NGC326 (fig. 4b) and 3C315 (Hogbom 1979). For these the projection of a more complicated precessional motion of the ejection axis is required (Ekers et al 1978a). The presence of nearby companions in most (all?) sources with strong inversion symmetry may provide a clue to the mechanism for swinging the ejection axis. Although they could not directly torque a nuclear engine deep inside the galaxy (such as a black hole), they could influence it by dumping fresh material with different angular momentum or by distorting the gas distribution which may be collimating the jet further out from the nucleus (Smarr private communication).

A different class of model which might be able to explain some inversion and mirror symmetry uses the effcts of bouyancy on jets or plasmons traversing a medium with strong density gradients. Ejection perpendicular to the gradient will produce mirror symmetric bending while ejection at an angle to the gradient can lead to an inversion symmetry. Recent support for this model is given by the observation of the inversion symmetric source 3C293 (Bridle et al 1981). This has a jet which curves towards the minor axis as it comes out of a very flattened galaxy at an oblique angle.

IV. Other Axes

When the radio data is combined with other information on the parent galaxy we can investigate more of the symmetry axes in the system.

Various investigations now show that the major radio axis is preferentially aligned with the minor optical axis of the galaxy (Palimaka et al 1979, Guthrie 1979, Shaver et al this symposium). However many clear exceptions to this correlation occur. Probably the unexpected complexity of the optical morphology of the elliptical galaxies (oblate, prolate or triaxial and some twisted isophotes!) complicate this issue and if we could pin down the physical meaning of the radio axis it might help unscramble the three dimensional optical structure. A clearer correlation is found with the axes defined by dust lanes in elliptical galaxies. The radio axis is usually nearly perpendicular to the dust lane and hence normal to a gaseous disk in these systems (Kotanyi and Ekers 1979), but again there are some clear exceptions.

Finally, we now have enough optical data to make a statistical comparision between the radio ejection axis and the optical rotation axis. The distribution of differences in fig. 6 is taken from the compilation in Ekers and Simkin (1981). This distribution clearly

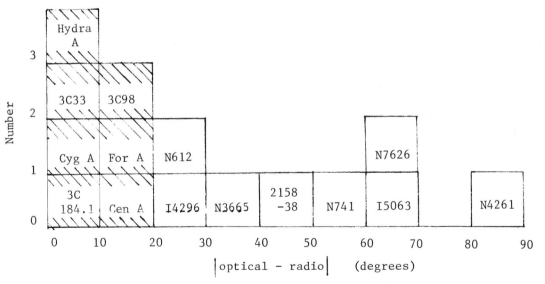

Figure 6. The distribution of differences between the position angle of the optical rotation axis and the inner radio axis of 15 radio galaxies. Objects with radio power 10^{24} WHz^{-1} Ster^{-1} at 1.4 GHz are cross hatched. (from the compilation in Ekers and Simkin 1981).

favours radio ejection along the optical axis but again with many exceptions. However, this also shows that the higher power radio galaxies (shaded) are much more accurately aligned with the rotation axis than those of lower power. This result in combination with the concepts of varying ejection axis direction suggested by the inversion symmetric sources and a general tendency for highly distorted sources to have lower luminosity suggests the following scenario. If the central engine in a radio galaxy happens to be well aligned with the rotation axis a long lived well collimated radio source results and the magnetic field and particle energy from the beam will build up in the lobes. On the other hand if the engine (or whatever does the collimation) is waving the beam about no well collimated structure will form and the energy from the nucleus will be spread over a much larger volume and a much smaller fraction of it will come out of the synchrotron window.

REFERENCES

Bridle, A.H., Fomalont, E.B., Cornwell, T.J., 1981, Astron.J. 86, 1294
Blandford, R.D., Icke, V., 1978, Mon.Not.Roy.astr.Soc. 185, 527
Cooper, B.F.C., Price, R.M., Cole, D.J., 1965, Aust. J. Phys. 18, 589
Ekers, R.D., Fanti, R., Lari, C., Parma, P., 1978a, Nature 276, 588
Ekers, R.D., Goss, W.M., Kotanyi, C.G., Skellern, D.J., 1978b, Astr. Ap. 69, 21
Ekers, R.D., Simkin, S.M., 1981, Astrophys. J. (submitted)
Ekers, R.D., Fanti, R., Lari, C., Parma, P., 1981, Astr. Ap. 101, 194
Fomalont, E.B., 1969, Astrophys. J. 157, 1027
Guthrie, B.N.G., 1979, Mon.Not.Roy.astr.Soc. 187, 581
Hargrave, P.J., Ryle, M., 1974, Mon.Not.Roy.astr.Soc. 166, 305
Hogbom, J.A., 1979, Astr. Ap. Suppl. 36, 173
Ingham, W., Morrison, P., 1975, Mon.Not.Roy.astr.Soc. 173, 569
Kotanyi, C.G., Ekers, R.D., 1979, Astr. Ap. 73, L1
Longair, M.S., Riley, J.M., 1979, Mon.Not.Roy.astr.Soc. 188, 625
Mackay, C.D., 1971, Mon.Not.Roy.astr.Soc. 154, 209
Mackay, C.D., 1973, Mon.Not.Roy.astr.Soc. 162, 1
Miley, G.K., 1980, Ann. Rev. Astron. Astrophys. 18, 165
Miley, G.K., Perola, G.C., van der Kruit, P.C., van der Laan, H., 1972, Nature 237, 269
Owen, F.N., Burns, J.O., Rudnick, L., 1978, Astrophys.J. Lett. 226, L119
Palimaka, J.J., Bridle, A.H., Fomalont, E.B., Brandie, G.W., 1979, Astrophys. J. Lett. 231, L7
Perley, R.A., Willis, A.G., Scott, J.S., 1979, Nature 281, 437
Ryle, M., Longair, M.S., 1967, Mon.Not.Roy.astr.Soc. 136, 123

A CONSEQUENCE OF THE ASYMMETRY OF JETS IN QUASARS AND ACTIVE NUCLEI OF GALAXIES

I. S. Shklovsky
Institute for Space Research, USSR

ABSTRACT

It is concluded that many, if not most, jets are truly one-sided. The hypothesis that the powerful radio emission of quasars and radio galaxies is caused by ejections of "plasmoids" originating in super-critical accretion on massive black holes is discussed. Because of asymmetry in the ejection of plasmoids from the thick accretion disks which form around massive black holes, the latter acquire considerable recoil momentum and should escape from the nuclei of the galaxies with large velocities. This provides a possibility for explaining a number of evolutionary effects and an approach to solving the problem of "dead" quasars.

THE REALITY OF ONE-SIDED JETS

Although over 80% of extragalactic sources--radio galaxies and quasars--are double, typically with an optical object lying between two extended radio-emitting clouds, the so-called "jets" emerging from active nuclei usually are "one-sided". The classical example is the famous jet in NGC 4486. In 1968, I suggested an interpretation of this feature as an ejection of compact plasma clouds ("plasmoids") from the nucleus of the galaxy at relativistic velocity and at a small angle to the line of sight (Shklovsky, 1968). I subsequently developed this idea further and applied it to several other objects (Shklovsky 1977, 1980). Scheuer and Readhead (1979) used it in interpreting the superluminal transverse velocities of compact components of several extragalactic radio sources observed by intercontinental interferometry. They advanced the interesting hypothesis that the so-called "radio-loud" quasars differ from the "radio-quiet" quasars only in having a favorable orientation for the direction of ejection of plasmoids moving at relativistic velocities.

It is assumed in all of the above mentioned papers that the plasmoids are ejected in two diametrically opposed directions. Owing to the Doppler effect, at relativistic velocities the flux from a receding component will be sharply reduced and can become unobservable, while the flux from a source approaching the observer is strongly enhanced. There can be no doubt that this effect does occur in nature. Recently, VLBI observations have shown reliably that a plasmoid ejected from the nucleus of 3C 273 has a transverse velocity $V_t \sim 10c$ (Pearson et al., 1981). From this it follows that the ejection occurred at an angle $\sim 6°$ to the line of sight and that its flux was enhanced by several thousand times owing to the Doppler effect. The question then arises: Is the absence of a "counter-jet" always due to a redshift which strongly attenuates the radiation? Do truly one-sided jets exist?

Analysis of the observational data leads us to conclude that one-sided jets do in fact exist.

(a) Cygnus A. A jet ~ 5 pc in length has recently been discovered; its direction differs only 6° from that of the radio axis of the source (Linfield, 1980). No counter-jet as strong as 5% of the jet was found. According to Hargrave and Ryle (1974), the radio axis of Cygnus A lies within 25° of the sky plane. Moreover, Simkin (1977) found that the rotation axis of the Cygnus A galaxy is inclined 6° with respect to the sky plane. This excludes the possibility of explaining the absence of a counter-jet by the relativistic Doppler effect. Linfield (1980) also studied the jets in the radio galaxies 3C 111, 3C 390.3, and 0055+30. One-sided jets were observed in all three cases. The orientations of the radio axes of these galaxies are not known, although in the case of 3C 111 it probably lies close to the sky plane.

(b) A one-sided jet was recently found in the nearest radio galaxy, Centaurus A (Feigelson et al., 1981). It is hard to imagine that the radio axis of NGC 5128 is close to the line of sight, particularly in view of the fact that the second nearest radio galaxy, NGC 4486, should also be "favorably oriented". The probability that the two nearest radio galaxies independently have a special orientation is about 10^{-2}.

(c) Two-sided as well as one-sided jets are observed, as in the radio galaxy Fornax A, where the jets are ~ 1 pc long (Fomalont and Geldzahler, 1980).

Thus, we can say that both one-sided and two-sided jets occur near the active galactic nuclei, and that the one-sided jets appear to occur several times more frequently than the two-sided jets. It can no longer be doubted that the jets are the source of the radio-emitting matter (i.e., relativistic electrons) in the extended clouds symmetrically flanking the optical galaxy. The fact that two such clouds are usually seen, while the nuclear jets are most often one-sided, means that the jets occur sometimes in one direction and sometimes in the diametrically opposite direction.

The time through which the ejection of plasmoids continues in a given direction can be rather long. For example, the one-sided jet from the nucleus of NGC 5128, consisting of several condensations, extends for some 5 kpc, so the duration of the one-sided ejection is at least 15000 years and probably much longer. Near this galaxy two quite asymmetrical intermediate maxima in radio brightness are observed. The center of the northeastern maximum is 23' from the nucleus of the galaxy, while the center of the southwestern maximum is fully 115' away. On the other hand, the "innermost" and "outermost" brightness maxima are very symmetrically placed; their distances from the nucleus are the same, 4' and 190', respectively. It seems most natural to regard the intermediate maxima as having been caused by one-sided jets which occurred at different epochs in the evolution of NGC 5128. With this interpretation of the intermediate maxima, the duration of the one-sided ejections must be reckoned in millions of years.

Finally, let us consider radio component "A" in 3C 273. It lies at the very end of the optical jet and has a diffuse structure; its angular radius at 20 cm wavelength is ~6", which corresponds to a linear size ~50 kpc (Perley et al., 1980). The magnetic field at its periphery is perpendicular to the direction of the jet, as in Cygnus A.

It is natural to regard this component not as an extension of the jet but rather as a feature "fed" by the jet, wholly analogous to the clouds in the double radio galaxies (see Perley et al., 1980). If this is so, the velocity of this component must be definitely non-relativistic. This interpretation raises a question: Where is the second, symmetrically placed, radio cloud 3C 273? According to Perley et al. (1980), some 30% of compact sources (mostly quasars) have components at distances of several arcseconds. These components are optically thin and are distinguished by steep radio spectra ($\alpha \sim 1$). The above authors suggest that in all such cases the objects are like Cygnus A, but with their radio axes lying fairly close to the line of sight. They explain the absence of symmetrical components by projection on the compact components. In this case, however, the spectrum of the well-studied component 3C 273B should show a strong excess of radiation at low frequencies, which certainly is not observed. The absence of the symmetrical component simply means that the jet in 3C 273 is really one-sided (although with relativistic velocity) and that it has remained one-sided throughout the evolution of this quasar, i.e., for several million years.

Thus the ejection of plasmoids is an inherently non-symmetrical event. Since the characteristic time of "one-sidedness" of the activity of nuclei can exceed several million years, and the duration of the radio-emitting phase in galaxies and quasars is of the same order, the overall activity of galactic nuclei (i.e., integrated over the whole time of evolution) is probably an asymmetrical phenomenon.

CAN MASSIVE BLACK HOLES LONG REMAIN IN THE NUCLEI OF GALAXIES?

In §1, we concluded from an analysis of the observational data that many and perhaps most jets are truly one-sided, and that this property is preserved for millions of years. This implies a very important result: Because of the asymmetrical character of the activity, a supermassive black hole can acquire momentum in the direction opposite to the ejection of the plasmoids, i.e., along the axis of the nucleus. With each ejection from the disk of a plasmoid of mass Δm at a velocity $v \sim c$, the black hole receives a small velocity increment

$$\Delta v_1 \sim \frac{\Delta m \cdot c}{M_H} .$$

During the lifetime of the one-sided relativistic ejection, the black hole acquires a recoil velocity

$$v_H \sim \frac{M}{M_H} \cdot c .$$

This effect is entirely similar to the acquisition of added velocity by the center of mass of a binary system where one component has undergone a supernova explosion, when the explosion is not spherically symmetrical. The only difference is that in our case the momentum is acquired gradually. In principal, the effect should be impeded by dynamical friction from all of the stars of the galaxy. This process has been investigated by White (1976), who calculated its characteristic time as

$$t_{df} \sim \frac{1}{30} \frac{M(R)}{M_H} \left[\frac{GM(R)}{R^3} \right]^{-1/2} ,$$

where $M(R)$ is the galactic mass interior to radius R and M_H is the mass of the black hole. In all cases of real interest, however,

$$t_{df} \gg \tau = \frac{R}{V_H} ,$$

where τ is the time which the recoil-accelerated black hole takes to move out of the galaxy.

Consider the specific case of 3C 273, whose jet must be one-sided (see §1). We can estimate the energy of the relativistic particles and the field in component A in the usual way. The calculations can be made by reference to Cygnus A, which has a spectral index similar to that of 3C 273A. The effective angular diameter of each component of Cygnus A is $\sim 30"$, five times greater than for 3C 273A, while the radio flux from each component of Cygnus A is also 20 times greater. Then the energy in 3C 273A is about two times less than in either component of Cygnus A. We adopt $W \sim 3 \times 10^{59}$ erg. The sole source of this energy is the total kinetic energy E of the jets, so $W \sim \alpha E$. We found from an analysis of the jets in M87 that $\alpha \sim 10^{-2} - 10^{-1}$. The total momentum is

$$P = \frac{W}{\alpha \cdot c} \sim 10^{51} \text{ g cm sec}^{-1}.$$

We assume (rather arbitrarily) that the mass of the black hole in 3C 273 is $\sim 5 \times 10^8$ $M_\odot \sim 10^{49}$ g. It follows that the velocity ultimately attained by the black hole during the course of the activity leading to the formation of 3C 273A is $\sim 10^9$ cm sec^{-1}. Moving with this speed for the time which has elapsed since the formation of component A ($\tau \sim 3 \times 10^6$ years), the black hole should have moved ~ 30 kpc from its original place. This means, however, that the black hole has moved far outside the relatively dense part of the parent galaxy (probably spheroidal), and it is not clear how the observed supercritical accretion on it can be continuing!

The contradiction can be circumvented in the following way. We do not doubt the law of conservation of momentum so, if all of the premises we have adopted are correct (one-sided jet, black hole mass $M_H \sim 5 \times 10^8$ $M_\odot$, interpretation of 3C 273A as an object like the components of Cygnus A), black holes must escape from their parent galaxies and accretion on them should drop far below the critical level. Simply stated, they will cease to be compact radio sources. Since in the case of 3C 273B (like in other similar cases) we still observe such a source, the only possible explanation is that it is <u>not the same</u> black hole that gave birth to 3C 273A.

The most general picture that can be drawn at present for the structure of an active galactic nucleus is as follows: In a small (1 - 10 pc) region there is an exceptionally dense ($10^8 - 10^{11}$ stars) super-cluster which contains interstellar gas. In the course of its evolution a massive black hole is formed, around which an accretion disk forms. The plane of this disk is approximately parallel to the symmetry plane of the cluster, which as a rule coincides with the symmetry plane of the galaxy. It is tacitly assumed that a single massive black hole forms in a nucleus. Even with a black hole mass $\sim 10^8$ $M_\odot$ and a velocity of $\sim 3 \times 10^7$ cm sec^{-1} for macroscopic motions in the nucleus, however, the sphere of gravitational action of the black hole extends only to ~ 0.1 pc, much less than the typical dimensions of a nucleus. Therefore, there is no reason why several or even many massive black holes should not simultaneously be present in a nucleus, with approximately parallel accretion disks. At a given time in the active state, for example, there can be one black hole, or there might be more.

In some radio galaxies several more or less symmetrically placed pairs of radio-emitting clouds are observed. The classical example is NGC 5128 (see §1). The usual interpretation of this postulates cycles of activity in a single compact object in the nucleus. One could equally well, however, hypothesize that different active compact objects are responsible for the various cycles.

The concept of massive black holes escaping from the nuclei of galaxies opens up some interesting new possibilities for interpreting long-known astronomical observations. Let us consider just the problem of the luminosity functions for different kinds of quasars and radio galaxies and features of their evolution. We shall draw on the summary of these problems given by Schmidt (1978).

There is a noteworthy difference between the luminosity functions for quasars with flat and steep spectra. The luminosity function for those with flat spectra is rather flat; for a 100-fold change in radio power, their space density changes by a factor of ~30, while the density changes by a factor of ~200 for those with steep spectra. We must interpret the quasars with flat spectra as being directly associated with the accretion disks around massive black holes (the prototype is 3C 273B), while those with steep spectra must be analogous to 3C 273A. The latter are structures which are fairly slowly diffusing into the intergalactic medium, their radio luminosities decreasing continually once the "pumping" has ended (compare Cygnus A and Centaurus A, whose energy content W differs by less than an order of magnitude, while their luminosities differ by a factor of several thousand). On the other hand, the power radiated by objects like 3C 273B remains more or less constant so long as supercritical accretion continues. It is determined by the mass of the black hole. Therefore, the luminosity function for sources of 3C 273B type should be similar to the mass function of the corresponding black holes, which must be rather flat.

The most important problem is the difference of the evolutionary effect for the different types of quasars. While the evolutionary effect is very large for the quasars with steep spectra and also for the strong radio galaxies and the radio-quiet quasars (at $z = 1$ their density is about 150 times the local value), it amounts only to a factor of 3 —4 for quasars with flat spectra (the densities are referred to a co-moving coordinate frame).

In principle, the difference can be explained by the different lifetimes of sources of the two kinds. As has been implied above, quasars with flat spectra must be short-lived. As soon as the super-critically accreting regime ends (as it must when a quasar moves out of the dense nuclear region), the object "goes out". On the other hand, the lifetimes of quasars with steep spectra are much longer, particularly at large z, where the density of the intergalactic medium was much higher than it is locally, making the spreading velocity significantly less than at present.

The idea that black holes escape from galaxies provides a possible approach to solving an important problem of long standing--that of the "dead quasars". It follows from the evolutionary effect that nearly every galactic nucleus should contain an "extinguished" quasar, which cannot be detected in any way. If in fact quasars escape from the nuclei of galaxies, the problem is decisively solved.

It is hardly necessary to point out that the above scenario for the evolution of quasars of different types is highly schematic. Our aim has been to call attention to the important astronomical consequences of the asymmetry of jets.

REFERENCES

Feigelson, E., Schreier, E., Delvaille, Y., Giacconi, R., Grindlay, Y., and Lightman, P. 1981, Center for Astrophysics preprint N1473.
Fomalont, E. B. and Geldzahler, B. 1980, Bull. Amer. Astr. Soc. 12, 804.
Hargrave, P. and Ryle, M. 1974, MNRAS 166, 305.
Linfield, R. 1980, Owens Valley Observatory preprint.
Pearson, Y., Unwin, S., Cohen, M., Linfield, R., Readhead, A., Seielstad, G., and Simon, R. 1981, Owens Valley Observatory reprint.
Perley, R. A., Fomalont, E. B., and Johnston, K. J. 1980, A. J. 85, 649.
Scheuer, P. and Readhead, A. 1979, Nature 277, 182.
Schmidt, M. 1978, Physica Scripta 17, 135.
Shklovsky, I. S. 1968, Astr. Zh. 45, 919.
Shklovsky, I. S. 1977, Astr. Zh. 54, 713.
Shklovsky, I. S. 1980, Pism. v Astr. Zh. 6, 131.
Simkin, S. 1977, Ap. J. 217, 45.
White, S. D. 1976, MNRAS 174, 19.

OTHER PAPERS PRESENTED AT THE SYMPOSIUM

Hot Spots in 3C 153, 3C 196, and 3C 268.4 / *I. Morison* (poster)
Strong Radio Sources in Bright Spiral Galaxies / *J. Condon, M. Condon, G. Gisler & J. Puschell* (poster)
Supersonic Jets / *M. Norman, L. Smarr, K. Winkler & M. Smith*
Einstein Observations of M87 / *E. Schreier, E. Feigelson & P. Gorenstein*
Flux Density Variations in Complete Samples of Radio Sources: Results of a Fourteen-Year Study / *G. Nicolson* (poster)
Superluminal Motions in 3C 120 / *R. Walker* (poster)
Optical Observations of QSOs from the UTRAO Survey / *D. Wills & B. Wills* (poster)
The Radio Structures of BL Lac Type Objects / *D. Stannard*
A Supernova Remnant in the Galaxy, NGC 4449 / *R. Bignell* (poster)
The Texas Survey / *J. Douglas* (poster)
Observations of a Selected Sample of Southern Radio Galaxies / *R. Smith* (poster)
Structures, Spectra and Optical Identifications of Radio Sources from the GB 20 cm Surveys and the NRAO/MPI 6 cm Surveys / *J. Machalski, J. Maslowski, J. Condon & M. Condon* (poster)
The Jodrell Bank Multi-Telescope Linked Interferometer / *J. Davis* (poster)
RADIO OBSERVATIONS OF THE DOUBLE QUASAR / *B. Burke* (invited talk)
Nobeyama mm-Wave Telescopes / *M. Inoue & H. Tanaka* (poster)
Deep CCD Photography of Radio Sources / *C. Mackay, D. Astill, M. Batty, D. Jauncey, A. Wright, R. Hunstead, D. Morton & J. Robertson* (poster)
A VLBI Search for Compact Components in Distant Extended Quasars / *P. Barthel* (poster)
New Approach to Detection of Transient Radio Sources: Phased Array Telescope with Large Field of View / *T. Daishido, R. Oka, T. Ohkawa, T. Maruyama, T. Yokoyama, K. Nagane & H. Hirabayashi* (poster)
Quasar Variability Data: An Alternate Presentation Mode / *E. Epstein & E. Schneider* (poster)

SUBJECT INDEX

Page references are to the first pages of the relevant papers.

Accretion 45, 197, 211, 235, 237, 247, 255, 263, 265, 369
Accretion - supercritical 255, 263, 465
Active galactic nuclei 107, 189, 247, 369, 475
Angular size distribution 393
 - flux density relation 393, 401
 - redshift relation 401, 411
Bautz-Morgan system 85, 87, 97
BL Lacertae objects 239, 311, 335, 363, 377, 383, 385
Black holes 1, 211, 247, 255, 263, 265, 475
 - precession 211, 255
Blank field radio sources 269, 435
Buoyant forces 13, 45, 77
cD galaxies 45, 97, 413
Clusters 77, 85, 87, 91, 97, 425, 453
Color-redshift diagram 413
Compact sources 1, 175, 295, 297, 317, 327, 329, 331, 335, 345, 369
Core-dominated sources 149, 169, 175, 363
Distortions 39, 45, 55, 465
Dust 239, 465
Evolution 1, 21, 35, 401, 427, 441, 451, 475
Galactic nuclei 1, 189, 239, 265, 387, 389
Galaxies - elliptical and S0 1, 25, 35, 65, 309, 377, 413, 427
 - spiral 1, 93, 145, 189
Gravitational lenses 345, 451, 461, 463

Hotspots 21, 25, 27, 43, 53, 59, 61, 135, 141, 149, 157, 161, 163, 177, 411
HI absorption 307, 311, 313
 - emission 309, 369
HII regions 107, 115, 195
In-situ acceleration 41, 107, 211, 229, 265
Intergalactic medium 93, 157, 453
Interstellar medium 25, 107, 179
Intracluster medium 13, 45, 91, 97
Jets 25, 43, 55, 61, 115, 121, 129, 133, 135, 139, 141, 167, 173, 175, 179, 189, 197, 207, 209, 211, 223, 227, 235, 237, 255, 263, 345
 - bends in 107, 129, 137, 145, 211, 345
 - collimation 107, 121, 129, 211, 223
 - confinement 25, 107, 129, 135, 145, 211, 229
 - entrainment 69, 121, 223
 - evolution 121, 197
 - gaps in 223, 231
 - magnetic configuration 121, 129, 141, 197, 211
 - morphology 121, 193, 229, 231, 279
 - one-sided 47, 51, 107, 121, 129, 149, 167, 197, 211, 265, 279, 289, 293, 345, 357, 465, 475
 - opening angle 25, 139, 223
 - optical 61, 65, 115
 - polarization 121, 135, 139, 141, 197
 precession 13, 133, 197, 255

- relativistic 75, 211, 345, 363
- stability 211, 229
- supersonic 121, 229
- twin 197, 211, 265
Jets - two-sided 121, 193, 265, 293, 475
- X-ray 107, 115, 129, 135
Jet-counterjet asymmetries 121, 141
Kelvin-Helmholtz instabilities 107, 157, 211, 229, 231
Linear size evolution 55, 393, 401, 411
Low-frequency variables 363
Line-locking 209
Luminosity-size relation 21, 401
Nuclear ejection 47, 179
Nuclear radio sources 13, 43, 191, 279, 291, 293
Number counts - quasars 269
- X-ray sources 269
Optical emission in radio lobes 43, 61, 69, 71
Optical emission lines 55, 65, 197, 369, 373
Optically violent variables 239, 311, 363
Particle reacceleration 107, 231, 233
Plasmoids 13, 41, 51, 121, 179, 475
Polarization - radio 43, 53, 61, 121, 139, 141, 173, 177, 179, 197, 239, 301, 335, 339
- optical 263, 341
- variability 331, 337
Quasars - distance scale 437
- luminosity evolution 437, 441
- nebulosity 375
- optical emission 263, 269, 341, 369, 373
- radio-quiet 305, 363, 369, 451
- X-rays 135, 263, 269, 359
Radiative acceleration 209, 365
Radio cores 1, 55, 65, 121, 149, 167, 169, 173, 175, 345, 363, 433
Radio galaxies 1, 13, 33, 35, 53, 55, 61, 65, 89, 239, 247, 279, 369, 413, 423, 427, 441, 465

Radio lobes 51, 77, 107, 157, 175, 177, 411, 475
Radio luminosity function 91, 393, 425, 427, 441
Ram pressure 45, 145, 179, 211
Relativistic beaming 149, 169, 265, 341, 363, 383, 433
- flow 209, 365
Scintillation 59, 325
Seyfert galaxies 1, 25, 55, 179, 189, 191, 239, 369, 453
Size-power distribution 91
Spectral curvature 27, 41
- energy distribution 377, 427
- index distribution 29, 33, 89, 433, 441
- index gradients 41, 121
Spinars 247, 265
Starburst models 179, 239
Superluminal motion 167, 211, 279, 317, 345, 355, 357, 359, 361, 363, 463, 475
Supernovae, extragalactic 391
Symmetry 13, 55, 161, 211, 255, 465
Tailed radio galaxies 13, 41, 45, 77
Turbulence 231, 233
Twin-exhaust model 211, 265
Vortex accretion funnel 211, 237
X-rays 77, 97, 107, 117, 269, 453

OBJECT INDEX

Page references are to the first pages of the relevant papers.

Abell Clusters 45, 77, 85, 89, 91, 97
Abell 2634 See 3C 465
BL Lacertae 239, 279, 317, 335, 337, 345, 385
Braccesi field objects 269
Centaurus A 1, 13, 61, 69, 107, 115, 117, 119, 121, 229, 231, 475
Coma A 61, 69, 211
Coma Cluster 97
CTA 21 279
CTA 102 317
CTD 93 See 1607+268
Cygnus A 13, 27, 29, 33, 157, 279, 465, 475
DA240 13, 177, 211, 401
DA344 See 1323+321
Double quasar See 0957+561
Fornax A 475
HB 13 13, 41
IC 708 13
IC 790 See NGC 4410
IC 4329 239
M81 1, 387
M82 239
M84 See 3C 272.1
M86 97
M87 See Virgo A
Markarian objects 179, 189, 195, 239, 335
Markarian 421 See 1101+384
Markarian 501 See 1652+398
NGC 253 239
NGC 315 121, 211, 227, 229, 231, 279, 475
NGC 326 13, 465
NGC 541 41
NGC 612 465
NGC 545/7 See 3C 40
NGC 1052 309

NGC 1068 179, 189
NGC 1265 1, 45, 465
NGC 1275 See Perseus A
NGC 3227 179, 239
NGC 3690 189
NGC 3801 121
NGC 4038/9 93
NGC 4151 179, 189, 239
NGC 4258 1, 145, 179
NGC 4278 309
NGC 4321 391
NGC 4410 93
NGC 4438 93
NGC 4486 See Virgo A
NGC 5128 See Centaurus A
NGC 5548 179, 191, 239
NGC 6251 13, 53, 121, 141, 163, 211, 229, 231, 279, 401
NGC 6764 179
NGC 6946 391
NGC 7318-20 See Stephan's Quintet
NGC 7385 See 2247+113
NGC 7469 179, 239
NGC 7720 See 3C 465
NRAO 140 See 0333+321
NRAO 150 See 0355+508
OC 457 See 0133+476
OI 090.4 239
OI 318 See 0711+356
OI 417 See 0710+439
OJ 287 See 0851+202
OJ 425 See 0814+425
OJ 508 See 0804+499
OS 562 See 1637+574
OV 591 See 1954+513
OY 091 See 2254+074
Perseus A 1, 13, 189, 279, 291, 307, 345, 453
Sagittarius A 1, 389
SC 0627-54 97

Sco X-1   211
SS433   197, 205, 207, 209, 211, 235, 265
Stephan's Quintet   95
UGC objects   89, 309
Virgo A   1, 13, 61, 97, 149, 211, 229, 231, 279, 293
W50   See SS 433
Zw 1615   77
I Zw 187   See 1727+502
III Zw 2   305
3C 9   157, 411
3C 13   423, 435
3C 20   53, 161
3C 31   13, 121, 211
3C 34   423
3C 40   13, 41
3C 61.1   413
3C 65   423
3C 66   121, 335
3C 68.2   413
3C 84   See Perseus A
3C 103   13
3C 105   13
3C 109   413
3C 111   279, 475
3C 119   279
3C 120   33, 279, 317, 331, 345
3C 123   413
3C 129   77
3C 129.1   77
3C 133   161
3C 147   279, 289
3C 179   345, 361
3C 184   423
3C 192   53
3C 196   149, 161
3C 216   279
3C 219   13, 163
3C 226   435
3C 228   435
3C 234   413
3C 236   13, 173, 279, 401
3C 247   435
3C 252   435
3C 254   413
3C 264   13, 89
3C 265   71, 413
3C 267   435
3C 268.1   413
3C 270.1   39
3C 272.1   97, 121
3C 273   33, 167, 175, 247, 279, 345, 355, 363, 475
3C 274   See Virgo A
3C 275.1   39
3C 277.3   See Coma A
3C 279   317, 331, 345
3C 280   435
3C 280.1   163, 411
3C 285   71, 163
3C 286   279
3C 289   423, 435
3C 293   61, 465
3C 294   413
3C 295   149, 423
3C 296   401
3C 299   413
3C 305   61, 69
3C 309.1   149
3C 310   13
3C 314.1   413
3C 315   13, 465
3C 318   413
3C 322   435
3C 324   435
3C 325   435
3C 326   401
3C 334   121, 129
3C 337   435
3C 340   413, 435
3C 345   121, 175, 279, 317, 329, 345, 357
3C 351   157, 163
3C 356   435
3C 368   423
3C 371   175, 279, 335
3C 380   149, 279
3C 382   413
3C 388   13, 231
3C 390.3   71, 279, 475
3C 395   279
3C 418   149
3C 427.1   423, 435
3C 430   53
3C 437   413
3C 446   345
3C 449   13, 121, 139, 211, 229, 231, 465
3C 454.3   149, 175, 279, 301, 317, 345
3C 465   13, 45

OBJECT INDEX 489

3C 469.1	413, 435
3C 470	413
4C 11.71	See 2247+113
4C 14.60	See 1538+149
4C 18.68	See 2305+187
4C 22.26	129
4C 24.02	129
4C 26.42	61, 69
4C 29.30	61
4C 29.68	121, 129
4C 31.63	See 2201+315
4C 32.69	129, 135, 211
4C 35.40	401
4C 38.41	See 1633+382
4C 39.25	See 0923+392
4C 39.27	129
4C 39.49	See 1652+398
4C 40.24	See 0945+408
4C 41.32	See 1624+416
4C 45.51	See 2351+456
4C 47.29	See 0859+470
4C 47.51	See 1919+479
4C 55.17	See 0954+55
4C 56.27	See 1823+568
4C 58.17	See 0850+581
4C 64.19	401
4C 69.21	See 1642+690
4C 73.08	401
4C 73.18	See 1928+738
4CT74.17.1	47
5C 7.190	393
0026+129	305
0048-097	383
0055+300	See NGC 315
0133+476	279
0202+31	297
0212+171	137
0212+735	279
0219+438	383
0235+164	211, 239, 311, 313, 317, 337, 383
0316+162	See CTA 21
0316+413	See Perseus A
0326+396	121, 193
0333+321	279, 345, 359
0349-27	65
0355+508	279, 345
0415+379	See 3C 111
0420-388	305
0428+205	279
0429+415	See 3C 119
0430+052	See 3C 120
0442-18	423
0454+844	295
0521-365	65, 335
0537-441	331
0538+498	See 3C 147
0548-322	377
0552+398	297
0607-157	337
0710+439	279
0711+356	279
0716+714	175
0727-115	337
0735+178	239, 317, 383
0736+01	317
0742+318	137
0754+100	383
0804+499	279
0812+020	43
0814+425	279
0818-128	383
0829+046	383
0836+710	279
0840+42	423
0850+581	279
0851+202	211, 317, 335, 345, 383
0859+470	279
0906+430	See 3C 216
0923+392	279, 329, 345
0932+02	411
0945+408	279
0954+55	295, 329
0957+561	461
1003+351	See 3C 236
1101+384	335, 383
1150+497	295
1217+023	137
1218+304	377
1219+28	377, 383
1226+023	See 3C 273
1228+126	See Virgo A
1244+32	411
1253-055	See 3C 279
1302-102	305
1308+326	383
1321+319	121, 193, 465
1323+321	279
1328+307	See 3C 286
1331-09	13
1400+162	335, 377, 383
1413+135	239, 335

1418+546	335
1504−16	317
1518+047	279
1524−13	317
1538+149	335, 383
1607+268	279
1624+416	279
1633+382	279
1636+473	149
1637+574	279
1637+826	See NGC 6251
1641+399	See 3C 345
1642+690	279
1647+43	423
1652+398	279, 335, 383
1807+698	See 3C 371
1823+568	279
1727+502	335, 377
1823+564	149
1828+487	See 3C 380
1845+797	See 3C 390.3
1901+319	See 3C 395
1919+479	77
1928+738	279
1954+513	279
1957+405	See Cygnus A
2021+614	279
2050+364	279
2134+004	297, 331, 345
2155−304	335, 363
2158−380	65
2200+420	See BL Lacertae
2201+04	383
2201+315	137
2247+113	13, 41, 61, 401
2251+158	See 3C 454.3
2254+074	335, 383
2305+187	133
2351+456	279